2025 10th IEEE International Conference on Integrated Circuits, Design, and Verification (ICDV 2025)

AA001044

Ho Chi Minh City, Vietnam
16-17 June 2025

IEEE Catalog Number: CFP25N19-POD
ISBN: 979-8-3315-1550-8

**Copyright © 2025 by the Institute of Electrical and Electronics Engineers, Inc.
All Rights Reserved**

Copyright and Reprint Permissions: Abstracting is permitted with credit to the source. Libraries are permitted to photocopy beyond the limit of U.S. copyright law for private use of patrons those articles in this volume that carry a code at the bottom of the first page, provided the per-copy fee indicated in the code is paid through Copyright Clearance Center, 222 Rosewood Drive, Danvers, MA 01923.

For other copying, reprint or republication permission, write to IEEE Copyrights Manager, IEEE Service Center, 445 Hoes Lane, Piscataway, NJ 08854. All rights reserved.

*** This is a print representation of what appears in the IEEE Digital Library. Some format issues inherent in the e-media version may also appear in this print version.

IEEE Catalog Number: CFP25N19-POD
ISBN (Print-On-Demand): 979-8-3315-1550-8
ISBN (Online):   979-8-3315-1549-2

Additional Copies of This Publication Are Available From:

Curran Associates, Inc
57 Morehouse Lane
Red Hook, NY 12571 USA
Phone: (845) 758-0400
Fax: (845) 758-2633
E-mail: curran@proceedings.com
Web: www.proceedings.com

Proceedings of

The 2025 10th International Conference on Integrated Circuits, Design, and Verification

ICDV 2025

June 16-17, 2025
Ho Chi Minh City, Vietnam

Table of Contents

2025 10th International Conference on Integrated Circuits, Design and Verification

Table of Contents _____ ii

Preface _____ v

Conference Committee _____ vi

Technical Program Committee _____ vii

Keynotes

Advanced Biomedical Imaging Technologies: Circuit Design and Techniques_____ x
    Yongfu Li

Photonics Integrated Circuits: Enabling the Next Era of High-Speed, Energy-Efficient Computing _ xi
    Le Quang Dam

Multi-core Multi-thread RISC-V-based System-on-Chip_____ xii
    Cong-Kha Pham

CASS Distinguished Lecture

Advanced Circuits and Systems for Navigation-Grade MEMS Accelerometers_____ xiii
    Jian Zhao

Technical Sessions

An Optimized Obstructive Sleep Apnea Detection Model Using Particle Swarm Optimization and
Machine Learning_____ 1
    Saroj Biswas, Atiya Khan, Chukhu Chunka

An Optimized Hybrid Quantum-Classical Neural Network Model for Handwritten Digit Classification
_____ 7
    Quoc Minh V. Nguyen, Trung-Khanh Le, Trong-Tu Bui, Duc-Hung Le

Harnessing TinyML for Accurate ECG Beat Detection _____ 13
    Dong Bui, Hoang Anh Vy Ngo, Dat Hoang Tran

FPGA-based Design and Implementation of Processing Element Array for Convolutional Neural
Networks_____ 19
    Chi Phuong Hoang, Nguyen D. Minh, Linh Nguyen-Thi-Thuy, Luu Nguyen-Van

Efficient AI Model and Hardware Architecture Based on CNN for Arrhythmia Prediction _____ 25
    Huy-Duc Pham, Thi-Minh-Tuyen Huynh, Tuan-Kiet Tran, Thanh-Dat Bui, Cong-Kha Pham,
    Huu-Thuan Huynh

High-PSR Capacitor-Less LDO with Enhanced Bandgap Reference in 65nm CMOS Technology _ 31
    Viet N. D Ngo, Cuong Huynh

Inductorless 5.405 GHz Fractional-N PLL for RF Synthesis with 5.6 mW Power Consumption ____ 37

Ha Thi Viet Nguyen, Cong-Kha Pham, Xuan Thanh-Pham, Kha Manh Hoang

Effect of Temperature on the Stability of SnSe Nanoribbons as a Channel Material for Field-Effect Transistors _____ 43
Nilüfer Ertekin, Wen Lei

A 12-bit 100MS/s SAR ADC with Sub-Radix and Optimize Digital Delay Path _____ 49
Long Pham Hoang Ho, Lam Thien Van, Cuong Huynh

QEA: An Accelerator for Quantum Circuit Simulation with Resources Efficiency and Flexibility _ 55
Van Duy Tran, Tuan Hai Vu, Vu Trung Duong Le, Hoai Luan Pham, Yasuhiko Nakashima

HW/SW Co-Design for a Variational AutoEncoder targeting Anomaly Detection on FPGA _____ 61
Tuan-Phong Tran, Thien-Duy Ho, Tung-Bach Nguyen, Xuan-Tu Tran, Duy-Hieu Bui

Efficient ECG Beat Classification Using Inception Network on Software and FPGA Platforms ___ 67
Diem Thi Tran, Le Nguyen Nhat Nam

Analysis of Plant Electrical Signals on an IoT Platform_____ 73
Xuan Bach Duy Nguyen, Bao Chau Pham Ngoc, Anh-Vu Dinh-Duc

A Transformer Feedback Oscillator _____ 79
Weiwen Lin, Zhiqun Li, Zhennan Li, Yan Yao, Bofan Chen, Muhammad Hashim, Yassin Abdullah

Synthesis of cosecant squared pattern antenna arrays using the methods of stacked beams _____ 83
Nhu Thai Le, Thanh Cong Vu, Tuan Anh La, Hoai Son Nguyen, Hang Le Thi

Nonlinear Capacitance Compensation Low Noise Amplifier and Mixer with UWB Anchor Antenna 87
Wen Cheng Lai

Dynamic Queue Management and Packet Loss Mitigation in P4-Enabled Data Planes _____ 91
Bui Ngoc Thanh Binh, Tran Nguyen Tuan Kiet, Nguyen Viet Ha

A Data Labeling Method in Deep Learning Model for User Clustering in the NOMA Systems ____ 97
Ngo Minh Nghia, Nguyen Thi Xuan Uyen, Nguyen Dung, Kha Duy Thai Ngoc, Dang Le Khoa

RTL Design of Convolution for CNN Using Baugh Wooley and Wallace Tree Multipliers_____ 103
Vinh Truong Quang, Quan Doan Duy, Khang Nguyen Minh

High-Efficiency 4:2 Compressor Designs: A Comparative Study on Hardware Cost and Error Trade-Offs _____ 109
Vishnu Padmakumar, Adhiraj Nandy, Sourav Nath, Koushik Guha, Krishna Baishnab, Saroj Biswas

SDR Implemented Algorithm for Real Time Intra-pulse Modulated Radar Signal Analysis_____ 115
Duong Van Minh, Duy-Cong Nguyen, Phuong Nguyen, Hoa Quang Nguyen, Giang Phan, Tan Phat Huynh, Manh Long Nguyen

DDoS Attack Detection for Software-Defined Network Architecture Based on Artificial Intelligence _____ 121
Thai-Bao Pham, My Nguyen-Le-Ha, Luan Van-Thien, Thuat Nguyen-Khanh, Quan Le-Trung

A Solution for Built-in On-chip Hardware Integrity Protection Adopting Resource-optimized RO-PUF _____ 127
Hoa Quang Nguyen, Hoang-Long Nguyen, Tri-Hieu Le, Van-Toan Tran, Duy-Cong Nguyen, Quang-Kien Trinh

Correlation Power Analysis of Pipelined and Multi-Threaded Coarse-Grained Reconfigurable Cryptographic Accelerator _____ 133
Van-Tuan Luu, Hoai Luan Pham, Van-Tinh Nguyen, Van-Phuc Hoang, Nguyen Van Trung, Vu Trung Duong Le, Yasuhiko Nakashima

Data communication security for FANETs using Ascon lightweight cryptography _____ 139
Huyen-Trang Pham-Thi, Duy-Hieu Bui, Xuan-Tu Tran

Author Index _____ 145

PREFACE

It is our great pleasure to warmly welcome you to the 10th International Conference on Integrated Circuits, Design, and Verification (ICDV 2025).

Semiconductors form the foundation of modern technology, enabling a vast array of applications from smartphones and artificial intelligence to automotive systems and the Internet of Things. According to a recent Gartner report, the global semiconductor market is projected to reach 1,000 billion USD by 2030. In alignment with this growth, the Vietnamese government has taken strategic steps to develop domestic capabilities in integrated circuit (IC) design, advanced packaging, testing, and verification. While traditionally recognized for its role in backend assembly, packaging, and testing, Vietnam is now rapidly positioning itself as an emerging hub for front-end IC design — a high-value segment in the semiconductor supply chain — driven by robust government initiatives and growing foreign investment.

Since its inception in 2010, the ICDV conference has served as a vital platform for connecting academia and industry in Vietnam to foster innovation and collaboration in the semiconductor domain. The conference facilitates the exchange of ideas, the dissemination of research findings, and the presentation of novel chip designs and applications across solid-state and semiconductor technologies. As CMOS technology continues to scale — with the number of transistors on a VLSI chip approaching 100 billion — new opportunities are emerging for advanced applications and computational paradigms. The ongoing surge in artificial intelligence has further underscored the need for novel system architectures, design methodologies, and integrated circuit technologies.

This year, ICDV 2025 is proudly hosted in Ho Chi Minh City — Vietnam's most populous urban center and the nation's leading hub for IC design enterprises. The conference is co-organized by the University of Science, Vietnam National University Ho Chi Minh City, and the Information Technology Institute, Vietnam National University Hanoi. We are honored to have the patronage of both Vietnam National Universities, and the technical sponsorship of the IEEE Vietnam Section (IEEE CASS, IEEE SSCS chapters), and IEICE Japan. We also extend our gratitude to Marvell Inc. for their generous support.

For ICDV 2025, we received 49 paper submissions, of which 25 were selected for presentation, resulting in an acceptance rate of 51%. The program includes 3 keynote addresses by distinguished leaders in the field and 4 invited talks from renowned experts in both academia and industry.

On behalf of the Organizing Committee, we would like to extend our sincere thanks to all sponsors, keynote and invited speakers, session chairs, reviewers, authors, and participants. Your valuable contributions have made this conference both meaningful and memorable. We hope that you find ICDV 2025 intellectually enriching and that you enjoy your time in vibrant Ho Chi Minh City.

General Co-Chairs
Minh-Triet Tran
Ho Chi Minh City University of Science, Vietnam
Xuan-Tu Tran
Vietnam National University, Hanoi, Vietnam

Conference Committee

General Co-Chairs: Minh-Triet Tran, *Ho Chi Minh City University of Science, Vietnam*
Xuan-Tu Tran, *Vietnam National University, Hanoi, Vietnam*
Kunio Uchiyama, *AIST, Japan*
Makoto Ikeda, *The University of Tokyo, Japan*

Technical Conference Chairs:
Duc-Hung Le, *Ho Chi Minh City University of Science, Vietnam*
Cong-Kha Pham, *University of Electro-Communications, Japan*
Van-Phuc Hoang, *Le Quy Don Technical University, Vietnam*

Local Arrangement Chairs:
Trong-Tu Bui, *Ho Chi Minh City University of Science, Vietnam*
Le-Khoa Dang, *Ho Chi Minh City University of Science, Vietnam*

Publication Chairs: Duy-Hieu Bui, *VNU Information Technology Institute, Vietnam*

Special Session Chairs:
Huynh Phu Minh Cuong, *Ho Chi Minh City Univ. of Technology, Vietnam*

Publicity Co-Chairs: Minh-Tri Nguyen, *Ho Chi Minh City University of Science, Vietnam*

Secretary: Lan-Huong Dang, *Ho Chi Minh City University of Science, Vietnam*

Technical Program Committee

Technical Program Chairs:

Duc-Hung Le, *Ho Chi Minh City University of Science, Vietnam*
Cong-Kha Pham, *University of Electro-Communications, Japan*
Van-Phuc Hoang, *Le Quy Don Technical University, Vietnam*

TPC members:

Duy-Hieu Bui, *VNU Information Technology Institute, Vietnam*
Minh-Cuong Huynh Phu, *Ho Chi Minh City University of Technology, Vietnam*
Dang Le Khoa, *Ho Chi Minh City University of Science, Vietnam*
Thuy Minh Le, *Hanoi University of Science and Technology, Vietnam*
Nguyen Minh Tri, *Ho Chi Minh City University of Science, Vietnam*
Xuan-Tu Tran, *Vietnam National University, Hanoi*
Tuan Minh Vo, *Da Nang University of Science and Technology, Vietnam*
Vassilis Alimisis, *National Technical University of Athens, Greece*
Mohammad Arif Sobhan Bhuiyan, *Xiamen University Malaysia, Malaysia*
Trong-Tu Bui, *Ho Chi Minh City University of Science, Vietnam*
Van Loi Cao, *Le Quy Don Technical University, Vietnam*
Ramon Carvajal, *University of Sevilla, Spain*
Nga Dao, *Le Quy Don Technical University, Vietnam*
Nguyen Ngo Doanh, *VNU Information Technology Institute, Vietnam*
Thị Thùy Dương Đinh, *Academy of Military Science and Technology, Vietnam*
Chau Hoang, *Hanoi University of Science and Technology (HUST), Vietnam*
Huu-Thuan Huynh, *Ho Chi Minh City University of Science, Vietnam*
Tung Huynh, *Danang University of Science and Technology, Vietnam*
Abumoslem Jannesari, *University of Tarbiat Modares of Tehran, Iran*
Fabian Khateb, *Brno University of Technology, Czech Republic*
Ah-Reum Kim, *Sungkyunkwan University, Korea*
Anshul Kundra, *Kurukshetra & HCL Technologies, India*
Trong-Hieu Le, *Electric Power University, Vietnam*
Trung-Khanh Le, *Ho Chi Minh City University of Science, Vietnam*
Vu Trung Duong Le, *Nara Institute of Science and Technology, Japan*
Antonio López-Martín, *Public University of Navarra, Spain*
Quang-Trung Luu, *Hanoi University of Science and Technology, Vietnam*
Duc-Tho Mai, *Academy of Cryptography Techniques, Vietnam*
Khairun Nisa' Minhad, *Xiamen University Malaysia*
Huy Cu Ngo, *Sony Semiconductor, Japan*
Duong Nguyen, *International University, Vietnam*
Hung Nguyen, *VNU University of Engineering and Technology, Vietnam*
Huu Muon Nguyen, *Synopsys Inc.*
Nhu Huan Nguyen, *CROMA Laboratory, France*
Tan Nguyen, *Ton Duc Thang University, Vietnam*
Thành Long Nguyễn, *Academy of Military Science and Technology, Vietnam*
Tran Thi Thao Nguyen, *Ho Chi Minh City University of Science, Vietnam*
Tri-Hai Nguyen, *Van Lang University, Vietnam*

Truong Khang Nguyen, *Van Lang University, Vietnam*

Van Tinh Nguyen, *Le Quy Don Technical University, Vietnam*

Quang Nguyen-The, *Le Quy Don Technical University, Vietnam*

Hesham Omran, *Ain Shams University, Egypt*

Chi Dat Pham, *Chiba University, Japan*

Hoai Luan Pham, *NARA Institute of Science and Technology, Japan*

Loan Pham-Nguyen, *Hanoi University of Science and Technology, Vietnam*

Hong Phuong Phan, *ARELIS SAS*

Tran Dang Khoa Phan, *Danang University of Science and Technology, Vietnam*

Fadi Shahroury, *Princess Sumaya University for Technology, Jordan*

Paul Sotiriadis, *Johns Hopkins University EPP & Sotecko Electronics LLC, USA*

PS Sushma, *NMAM Institute of Technology, Nitte, India*

Hoang Thi Yen, *Le Quy Don Technical University*

Tran Thinh, *Ho Chi Minh City University of Technology, Vietnam*

Intissar Toihria, *Laboratory of Electronics-Telecommunication-Computer Lyon, Tunisia*

Lan Thi Tran, *University of Transport and Communications, Vietnam*

Van-Toan Tran, *Le Quy Don Technical University, Vietnam*

Dung Trinh, *Ho Chi Minh City University of Technology, Vietnam*

Manh Cuong Truong, *Marvell Technology, Vietnam*

Vinh Truong Quang, *Ho Chi Minh City University of Technology, Vietnam*

Nguyễn Vinh, *Ho Chi Minh City University of Science, Vietnam*

Tan Phu Vuong, *Grenoble INP-MINATEC, France*

Zunsong Yang, *The University of Tokyo, Japan*

Zhao Zhang, *Chinese Academy of Sciences, China*

Xiao Zhao, *China University of Geosciences, China*

Yuanjing Zheng, *Nanyang Technological University, Singapore*

Conference Sponsors

Patrons

Vietnam National University, Hanoi

Vietnam National University, Ho Chi Minh City

Host

Ho Chi Minh City University of Science

Co-Organized by

VNU Information Technology Institute

Technical sponsors

IEEE Vietnam Section

IEEE SSCS Vietnam Section Chapter

IEEE CASS Vietnam Section Chapter

IEICE Vietnam Section

Financial Sponsors

Marvell Technology, Vietnam

Keynote #1

Advanced Biomedical Imaging Technologies: Circuit Design and Techniques

Yongfu Li
Shanghai Jiaotong University, China

Abstract— **Electrical Impedance Tomography (EIT) is transforming biomedical imaging, offering non-invasive, cost-effective solutions for critical healthcare diagnostics. This keynote explores state-of-the-art EIT technologies, emphasizing advanced circuit design and innovative imaging methods. I will present the latest developments in circuit design and imaging techniques, highlighting their impact on diagnostic accuracy and system performance. We will discuss the rising demand for wearable EIT sensors, driven by the urgent need for continuous, real-time monitoring of patient health beyond traditional clinical settings. Attendees will gain insights into practical applications of EIT, equipping them to drive innovation, enhance diagnostics, and significantly improve patient outcomes.**

Biography

Yongfu Li (S'09-M'14-SM'18) received his B.Eng. and Ph.D. in Electrical and Computing Engineering from the National University of Singapore in 2009 and 2014. Following his studies, he worked at NUS as a research engineer and then at GLOBALFOUNDRIES, progressing from senior to principal engineer and ultimately serving as a member of the technical staff. He is currently a tenured Associate Professor at Shanghai Jiao Tong University, focusing on analog and mixed-signal circuits, biomedical signal processing, and circuit automation.

Dr. Li has garnered prestigious awards throughout his career, such as the International Excellent Young Scientists Award (2023) and the IEEE MGA Young Professionals Achievement Award (2022). His work in IEEE includes roles as Board of Governors member in the IEEE Circuits and Systems Society and membership in several committees. He has also been an editor for publications like the IEEE Open Journal of Circuits and Systems and IEEE Transactions on Biomedical Circuits and Systems.

An active contributor to the academic community, Dr. Li founded IEEE YP Affinity Groups in Singapore and Shanghai, guiding numerous students to win over 40 international and national awards. He has served in various leadership roles in major IEEE conferences, further establishing his influence and commitment to advancing engineering education and research globally.

Keynote #2

Photonics Integrated Circuits: Enabling the Next Era of High-Speed, Energy-Efficient Computing

Le Quang Dam
General Director, Marvell Technology, Vietnam

Abstract— As the limits of traditional electronic integrated circuits become increasingly apparent in the face of exponential data growth and power demands, photonic integrated circuits (PICs) have emerged as a transformative technology to address the challenges of speed, bandwidth, and energy efficiency. In this keynote, we explore the evolution, current breakthroughs, and future directions of PICs in both data communication and computing applications. We delve into the integration of photonic elements with existing CMOS platforms, discuss recent advances in silicon photonics, and highlight how heterogeneous integration is enabling compact, scalable, and cost-effective PIC solutions. The talk also addresses key design challenges such as thermal management, coupling efficiency, and standardization, while emphasizing the critical role of advanced design automation and simulation tools. Looking ahead, we envision a future where photonic and electronic co-design becomes a cornerstone of next-generation IC systems, propelling innovations in AI, quantum computing, and beyond. This session aims to inspire new approaches and collaborations at the intersection of photonics and electronics, laying the groundwork for the IC industry's next leap forward.

Biography

Quang-Dam Le joined Marvell in 2011, and held different positions such as Technical Director, Senior Director, Associate Vice President, and presently, he is the General Director of Marvell Technology Vietnam. He was graduated with a Bachelor of Science (B.Sc.) degree from Ho Chi Minh City University of Science (HCMUS) in 1988, a Master (MSc.) degree in Physics in Canada in 1993 , and a Doctorate (Ph.D.) degree in Signal Processing (Artificial Intelligence) in Canada in 1996, before starting his career in the semiconductor industry.

QD began his career as an algorithm designer for all Digital Signal Processing (DSP) IPs for Miranda Technologies, thereafter Gennum Corporation as Senior Video System Architect, and went on to join ATI Technologies Inc (acquired later by Advanced Micro Devices - AMD) as Senior Manager to manage the DSP teams, responsible for Markham (Canada), Bangalore (India), Shanghai (China) and Munich (Germany). Just prior to Marvell, he was a Senior Principal Scientist with Broadcom. He has strong interest in System Architecture, Signal Processing and Artificial Intelligence..

Keynote #3

Multi-core Multi-thread RISC-V-based System-on-Chip

Cong-Kha Pham
University of Electro-Communications (UEC), Tokyo, Japan

Abstract— The relentless demand for computational power has driven the development of diverse parallel architectures, fundamentally categorized as general-purpose and special-purpose systems. General- purpose systems, characterized by programmable controllers, provide versatility across a wide range of applications. Conversely, special-purpose systems, utilizing fixed hardware controllers, prioritize efficiency for specific tasks. Multicore processors, a cornerstone of modern computing, are employed in both categories, leveraging interconnected functional modules to enable concurrent execution. General-purpose multicore architectures, such as linear arrays and trees, are adept at handling varied workloads, while specialized multiprocessors, including systolic arrays and hypercubes, are tailored to specific computational patterns. Designing these complex systems requires significant expertise to optimize performance and overcome the inherent memory bottlenecks associated with traditional architectures. The advent of the big-data era has brought about a paradigm shift towards data-centric design, placing a strong emphasis on the importance of high-quality, structured data. This shift has also blurred the traditional boundaries between general-purpose and special-purpose systems, leading to the emergence of hybrid designs that integrate features from both categories. Notable examples include NVIDIA's GH200 and GB200 architectures, which effectively combine the flexibility of programmability with the efficiency of specialized hardware accelerators. Furthermore, the importance of hardware/software co-design has become increasingly evident. Modern computing systems are highly complex, requiring a holistic approach where hardware and software are developed in tandem to achieve optimal performance and efficiency. Software frameworks developed by companies like Tenstorrent and Cerebras, alongside custom chips designed by companies like Meta and Microsoft, underscore the significance of integrated design methodologies. These efforts aim to strike a balance between the flexibility of general-purpose systems and the efficiency of application-specific designs.

Our research leverages the RISC-V open-source Instruction Set Architecture (ISA) to develop advanced multicore systems, focusing on both hardware design and multithreaded software. A high- performance core initiates thread generation for non-parallel tasks, storing thread information in a queue accessible by other cores. Upon queue writes, idle cores are activated to fetch and store thread data, mitigating shared resource bottlenecks. Bidirectional private buses are employed to further minimize data movement overhead. We have explored the integration of near-cache processing capabilities and tightly coupled accelerators, utilizing a hybrid Level 1 (L1) cache. This cache can function as both a traditional cache and local memory, enabling efficient data access for accelerators such as Matrix Processors. Techniques like divide-and-conquer and task splitting are utilized to optimize performance, with minimal hardware overhead. Software benchmarks, including matrix multiplication and convolution, have demonstrated significant performance gains. Four-core configurations achieve speed-up factors ranging from 4 to 6 times, while eight-core configurations yield speed-ups ranging from 6 to 18 times. The hybrid L1 cache, coupled with tightly coupled accelerators, achieves speed-ups ranging from 6 to over 1,000 times compared to a single core, highlighting the effectiveness of this integrated design approach..

Biography

Cong-Kha Pham (Senior Member, IEEE) received the B.S., M.S., and Ph.D. degrees in electronics engineering from Sophia University, Tokyo, Japan. He is currently a Professor with the Department of Information and Network Engineering, The University of Electro-Communications (UEC), Tokyo. His research interests include hardware system design implementation by FPGA and integrated circuits. His recent projects include research on energy harvest power supply and low-power data-centric sensor network system utilizing it, development of long-distance transmission/miniaturization equipment of sensor network by low power wireless, super low-voltage device project, research on memory-based information detection systems, hardware implementation of hardware system by FPGA, and integrated circuit. He is teaching many undergraduate and postgraduate students and has received numerous awards for dissertations. The University of Electro-Communications Integrated Circuit Design Laboratory (Pham Laboratory) educates the design, implementation, and evaluation of hardware systems and VLSI, aims to design a system-on-chip by integrating various information processing hardware, and develops a high-performance computational circuit realized with a small number of elements.

CASS Distinguished Lecture

Advanced Circuits and Systems for Navigation-Grade MEMS Accelerometers

Jian Zhao
Shanghai Jiao Tong University

Abstract— Silicon oscillating accelerometers have the potential to achieve navigation-grade precision due to their frequency-modulation nature. As a result, they present a promising alternative to quartz accelerometers in smart unmanned platforms that require satellite-free navigation. This tutorial will begin with an introduction to the fundamental principles and specification requirements of MEMS resonant sensors. It will then delve into the noise modelling and agile design methodologies essential for optimizing performance. Advancements in low-noise, high-efficiency readout circuits will also be explored. Finally, the impact of long-term drift errors caused by temperature fluctuations, process variations, and environmental changes will be discussed, along with advanced compensation techniques. By addressing key challenges and solutions, this tutorial aims to provide valuable insights for sensor and system designers working on high-precision MEMS accelerometers and navigation applications.

This talk is sponsored by IEEE Circuits and Systems Society.

Biography

Jian Zhao, (S'14, M'17, SM'21) received his Ph.D. degree from School of Mechanical Engineering, Nanjing University of Science and Technology, China, in 2017.

From 2012 to 2015, he worked as a visiting scholar in the VLSI and Signal Processing Lab, National University of Singapore, where he designed CMOS readout circuits for MEMS sensors. In 2017-2019, he joined the Department of Electronic Engineering, Tsinghua University as a post-doctoral researcher developing ICs for wireless body area networks. He is currently an Associate Professor in the Department of Micro/Nano Electronics, Shanghai Jiao Tong University, China. He has authored and co-authored over 50 technical papers and 2 book chapters. His current research interests include biomedical & bio-inspired circuits and systems.

Since 2019, he has also served as an organization committee/technical program committee/review committee for many prestigious IEEE conferences. These conferences include ISCAS, ISICAS, AICAS, IFETC, ICTA and APCCAS. He also serves as an Associate Editor for IEEE Transactions on Biomedical Circuits and Systems, IEEE Transactions on Circuits and Systems I: Regular Papers (TCAS-I) and SPJ Cyborg and Bionic Systems. He is the receipt of IEEE BioCAS 20th Anniversary Top WiCAS/YP Contributor Award. He served as IEEE CASS Distinguished Lecturers from 2025-2026. He is the Co-founder and Past-chair of IEEE Shanghai Section Young Professional Affinity Group.

An Optimized Obstructive Sleep Apnea Detection Model Using Particle Swarm Optimization and Machine Learning

1st Atiya Khan
Dept. of Computer Science & Engg.
National Institute of Technology Silchar
Silchar, India - 788010
atiya21_rs@cse.nits.ac.in

2nd Saroj K. Biswas
Dept. of Computer Science & Engg.
National Institute of Technology Silchar
Silchar, India - 788010
saroj@cse.nits.ac.in

3rd Chukhu Chunka
Dept. of Computer Science & Engg.
National Institute of Technology Silchar
Silchar, India - 788010
chukhu@cse.nits.ac.in

Abstract—Obstructive Sleep Apnea (OSA) is a severe health issue all over the world, characterized by repeated interruption in breathing during sleep. Traditional modes of diagnosis, like polysomnography, require huge costs and time for evaluation, due to which there is a need for efficient and automated diagnostic systems. This paper presents an enhanced model named Optimized Intelligent System for Obstructive Sleep Apnea (OISOSA) for efficient OSA detection using single-lead EEG data. The proposed model utilizes the discrete wavelet transform with db8 for the decomposition of the EEG signal into the sub-bands and extraction of features from each of the sub-bands. Further, the model incorporates Gaussian filter for feature smoothing, Isolation Forest algorithm for the removal of outliers, and Particle Swarm Optimization (PSO) algorithm for feature optimization. Finally, the classification of apnea and non-apnea is done using the extra tree classifier. Performance evaluation using metrics such as accuracy, precision, recall and F1-score have been used to demonstrate the effectiveness of the proposed system. OISOSA achieved the highest accuracy of 86% with 10-fold cross-validation and 85% with holdout validation, outperforming the benchmark models. This study highlights the potential of PSO-enhanced ensemble analytics for reliable and accessible OSA detection.

Index Terms—Obstructive Sleep Apnea, discrete wavelet transform, Gaussian filter, Isolation Forest, Particle Swarm Optimization.

I. INTRODUCTION

Sleep plays a crucial role tha hepls in maintaining overall health and well-being. It enables the processes of restoration and repair of a person's nervous, muscular, skeletal, and immune systems. High-quality sleep enhances cognitive function, emotional stability, and cardiovascular health, whereas irregular or improper sleep can lead to high-stress levels, impaired metabolic functions, and numerous health complications. Among various sleep-related disorders such as insomnia, parasomnia, narcolepsy, and restless leg syndrome, Obstructive Sleep Apnea (OSA) is one of the most prevalent and severe conditions. OSA is characterized by repeated episodes of partial or complete airway obstruction during sleep, leading to interrupted breathing during sleep. These interruptions can result in poor sleep quality and cause significant health issues, including cardiovascular disorders, diabetes, stroke, impaired cognitive function, and mood disturbances.

OSA has emerged as a global health challenge, affecting millions of individuals worldwide. It is estimated that around 200 million people suffer from OSA, yet up to 80% of moderate to severe cases remain undiagnosed due to a lack of awareness, misdiagnosis, and limited access to diagnostic facilities. Traditional diagnostic methods, such as polysomnography (PSG), involve overnight monitoring using multiple physiological sensors like Electroencephalography (EEG), Thoracic and abdominal belt. Electromyography (EMG), Electrooculography (EOG), Electrocardiography (ECG), and pulse oximetry to detect apnea events. Although PSG is very effective, it is time-consuming, complex, expensive, and requires specialized equipment and sleep specialists, making it inaccessible for many patients. These limitations highlight the need for efficient, cost-effective, and widely accessible diagnostic alternatives.

The researchers around the globe are trying to develop intelligent systems that utilize one or multiple physiological signals for OSA detection. Among these physiological sensors, EEG signals are extremely valuable as they provide detailed insights into neural activity, sleep stages, and micro disruptions associated with apnea events. There have been several studies utilizing the single or multiple channels of EEG sensors for OSA detection. These existing studies have used machine learning (ML) models for OSA detection. However, the computational limitations and potential overfitting associated with these shallow learning models restrict their generalizability. Ensemble Learning (EL) models have shown promise in overcoming these limitations, improving accuracy by leveraging the strengths of multiple weak learners. The existing studies are also limited by the small dataset and limited preprocessing which often results in suboptimal performance and reduced robustness of the developed models. Small datasets can lead to model overfitting, where the model learns specific patterns present in the training data but fails to generalize to the test data.

979-8-3315-1550-8/25 $31.00 © 2025 IEEE

Limited preprocessing may leave noise or outliers unaddressed, which can obscure crucial patterns and sometimes negatively impact the classification accuracy of the model.

To address these challenges, this paper presents the Optimized Intelligent System for Obstructive Sleep Apnea (OISOSA), model that utilizes the single chanel EEG signal data for accurate OSA detection. The proposed system adopts a robust pipeline, incorporating signal processing, feature extraction, optimization, and classification. Initially, the EEG signals are decomposed into sub-bands using Discrete Wavelet Transform (DWT) with db8, which effectively captures both time and frequency domain information. From each sub-band, nine statistical features are extracted to represent key characteristics of the EEG signals. For preprocessing the proposed model incorporates such as Gaussian filtering (GF) for feature smoothing and Isolation Forest (IF) for outlier detection are applied. The GF and IF together reduces the impact of high-frequency noise and erroneous data, enhancing the robustness of the model. The study also employs Particle Swarm Optimization (PSO) for feature selection. PSO is a powerful meta-heuristic optimization algorithm inspired by swarm behavior observed in nature, to address the challenges of high-dimensionalilty of medical data. It effectively selects the most informative features, improving the performance and interpretability of the model. For classification, the Extra Tree (ET) classifier is employed, due to its ability to handle complex relationships and high-dimensional data with minimal overfitting. The OISOSA model is extensively evaluated using metrics such as accuracy (A), precision (P), recall (R), and F1-score (F1) on 10-fold cross validation (CV), 5-fold cross validation (CV) and holdout method. The proposed study highlights the effectiveness of EEG with advanced feature selection and ensemble learning techniques to create a robust and efficient solution for OSA detection.

Overall notabilities and Contributions of the paper are as follows - :

1) Development of a PSO-based optimized OSA detection system leveraging EEG signals and an ensemble classifier for improved accuracy.
2) Comprehensive preprocessing pipeline addressing noise and outliers to ensure data integrity.
3) Detailed evaluation of the model's performance against benchmark systems using key metrics to validate its reliability and applicability in real-world scenarios.

The paper is structured as follows: In Section II, an overview of the state-of-the-art OSA detection models, including their performance metrics, has been discussed. Section III details the proposed methodology, while Section IV presents comprehensive experimental results and comparisons. Finally, Section V concludes the paper and outlines future directions.

II. LITERATURE SURVEY

Taran et al. [1] in year 2020 proposed an adaptive Hermite decomposition method for detecting sleep apnea events using EEG signals. They used an artificial bee colony algorithm to select optimal Hermite functions for EEG signals. Features were extracted from the Hermite coefficients, and feature selection was done using Fisher score ranking. For classification, they used extreme learning machines and least squares support vector machine classifiers. Kumari et al. [2] in their paper used EEG signal data to detect OSA. They used DWT with db-2 for band separation. They extracted features such as energy, mean, median, and standard deviation from the decomposed signals. Then, they used an SVM classifier to classify apnea events and tested the classifier on different train-test splits and achieved the highest accuracy on a 90-10 split.

Saha et al. [3] presented an approach for automatic sleep apnea detection using features from multi-band EEG signals. They divided the EEG signal into five frequency bands and computed the entropy for each band. For classification, they used the K-nearest neighbor and compared it with the existing methods. The proposed approach surpassed the existing methods with high classification accuracy. Gholami et al. [4] proposed a CAD system to diagnose sleep apnea using complexity features extracted from EEG signals. First they decomposed the signal into six bands and extracted a total of 120 features from each sub-bands. The minimum redundancy maximum relevance (mRMR) algorithm was used to sort the features, and SVM and KNN classifiers were used to classify apnea and non-apnea events. The result indicated that the SVM classifier achieved the highest accuracy.

Almuhammadi et al. [5] presented an efficient method for classifying OSA using EEG signals. They used Infinite Impulse Response (IIR) to separate the EEG signal data into the delta, theta, alpha, beta, and gamma sub-bands. After that, they extracted the features like energy and variance from each sub-band. For classification, they compared the performance of ML algorithms such as SVM, ANN, Linear Discriminant Analysis (LDA) and Naive Bayes (NB). Among the various classifiers, SVM achieved the highest performance accuracy. They also discovered that the (90-10) train-test split achieved the highest accuracy. Jayaraj et al. [6] focussed on classifying sleep apnea using EEG signals. They used 159 subjects from ISRUC, Sleep-EDF, and CAP Sleep databases. They employed wavelet packet decomposition to decompose the EEG signal into five frequency bands. They calculate the features like entropy and energy from the frequency bands. They also included heart rate, brain perfusion, neural activity, and synchronization. Finally, they used SVM and RF to classify apnea and normal subjects. Among the SVM and RF, the SVM showed superior performance, achieving 90% of accuracy.

Wang et al. [7] presented an efficient method to detect sleep apnea-hypopnea events using EEG signal data. They

979-8-3315-1550-8/25 $31.00 © 2025 IEEE

used Tianjin Chest Hospital EEG data and decomposed it using an IIR Butterworth bandpass filter. Then, they employed DWT with db3 to extract features from different sub-bands of the EEG signal. They also utilized various machine learning algorithms, including SVM, KNN and RF. The results demonstrated the high accuracy of the proposed model, indicating the method's effectiveness in diagnosing sleep apnea-hypopnea syndrome. Vimala et al. [8] proposed an intelligent classification system using an EEG signal. The authors employed IIR Butterworth Band Pass Filter and Hilbert Huang Transform to preprocess the EEG signals. After filtering, the signals were decomposed into five frequency bands. The study utilized features such as energy, entropy, and variance from these bands for classification using machine learning algorithms like Support Vector Machine (SVM), K-Nearest Neighbors (KNN), and Artificial Neural Network (ANN). The results indicated that the SVM classifier yielded promising performance metrics, including accuracy, sensitivity, and specificity, demonstrating the potential of the proposed system in effectively classifying sleep apnea.

III. PROPOSED METHODOLOGY

The OISOSA model proposed in this paper, leveraging single-lead EEG signal data, consists of five primary stages crucial for OSA detection: III-A Dataset Preparation, B. Feature SmoothingIII-B, C. Outlier Removal III-C, D. Feature Selection III-D, and E. Classification III-E. The architecture of the proposed PSO-based model is depicted in Figure 1.

A. Data Preparation

The proposed model utilizes the Sleep Heart Health Study (SHHS 1) dataset [9], [10], available through the National Sleep Research Resource (NSRR). The dataset includes polysomnographic data from 6,441 subjects aged 40 years or older with no prior history of any treatment. In this paper, a single EEG electrode (C4-A1) is selected to simplify data acquisition while ensuring sufficient signal quality for OSA detection. The dataset is further refined, other instances are removed, and only apnea and non-apnea instances are left, resulting in 4,881 subjects (2,434 apnea and 2,447 non-apnea instances). The EEG signal data is then decomposed using Discrete Wavelet Transform (DWT) with the Daubechies 8 (db8) wavelet into five frequency sub-bands: gamma (γ) (40-100 Hz), beta (β) (12-40 Hz), alpha (α) (8-12 Hz), theta (θ) (4-8 Hz), and delta (δ) (0.5-4 Hz). From each sub-band, nine statistical features are extracted: Minimum, Maximum, Mean, Variance, Standard Deviation, Kurtosis, Energy, Power, and Root Mean Square (RMS). These features represent both the time-domain and frequency-domain characteristics of the EEG signals. The final feature set consists of 45 statistical features, nine features from each of the five sub-bands, ensuring a rich representation of the data for subsequent analysis.

B. Feature Smoothing

The raw EEG signals often contain high-frequency noise from several sources, such as muscle artifacts, environmental noise, and equipment malfunction. This noise can obscure the essential characteristics of the EEG signal that is important for OSA detection. To address these noises, the Gaussian filter has been incorporated into the proposed pipeline. The Gaussian filter is a noise removal technique that effectively attenuates high-frequency components while preserving the low-frequency element for analysis. The Gaussian filter is mathematically represented as: $G(x) = \frac{1}{\sqrt{2\pi}\sigma} e^{-\frac{x^2}{2\sigma^2}}$, Where $G(x)$ represents the Gaussian function, σ is the standard deviation of the Gaussian distribution which controls the level of smoothing, and x denotes the distance from the mean. The raw EEG signal $S(t)$ is smoothed by convolving it with the Gaussian filter $G(x)$, producing the smoothed signal $Y(t)$, which is computed as:

$$Y(t) = (S * G)(t) = \int_{-\infty}^{+\infty} S(\tau)G(t-\tau)\,d\tau,$$

The choice of σ is important to ensure a correct balance between effective noise suppression and the retention of essential signal features. The resulatant EEG signal after applying gaussian filter are free from high-frequency noise, and are well-suited for the subsequent steps of outlier removal, feature selection, and classification.

C. Outlier Romoval

In addition to the noise, outliers present in EEG data can also significantly affect the performance of ML models. These outliers may be caused by factors such as electrode movement, Sudden power fluctuations, or movements due to coughing, sneezing, or shifting position. Identification and removal of these outliers are essential for ensuring a reliable and robust pipeline for OSA detection.

To address the issue of outliers, the IF algorithm has been incorporated into the proposed model. The IF is an ensemble-based learning technique that works by constructing a forest of random binary trees, known as isolation trees, where each tree isolates a sample by randomly selecting a feature and splitting it at a random value. Anomalies are isolated more quickly due to their distinct and sparse characteristics, leading to shorter average path lengths in the tree structure.

The anomaly score $s(x, n)$ of a data point x is computed as: $s(x, n) = 2^{-\frac{E(h(x))}{c(n)}}$,

where $E(h(x))$ is the average path length of x across all isolation trees, n is the number of instances in the dataset, and $c(n)$ is the expected path length of a binary tree with n samples. Instances with $s(x, n)$ values close to 1 are classified as outliers, while instances with lower scores are considered normal. The IF enhances the overall reliability and generalizability of the proposed model by removing irrelevant samples. The cleaned dataset is then passed to the next stage for feature selection.

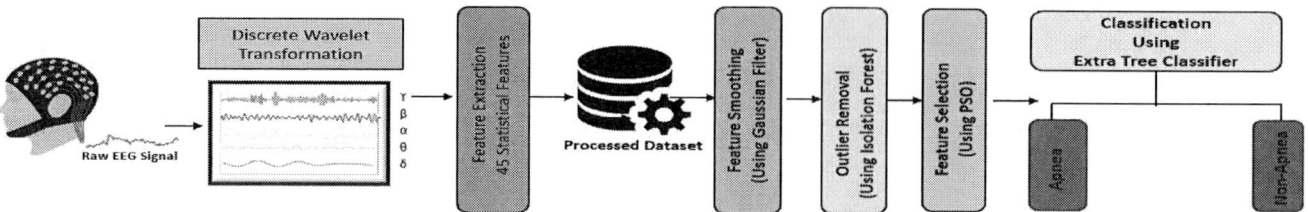

Fig. 1. Architecture of the proposed OISOSA model.

D. Feature Selection

The next step in the proposed pipeline is feature selection, which is performed using Particle Swarm Optimization (PSO). The PSO is an evolutionary algorithm inspired by the social behaviour of birds and fish. PSO is incorporated in the proposed model due to its ability to efficiently explore high-dimensional feature spaces, avoid local optima, and identify optimal or near-optimal subsets of features with reduced computational complexity. Its iterative nature and balance between exploration and exploitation make it particularly suitable for EEG data, where feature sets are often large and diverse. In this algorithm, each particle represents a subset of features, and its quality is evaluated based on classification accuracy. The particles iteratively adjust their positions and velocities in the feature space according to the equations:

$$v_i^{t+1} = \omega v_i^t + c_1 r_1 (p_i^t - x_i^t) + c_2 r_2 (g^t - x_i^t), \quad x_i^{t+1} = x_i^t + v_i^{t+1},$$

where v_i^t and x_i^t are the velocity and position of particle i at iteration t, ω is the inertia weight, c_1 and c_2 are acceleration coefficients, and r_1, r_2 are random values in $[0,1]$. The algorithm iteratively updates the personal best position (p_i^t) and global best position (g^t) until a stopping criterion is reached, which can be the number of iterations or convergence. This step solves the problem of high dimensionality of the dataset by selecting the most relevant features, thereby enhancing the classification accuracy and efficiency of the model.

E. Classification

After selecting the optimized feature subset using PSO, the classification is performed using the ET classifier. ET is an EL algorithm that can efficiently handle high-dimensional EEG signal data. The ET classifier combines the strength of multiple weak learners by constructing several decision trees, where each tree provides an independent prediction, and the final prediction is determined by aggregating these predictions. Each of these decision trees is trained on randomly selected subsets of features and data samples. The extra tree classifier also incorporates additional randomness by selecting a random split value rather than using criteria such as information gain or Gini impurity. This approach introduces the diversity in the decision trees that leads to improved generalization and reduces the risk of overfitting. This property is beneficial for analyzing complex and nonlinear EEG data. Integrating the ET classifier into the proposed pipeline complements the previous steps of noise reduction, outlier removal, and feature selection. Together, these steps create an effective framework for the accurate detection of OSA from EEG signals.

IV. RESULT AND DISCUSSION

The findings of this study provide important insights into OSA detection and have significant implications for developing intelligent systems for diagnosing sleep apnea. The result and discussion of this paper are divided into four sections. Section IV-A compares the performance of the proposed model against machine and ensemble learning models. Section IV-B compares the performance of PSO with other Feature selection techniques. Section IV-C discusses the ablation study of the proposed model. Section IV-D compares the proposed model with the state-of-the-art models present in the literature.

A. Comparative Assessment of OISOSA Model with Machine and Ensemble Learning Models

To comprehensively analyze the proposed OISOSA model, this section compares its performance with machine and ensemble learning models. The evaluation involves three validation schemes, including holdout, 10-fold, and 5-fold cross-validation (CV), using key metrics such as A, P, R, and F1. All the ensemble and machine learning models were implemented without any preprocessing. Among the considered machine learning models like SVM, KNN, and DT, SVM achieved the highest accuracy of 55%. However, the ensemble learning algorithms achieved better accuracy than shallow learning algorithms mainly because the ensemble models consist of multiple shallow learning algorithms, allowing them to capture a wider range of patterns and nonlinearities in the EEG data, causing improved generalization and higher predictive accuracy. The result of the analysis is shown in Table I.

From Table I, it can be seen that the Extra Trees algorithm achieved the highest accuracy of 57% among all the ensemble learning algorithms. The achieved performance has led to the selection of Extra Trees as the classifier in the proposed model, which, along with other elements of the pipeline, achieved an accuracy of 86% across all evaluation metrics on 10-fold CV.

979-8-3315-1550-8/25 $31.00 © 2025 IEEE

TABLE I
DETAILED PERFORMANCE COMPARISON OF OISOSA WITH DIFFERENT CLASSIFIERS.

Evaluation Metric → Classifiers ↓	10-fold Cross-Validation				5-fold Cross-Validation				Holdout Method			
	P(%)	R(%)	F1(%)	A(%)	P(%)	R(%)	F1(%)	A(%)	P(%)	R(%)	F1(%)	A(%)
SVM	55	56	55	55	56	56	55	55	55	55	55	55
KNN	54	54	54	53	53	52	52	52	52	52	52	52
DT	54	54	54	54	53	52	53	53	51	51	51	51
RF	56	56	56	56	55	55	55	55	56	56	56	56
AdaBoost	56	56	56	56	55	55	55	55	55	55	55	56
GB	53	53	53	53	54	54	54	54	55	55	55	55
Extra Tree	57	57	57	57	56	56	56	56	56	56	56	56
OISOSA	**86**	**86**	**86**	**86**	**85**	**85**	**85**	**85**	**86**	**86**	**86**	**86**

TABLE II
PERFORMANCE COMPARISON OF PSO WITH STATE-OF-THE-ART FEATURE SELECTION METHOD

Model		10-Fold CV				5-Fold CV				Holdout			
		A(%)	P(%)	R(%)	F1(%)	A(%)	P(%)	R(%)	F1(%)	A(%)	P(%)	R(%)	F1(%)
Non-Evolutionary Feature Selection	SFS	73	73	73	73	72	72	72	72	73	73	73	73
	SBS	83	83	83	83	81	81	81	81	82	82	82	82
	Boruta	83	84	83	83	82	82	82	82	83	83	83	83
Evolutionary Feature Selection	GA	82	82	82	82	80	80	80	80	80	80	80	80
	ACO	85	85	85	85	82	82	82	82	83	83	83	83
	PSO	**86**	**86**	**86**	**86**	**83**	**83**	**83**	**83**	**85**	**85**	**85**	**85**

B. Performance Comparison of PSO with Advanced Feature Selection Algorithms

The following section discusses the performance comparison of the proposed model with state-of-the-art feature selection algorithms. To effectively establish the efficiency of the PSO algorithm, it is compared with other evolutionary and non-evolutionary feature selection algorithms using the evaluation matrices such as P, A, R and F1. The benchmark feature selection methods incorporated for comparison are Sequential Backward Selection (SBS), Sequential Forward Selection (SFS), Boruta, Ant Colony Optimization (ACO) and Genetic Algorithm (GA). To provide a thorough assessment, the experimentation has been performed using the 10-fold CV, 5-fold CV and holdout methods. The results of this experimentation are shown in Table II. All the elements of the pipeline have been incorporated in all the cases.

The results demonstrate that PSO achieved the highest accuracy of 86% on 10-fold CV, outperforming Boruta, SBS, GA, ACO and SFS, which achieved accuracies of 83%, 83%, 82%, 85% and 73%, respectively. These findings underscore the effectiveness of PSO in optimizing the feature set for accurate and reliable OSA detection while also providing insights into the strengths and limitations of different feature selection techniques.

C. Detailed Ablation Study of the Proposed Model

This section discusses the efficacy of the proposed OISOSA model. By integrating advanced signal processing techniques, efficient noise and outlier removal methods, an optimal feature selection algorithm, and an Extra Trees classifier, the proposed model achieves optimal performance across all the metrics. Each of these components plays a significant role in enhancing the performance of the proposed model. To assess the effectiveness of each element, they have been removed from the pipeline one by one, and

the performance has been recorded as shown in Table III. Initially, omitting IF and PSO led to a drop in accuracy from 86% to 83%, then omitting GF and PSO resulted in a decrease to 61%. The ROC-AUC curve analysis shows OISOSA's superior performance, with a true positive rate and an average AUC of 91%, as illustrated in Fig. 2(a) compared to the baseline model without IF, GF and PSO as illustrated in Fig. 2(b).

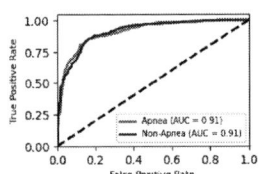 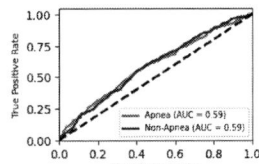

(a) ROC-AUC curve of OISOSA. (b) ROC-AUC curve of OISOSA without IF, GF and PSO.

Fig. 2. ROC-AUC curve of the proposed OISOSA with and without Preprocessing.

TABLE III
EFFICACY OF THE PROPOSED MODEL USING ABLATION STUDY

Model's Ablation	10-Fold CV			
	P(%)	R(%)	Acc.(%)	F1(%)
OISOSAEA without GF, IF and PSO	58	58	58	58
OISOSAEA without IF and GF	59	59	59	59
OISOSAEA without GF and PSO	61	61	61	61
OISOSAEA without IF and PSO	83	83	83	83
OISOSAEA	**86**	**86**	**86**	**86**

Again, omitting IF and GF causes the accuracy to decline to 59%, and excluding all three components reduces the accuracy to 58%. This analysis highlights the crucial role of each element in optimizing the model's performance for accurate OSA detection.

D. Benchmark comparison of the Proposed Model Against State-of-the-Art Models

This section provides a comparative analysis of the proposed OISOSA model with benchmark models from the literature. The comparison has been performed using the SHHS dataset, following the same methodology as described in their respective papers. Key performance metrics such as P, R, F1, and A have been evaluated using the 80-20 holdout and 10-fold CV methods. The results of the holdout method are shown in Table IV, and a graphical representation of a 10-fold CV is provided in Figure 3.

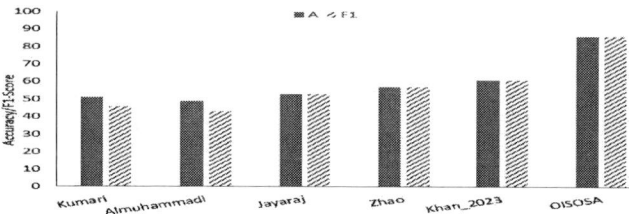

Fig. 3. Performance comparison of the proposed OISOSA model with benchmark models using 10-fold CV.

The results demonstrate that the proposed OISOSA model significantly outperforms the benchmark models in all evaluation metrics. Specifically, [5] achieved an accuracy of 49%, [2] achieved 51%, [6] achieved 53%, [11] achieved 57%, and [12] achieved 61%. The suboptimal performance of the benchmark models is attributed to smaller datasets, limited preprocessing, and the use of shallow learning classifiers. In contrast, the OISOSA model achieved a high accuracy of 86% with 10-fold CV and 85% with the 80-20 holdout method, highlighting its capability in accurately detecting obstructive sleep apnea. This highlights the potential of the OISOSA model for clinical applications and research in sleep medicine.

TABLE IV
PERFORMANCE COMPARISON OF OISOSA WITH BENCHMARK
MODELS USING 80-20 HOLDOUT METHOD

Paper	P	R	A	F1
Kumari *et al.* [2]	53	52	51	46
Almuhammadi *et al.* [5]	51	50	49	43
Jayaraj *et al.* [6]	54	54	53	53
Zhao *et al.* [11]	57	57	57	57
Khan *et al.* [12]	61	61	61	61
OISOSA	**85**	**85**	**85**	**85**

V. CONCLUSION

This study presents an optimized intelligent system for obstructive sleep apnea model for OSA detection, integrating advanced signal processing techniques with state-of-the-art feature selection algorithms and machine learning classifiers. The proposed model has utilized EEG signal data from the SHHS dataset. By using PSO for feature selection, combined with robust preprocessing techniques and the Extra Trees

classifier, the proposed model has shown an excellent performance in terms of A, P, R, and F1 across various evaluation strategies, including 10-fold CV, 5-fold CV, and holdout. The experimental results have shown that the proposed OISOSA model has achieved the highest accuracy of 86% on 10-fold CV, highlighting its reliability for applications in healthcare. The detailed comparative analysis reveals that the proposed model outperforms other evolutionary and non-evolutionary feature selection techniques. Also, the ablation study highlights the importance of each component, like the GF, IF, and PSO, in enhancing the overall performance of the model. The integration of these components has improved the robustness and generalizability of the proposed model.

Future work will focus on exploring hybrid optimization approaches for feature selection and incorporating deep learning techniques to enhance the OSA detection model further.

REFERENCES

[1] S. Taran and V. Bajaj, "Sleep apnea detection using artificial bee colony optimize hermite basis functions for eeg signals," *IEEE Transactions on Instrumentation and Measurement*, vol. 69, no. 2, pp. 608–616, 2020.

[2] C. Usha Kumari, P. Kora, K. Meenakshi, K. Swaraja, T. Padma, A. K. Panigrahy, and N. Arun Vignesh, "Feature extraction and detection of obstructive sleep apnea from raw eeg signal," in *International Conference on Innovative Computing and Communications*, A. Khanna, D. Gupta, S. Bhattacharyya, V. Snasel, J. Platos, and A. E. Hassanien, Eds. Singapore: Springer Singapore, 2020, pp. 425–433.

[3] S. Saha, A. Bhattacharjee, M. A. A. Ansary, and S. Fattah, "An approach for automatic sleep apnea detection based on entropy of multi-band eeg signal," in *2016 IEEE Region 10 Conference (TENCON)*, 2016, pp. 420–423.

[4] B. Gholami, M. H. Behboudi, A. Khadem, A. Shoeibi, and J. Gorriz, *Sleep Apnea Diagnosis Using Complexity Features of EEG Signals*, 05 2022, pp. 74–83.

[5] W. S. Almuhammadi, K. A. I. Aboalayon, and M. Faezipour, "Efficient obstructive sleep apnea classification based on eeg signals," in *2015 Long Island Systems, Applications and Technology*, 2015, pp. 1–6.

[6] R. Jayaraj and J. Mohan, "Classification of sleep apnea based on sub-band decomposition of eeg signals," *Diagnostics*, vol. 11, p. 1571, 08 2021.

[7] Y. Wang, S. Ji, T. Yang, X. Wang, H. Wang, and X. Zhao, "An efficient method to detect sleep hypopnea- apnea events based on eeg signals," *IEEE Access*, vol. 9, pp. 641–650, 2021.

[8] V. Vimala, K. Ramar, and M. Ettappan, "An intelligent sleep apnea classification system based on eeg signals," *Journal of Medical Systems*, vol. 43, 01 2019.

[9] G.-Q. Zhang, L. Cui, R. Mueller, S. Tao, M. Kim, M. Rueschman, S. Mariani, D. Mobley, and S. Redline, "The National Sleep Research Resource: towards a sleep data commons," *Journal of the American Medical Informatics Association*, vol. 25, no. 10, pp. 1351–1358, 05 2018.

[10] S. Quan, B. Howard, C. Iber, J. Kiley, F. Nieto, G. O'Connor, D. Rapoport, S. Redline, J. Robbins, J. Samet, and P. Wahl, "The sleep heart health study: design, rationale, and methods," *Sleep*, vol. 20, no. 12, pp. 1077–1085, 1997.

[11] X. Zhao, X. Wang, T. Yang, S. Ji, H. Wang, J. Wang, Y. Wang, and Q. Wu, "Classification of sleep apnea based on eeg sub-band signal characteristics," *Scientific Reports*, vol. 11, no. 1, p. 5824, Mar 2021.

[12] A. Khan, S. K. Biswas, and C. Chunka, "Esad: Expert system for apnea detection using enhanced dwt feature extraction and machine learning algorithms," in *2023 14th International Conference on Computing Communication and Networking Technologies (ICCCNT)*, 2023, pp. 1–6.

An Optimized Hybrid Quantum-Classical Neural Network Model for Handwritten Digit Classification

Vu-Quoc-Minh Nguyen, Trung-Khanh Le *Graduated Member, IEEE,*
Trong-Tu Bui *Member, IEEE,* and Duc-Hung Le[†], *Member, IEEE*
Faculty of Electronics and Telecommunications, the University of Science,
Vietnam National University, Ho Chi Minh City, Vietnam
[†]Email: ldhung@hcmus.edu.vn

Abstract—**Quantum computers are the latest dream of humans in solving nondeterministic polynomial problems. These problems include machine learning and object classification. However, commercial quantum computers are still not available. In the meantime, scientists developed hybrid models by combining Convolutional Neural Networks (CNN) and Quantum Neural Networks (QNN) to apply quantum computing to artificial intelligence (AI) applications. The hybrid model accelerates the processing of an AI machine by combining the advantage of CNN in feature extraction with the parallel computing capability of QNN. In this study, we optimized the existing hybrid models with a fewer number of qubits to improve the performance of QNN on digit recognition. The study also presents the implementation of the model using Qiskit, describes the training process, and analyzes the impact of the number of qubits and quantum gates on the classification results. Experiments show that Hybrid Quantum-Classical Neural Networks (H-QNN) have the potential to outperform pure CNN. Finally, the study suggests future improvements, including the optimization of quantum algorithms and extending experiments to more complex data sets.**

Index Terms—**Quantum, convolution, neural network, hybrid, classical.**

I. INTRODUCTION

Quantum computing is gradually becoming a breakthrough in human computation. It operates on two most important principles, namely Entanglement and Quantum Superposition, which has great potential to perform complex calculations quickly, optimizing performance and time with many difficult and tricky problems. Quantum Machine Learning (QML) is one important research direction in the field of quantum computing, incorporating machine learning principles with the stable computing power of electronic systems, opening up opportunities to solve extremely complex problems but solving them using classical models is disadvantageous and cumbersome [1].

With the explosive development of artificial intelligence, one of its most famous applications is image-based object classification. Previously, feature extraction methods were all manually implemented for image classification models, but since the advent and rapid development of deep learning, especially the CNN model, it has helped automate the feature extraction process and significantly improved the accuracy [2]. Despite such significant improvements, as the volume of data increases, even powerful CNN models are limited in terms of computational resources and processing time. This leads to the research and application of QML targets to achieve optimal performance, taking advantage of the multidimensional data processing capabilities of operational algorithms to increase the speed and accuracy of image recognition. Although QML holds great promise, current quantum hardware is still limited, making practical implementation difficult. To solve this problem, hybrid models of classical deep learning and Quantum Neural Networks (QNN) have emerged, typically the H-QNN [3]. This model combines the power of CNN in feature extraction with the parallel data processing ability of QNN, which helps to improve the efficiency of image classification systems.

In this paper, we focus on improving the accuracy performance and also extending the classification capability of the existing H-QNN model [4] to classify MNIST images from '0' to '9' in which the classical CNN model plays a bridge role because current quantum computers are not powerful enough to fully utilize QNNs to replace current classical models. Combining classical and quantum computing models not only saves computational resources and improves accuracy, but also opens up the potential for applications in areas such as medicine, materials development, and real-time image recognition.

979-8-3315-1550-8/25 $31.00 © 2025 IEEE

As quantum hardware advances, hybrid models like H-QNN promise to become an important part of the future of computer vision and machine learning.

II. QUANTUM CONVOLUTIONAL NEURAL NETWORKS AND HYBRID QUANTUM-CLASSICAL NEURAL NETWORKS

A. Quantum Convolutional Neural Network

The current Quantum Convolutional Neural Networks (QCNN) model still has the typical structural characteristics of CNN in general. Before being processed by quantum algorithms, data must be encoded with qubits, and the larger the data, the more qubits are required. To solve this problem, the CNN structural model is applied. And it is basically structured as follows [5]:

1) The convolution circuit works in the same way as CNN, except that the convolution circuit of QCNN must encode the data first and then use quantum logic gates to find the hidden state of the data.

2) The pooling circuit of QCNN reduces the size of the system by processing the data and then measuring a portion of the qubits using 2-qubit logic gates.

3) Repeat the convolution and pooling process using circuits (1) and (2) until the system size is small enough, the fully connected circuit are used to predict the classification result.

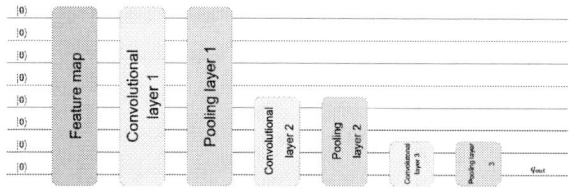

Fig. 1. A specific QCNN model contains 8 qubits. Where Convolutional layer processes data to get features, Pooling layer is used to reduce the data size by half. This process is repeated until there is only 1 output qubit with the σ_z operator to get the eigenvalue(can be -1 or 1), this result is used for prediction in classification applications [6].

The Fig.1 above is a concrete example of a QCNN model. It is implemented on quantum computers at very fast speeds. For example, Shor's algorithm [7], Grover's algorithm [8] or the Harrow–Hassidim–Lloyd algorithm (HHL) [9]. Although they process and search in much shorter time than traditional methods, they have shortcomings in accuracy and stability. Because the way quantum computing works is based on the entanglement and superposition of quantum states and the results are

based on the probability of those states, which means that quantum algorithms must measure the results many times to determine the correct results. In addition, to encode data into qubits, we must physically isolate these qubits and to achieve this isolated state, there must be very strict conditions. Otherwise, it is very susceptible to external influences that lead to errors. As the number of qubits increases, this problem also become many times larger, causing quantum algorithms to collapse [10] .

B. Hybrid Quantum–Classical Neural Network

H-QNN are gaining attention due to their potential to outperform classical methods in a variety of complex domains. H-QNN leverage the powerful and fast computational power of quantum algorithms while maintaining the high stability and reliability of classical architectures, resulting in improved performance in areas [11]. To visualize the structure of the H-QNN model, see the Fig.2 below:

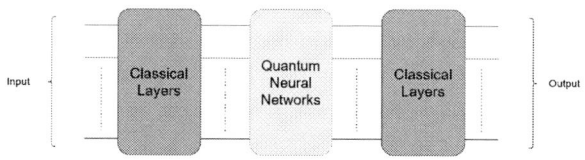

Fig. 2. A case for a general model of one of the many methods for constructing H-QNN [12]. Here we build a test model based on such a general structure for image classification.

Although they possess the superior characteristics of both quantum algorithms and classical models, they still cannot taking full advantage of those two parts, such as not being able to calculate (simulate) as quickly as QCNN, the stability and accuracy are still not as good as CNN [13].

These models can be deployed on classical computer systems. And each part of H-QNN structure has many different ways of construction, leading to many different research directions and applications to manipulate and solve problems that challenge classical computers.

III. MNIST CLASSIFICATION USING HYBRID QUANTUM-CLASSICAL NEURAL NETWORK MODEL

A. Quantum Neural Network Construction

Regarding the structure of the Hybrid Quantum-Classical Neural Networks, there are two directions: Feature map, Ansatz A and an Observable O as in the Fig.3. On the Feature map, Information is encoded into the quantum states of qubits:

979-8-3315-1550-8/25 $31.00 © 2025 IEEE

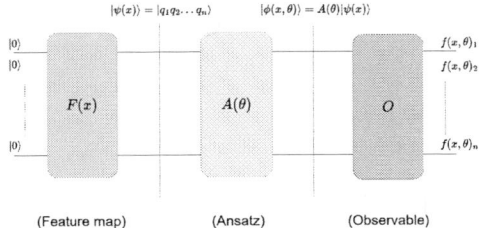

Fig. 3. The structure of QNN consists of three parts: Feature map, an Ansatz and an Observable.

$$|\psi(x)\rangle = F(x)|0\rangle^{\otimes n} \tag{1}$$

In which, x: Classical input data (can be a vector or a decimal value,...); $|0\rangle^{\otimes n}$: Initial state of n qubits, with each qubit being in state $|0\rangle$; $F(x)$: Feature map — a quantum circuit that encodes data x into quantum state; $|\psi(x)\rangle$: Quantum state after encoding data x. There are three popular ways to encode data: amplitude encoding, angle encoding and basis encoding. In this article, we have used Qiskit's ZZFeatureMap (a variant of angle encoding) in the Feature map because they require fewer qubits, obtain nonlinear relationships of data and are also less error-prone, which is more suitable for simulations of current quantum computing systems [14].

The Ansatz A circuit [15] is used to train the model with parameters and this is done by rotation gates and quantum entangled states. An Ansatz circuit optimizes and refines the model with parameters θ. As more qubits are needed, more such parameters are needed and their values can be obtained by running model training loops:

$$|\phi(x,\theta)\rangle = A(\theta)|\psi(x)\rangle \tag{2}$$

With: $A(\theta)$: Ansatz quantum circuits consist of parameterized gates, used to transform $|\psi(x)\rangle$; θ: Training parameters adjusted during model optimization; $|\phi(x,\theta)\rangle$: The output quantum state after applying $A(\theta)$, carries the information needed to measure and generate the model output.

The Real Amplitudes Ansatz used in the experiments of this paper consists of R_Y rotation gates applied to each qubit individually with from parameters θ_1 to θ_d and two qubit C_X entanglement gates applied to each pair of qubits and at the output there are also R_Y rotation gates applied to each qubit individually with parameters θ_{d+1} to θ_{2n}. Note that the number of parameters used is twice the number of qubits. Moreover, we come to an Observation O. It is used to measure the eigenvalues of the quantum states of qubits. That is, we obtain the expected results of the operators $f(x,\theta)$ acting on the qubits, which are characterized by quantum logic gates, and want to obtain that expected result by a variational quantum circuit [16] many times to get the highest accuracy:

$$f(x,\theta) = \langle \phi(x,\theta)|O|\phi(x,\theta)\rangle \tag{3}$$

Finally, with the $f(x,\theta)$ value just obtained, we use them to calculate the Hybrid Gradient Backpropagation. Estimator() of Qiskit (can be simulated on a classical computer or using a quantum computer) do this job. In our experiment, we use observable σ_z which is a spin measurement along the Oz coordinate axis in three-dimensional space, and use the Bloch sphere corresponding to each qubit to measure Observation O Meaning:

$$O = \sigma_z^{\otimes n} \tag{4}$$

B. CNN model used for MNIST classification

For the Classical Neural Network, as part of the Hybrid model, we use the CNN structure to calculate the input. In CNN, Convolutional layers are often used with image data to extract features from the input data. This is a very important process of the model because we deploy the feature extractions of these Convolutional layers to classify the image. We use two 5×5 Convolutional layers with a stride of 1x1, for both cases $n = 2$ and $n = 3$ qubits.

This is followed by a 2D Max pooling layer that reduces the data size by taking the maximum value in its 2x2 filter. Use the Relu activation function for each of these Convolutional layers:

$$f(x) = \text{ReLU}(x) = \max(0, x) \tag{5}$$

The ReLU activation function is present in the first two Convolutional layers and the Fully connected layer. This function simply removes negative values to help the model learn better, and is used to activate nonlinear functions for them. Next, the Fully connected layer helps reduce the amount of information and retain important features:

$$Y = XW^T + b \tag{6}$$

Where X is the input data matrix, W is the weight matrix, b is the bias vector that helps the model learn better by shifting the output, and Y is the output data matrix. As for the loss function and optimizer, this

979-8-3315-1550-8/25 $31.00 © 2025 IEEE

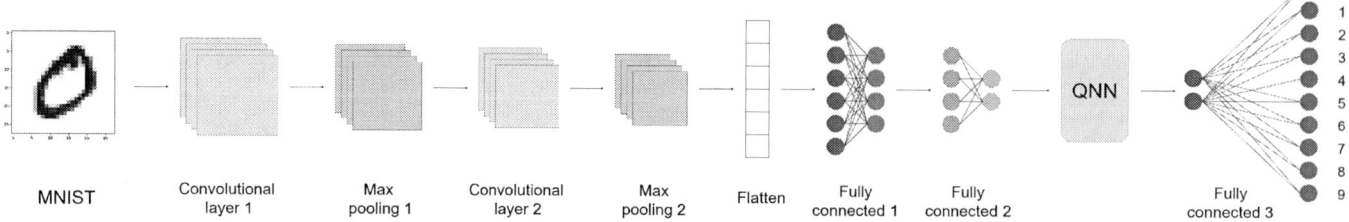

Fig. 4. The H-QNN model used in this experiment consists of CNN and QNN, CNN consists of 2 Convolutional layers and 2 Max pooling layers alternating with each other, these data points are flattened and then connected to 2 Fully connected layers in turn. Next is QNN, the output of QNN is fed into the 3rd Fully connected layer to give the final result which is the probability of the digits.

model uses Negative Log-Likelihood Loss (NLLLoss) and Adaptive Moment Estimation (Adam Optimizer). Negative Log-Likelihood Loss (NLLLoss) has the advantage of being suitable for log-probability inputs, balancing the weights in the model, and being computationally efficient:

$$\text{NLLLoss} = -\sum_{i=1}^{N} y_i \log(\hat{y}_i) \qquad (7)$$

The quantities in this equation include N: Total number of samples in the dataset; y_i: The actual label of the i-th sample; $\hat{y}_i$: Prediction probability of the model for the correct label of the i-th sample.

The reason we chose ADAM optimizer is because of its ease of implementation, low memory requirements, and self-tuning learning rate for faster convergence [17]:

$$\begin{cases} m_t = \beta_1 m_{t-1} + (1 - \beta_1)g_t \\ v_t = \beta_2 v_{t-1} + (1 - \beta_2)g_t^2 \\ \hat{m}_t = \dfrac{m_t}{1 - \beta_1^t}, \quad \hat{v}_t = \dfrac{v_t}{1 - \beta_2^t} \\ \theta_t = \theta_{t-1} - \dfrac{\alpha \hat{m}_t}{\sqrt{\hat{v}_t} + \epsilon} \end{cases} \qquad (8)$$

Where: g_t is the gradient of the loss function at step t; m_t and v_t are the moving average (momentum) estimates of the gradient and the squared gradient (velocity), respectively; β_1 and β_2 are parameters characterizing the decay rate for the gradients and the decay rate for the squared gradients; α is the learning rate.In this experiment α=0.001; ϵ is an arbitrary small value to avoid division by zero; θ_t is the model parameter at step t.

C. Combining QNN and CNN to form H-QNN model for MNIST classification

After introducing the structures and properties of QNN and CNN, we combined them into a Hybrid

Quantum-Classical Neural Network for MNIST classification. Input data passes through CNN layers—two Convolutional layers each followed by Max pooling layers—then through two Fully connected layers. The output is connected to the QNN using PyTorch's Torch-Connector(), followed by a final Fully connected layer with 10 outputs representing log-probabilities for digits 0–9, fed into the Estimator simulation layer and used to calculate the NLLLoss function using the ADAM optimizer. The combination of QNN and CNN is shown as in Fig.4.

TABLE I

COMPARISON ON H-QNN STRUCTURAL COMPONENTS WITH TWO CASES $n = 2$ AND $n = 3$ QUBITS.

Resources	$n = 2$	$n = 3$
Convolutional layer 1	(1,2)	(1,2)
Max pooling 1	(2,2)	(2,2)
Convolutional layer 2	(2,32)	(2,64)
Max pooling 2	(2,2)	(2,2)
Flatten	1024	512
Fully connected 1	(1024,512)	(512,128)
Fully connected 2	(512,2)	(128,3)
Fully connected 3	(2,10)	(3,10)

IV. EXPERIMENT

This section presents the core of the article, with the model structure shown in Fig.4. The experiment considers two cases: QNN with 2 and 3 qubits. The dataset is PyTorch's MNIST, using 1000 random training images and 1000 test images, randomly shuffled for each training with a certain number of epochs. The model is run on a 12th Gen Intel(R) Core(TM) i7-12700H CPU. Table I summarizes the results for both qubit settings in the image classification simulation.

After performing and displaying the results (as shown in Fig.5), we tested two batch sizes (1 and 5) for both 2-qubit and 3-qubit cases. Batch size is the number

979-8-3315-1550-8/25 $31.00 © 2025 IEEE

TABLE II

COMPARISON OF PARAMETERS OF DIFFERENT ARTICLES ABOUT THE H-QNN MODEL.

Resources	[18]	[3]	[13]	[19]	This work
Number of qubits	N/A	N/A	9	4	3
Accuracy	avg. 94.5%	≃85%	≃80%	96.5%	≃95%
Loss	avg. 0.33	0.4086	N/A	0.148	0.1971
Training/testing data type	Synthetic Dataset	MNIST(28x28)	MNIST(28x28)	MRI(250x250)	MNIST(28x28)
Number of epochs	20	50	60	20	20
Class	2	10 [0,9]	4 [0,3]	2	10 [0,9]
Layers of QNN	N/A	N/A	6	N/A	1

of images divided from the original data set and then fed into the model for training. With 1 batch size, although training time is slightly longer, accuracy is higher and loss is lower compared to 5 batch size. For 3 qubits, 1 batch size shows faster loss decay but has more fluctuation and 15–20% longer training time. The 3 qubit case requires fewer iterations (almost half) to achieve high accuracy than the 2 qubit case, but the processing time is about 3 times longer even though the CNN layer for the 2 qubit case has more weights which increases the computation time. H-QNN processing time is still slower and accuracy is not superior to CNN. To explain this, it is because we simulate QNN on a classical computer, so this model also spends resources simulating the Quantum Network and then uses the Quantum Neural Network to calculate, when the model uses more qubits, the impact of noise is also more obvious and also makes it more difficult to simulate on a classical computer. More importantly, the algorithms for computing the loss function and optimizing the CNN part are still classical algorithms, not designed to run for models using QNN. Besides, when measuring the output of QNN, it takes many measurements to get an accurate result based on probability. If this model has a loss function algorithm along with an optimization function suitable for QNN and it is implemented on a real quantum computer, the processing time can be much faster. More qubits also introduce more noise and require more complex simulations. Additionally, batch size 1 needs more parameter updates (1000 per loop vs. 20 for batch size 5), leading to better accuracy but slightly longer processing time.

For both experimental cases about the number of qubits above, we can see that when it reaches a certain accuracy threshold (about 92% or more), it starts to oscillate and no longer increases (Plateau's problem). This phenomenon may result from the optimization algorithm failing to ensure proper convergence. A more effective optimization technique, which was not ex-

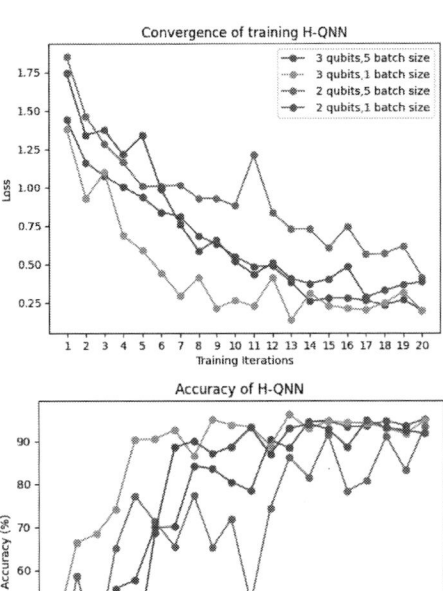

Fig. 5. Convergence of H-QNN training and loss with 2 and 3 qubits

plored in this paper, may be required [20]. Because of limited resources, we only calculate up to the 20th loop. If we run more iterations and more samples, the loss and accuracy results will be more accurate and stable. We take some typical experiments to compare some criteria with our model in Table II, most of the papers have image classifiers below 5 classes, very few papers consider higher classifiers. We decided to include paper [3] in the comparison table because it has many similarities to compare with our experiment. Our accuracy is about 10% higher than that paper and the cross-entropy loss is about 0.2 lower while running only 20 iterations and only a few qubits are needed (maybe our processing time is faster although this parameter is not mentioned in their research paper). But their number of test and validation samples is larger than ours, which may affect the accuracy of their classification model. Other models have very high accuracy, low loss, rich data files, and diverse model implementation methods, but the number of classifications is still modest, the reason can be attributed to limited resources and time.

In addition, parameters like throughput and power consumption are not included, as our main focus is on improving and comparing recognition accuracy with other studies. Most H-QNN research is still based on simulations, not real quantum computers, and has yet to outperform classical models. Moreover, such parameters vary across papers, which often use small and simple data sets to evaluated because hardware and algorithm limitations do not yet allow us to perform larger and more complex things in reality.

V. Conclusion

In this paper, we have implemented the H-QNN model by referring to and improving the models in other papers, the results are improved in terms of accuracy and loss is significantly reduced as well as the efficiency of the model. However, the problem of data processing time is prolonged due to the dependence on the simulation system and the optimization functions and loss functions. The field of QML and quantum algorithms applied with artificial intelligence to solve inherent problems related to artificial intelligence such as time, energy, ... has a lot of potential for research or development and application to image classification as our article or articles of other authors is a typical example of applying quantum computing to life. In addition, QML in general and QNN in particular can be applied to medicine to treat cancer, chronic diseases or genetic diseases. Or simulation to manufacture materials to serve human needs or algorithms to solve problems that classical computers are almost impossible to handle. If successfully applied, quantum computing will change the development of human science and technology in a profound and obvious way, which will be a stepping stone to conquer the heights that humans always aspire to explore.

Acknowledgment

This research was funded by Vingroup Innovation Foundation (VINIF) under project code VINIF.2024.DA085.

References

[1] David Peral García, Juan Cruz-Benito and Francisco José García-Peñalvo, "Systematic Literature Review: Quantum Machine Learning. and its applications," *arXiv:2201.04093v2*, 2022.

[2] A. Krizhevsky, I. Sutskever, and G. E. Hinton, "ImageNet Classification with Deep Convolutional Neural Networks," in *Advances in Neural Information Processing Systems 25 (NeurIPS)*, 2012, pp. 1097–1105.

[3] Mingrui Shi, Haozhen Situ and Cai Zhang, "Hybrid Quantum Neural Network Structures for Image Multi-classification," *arXiv:2308.16005v1*, 2023.

[4] Qiskit Machine Learning Tutorials, Torch Connector and Hybrid QNNs.

[5] Seunghyeok Oh,Jaeho Choi and Joongheon Kim, "A Tutorial on Quantum Convolutional Neural Networks (QCNN)," *arXiv:2009.09423v1*, 2020.

[6] Qiskit Machine Learning Tutorials, The Quantum Convolution Neural Network.

[7] Peter W. Shor, "Polynomial-Time Algorithms for Prime Factorization and Discrete Logarithms on a Quantum Computer ," *arXiv:quant-ph/9508027*, 1995.

[8] Lov K. Grover, "A fast quantum mechanical algorithm for database search," *arXiv:quant-ph/9605043v3*, 1996.

[9] Zaman, AMorrell, HWong H, "A Step-by-Step HHL Algorithm Walkthrough to Enhance Understanding of Critical Quantum Computing Concepts," *arXiv:2108.09004v4*, June 2023.

[10] Noson S. Yanofsky, "An Introduction to Quantum Computing," *arXiv:0708.0261*, 2007.

[11] Deepak Ranga,Sunil Prajapat,Zahid Akhtar,Pankaj Kumar and Athanasios V. Vasilakos, "Hybrid Quantum–Classical Neural Networks for Efficient MNIST Binary Image Classification," *DOI: 10.3390/math12233684*, Nov. 2024.

[12] Muhammad Kashif and Saif Al-Kuwari, "Design Space Exploration of Hybrid Quantum–Classical Neural Networks ," *DOI:10.3390/electronics10232980*, Nov. 2021.

[13] Kamila Zaman, Tasnim Ahmed, Muhammad Abdullah Hanif, Alberto Marchisio, and Muhammad Shafique, "A Comparative Analysis of Hybrid-Quantum Classical Neural Networks," *arXiv:2402.10540v2*, 2024.

[14] Minati Rath and Hema Date, "Quantum data encoding: a comparative analysis of classical-to-quantum mapping techniques and their impact on machine learning accuracy," *arXiv:2311.10375v1*, 2023.

[15] Xiaoyu Guo,Takahiro Muta, Jianjun Zhao, "Quantum Circuit Ansatz: Patterns of Abstraction and Reuse of Quantum Algorithm Design," *arXiv:2405.05021v3*, 2024.

[16] Georg Kruse, Theodora-Augustina Dragan, Robert Wille and Jeanette Miriam Lorenz, "Variational Quantum Circuit Design for Quantum Reinforcement Learning on Continuous Environments," *arXiv:2312.13798v1*, 2023.

[17] Diederik P. Kingma and Jimmy Ba, "Adam: A Method for Stochastic Optimization," *arXiv:1412.6980v9*, Jan. 2017.

[18] Davis Arthur and Prasanna Date, "A Hybrid Quantum-Classical Neural Network Architecture for Binary Classification," *arXiv:2201.01820v2*, 2022.

[19] Ryan Kim, "Implementing a Hybrid Quantum-Classical Neural Network by Utilizing a Variational Quantum Circuit for Detection of Dementia," *arXiv:2301.12505*, 2023.

[20] Andrew Arrasmith, M. Cerezo, Piotr Czarnik, Lukasz Cincio and Patrick J. Coles,"Effect of barren plateaus on gradient-free optimization," *arXiv:2011.12245v2*, Sep. 2021.

2025 10th IEEE International Conference on Integrated Circuits, Design, and Verification (ICDV)

Harnessing TinyML for Accurate ECG Beat Detection

1st Bui An Dong
HCMC University of Science
ITR VN CORPORATION
Ho Chi Minh City, Vietnam
badong@hcmus.edu.vn or dongbui@itrvn.com

2nd Tran Hoang Dat
ITR VN CORPORATION
Ho Chi Minh City, Vietnam
dattran@itrvn.com

3rd Ngo Hoang Anh Vy
ITR VN CORPORATION
Ho Chi Minh City, Vietnam
vyngo@itrvn.com

Abstract—**In this paper, we propose the use of a lightweight deep learning-based ECG beat detection model optimized for resource-constrained microcontrollers. Accurate R-peak detection is essential for analyzing heart rate variability and diagnosing arrhythmias. Despite achieving high performance, complex deep learning models often require high computational resources, limiting their applicability in real-time and wearable healthcare systems. To address this challenge, our study focuses on developing an efficient model that balances accuracy and computational efficiency. Our private collected ECG dataset is used for training. Because ECG signals are often noisy and vary across individuals, we implemented preprocessing techniques such as Butterworth band-pass filtering, baseline wander removal to enhance data quality. The proposed model combines traditional signal processing with deep learning to achieve robust QRS complex detection. Our model architecture is inspired and developed from the Fire module to balance high detection accuracy with a low number of parameters and inference time, making it suitable for wearable devices and real-time applications. Compared to a baseline approach, our method demonstrates improved detection accuracy in MIT-BIH dataset. Our model achieved 99.69% sensitivity and 99.69% PPV, outperforming [1] (99.47% sensitivity, 99.6% PPV) and [2] (98.92% sensitivity, 99.35% PPV), demonstrating superior accuracy in ECG beat detection. Experimental results demonstrate that our method effectively identifies R-peaks while maintaining efficiency, contributing to scalable and cost-effective continuous heart monitoring.**

Index Terms—**ECG beat detection, lightweight model, deep learning**

I. INTRODUCTION

ECG beat detection is a process of identifying individual heartbeats within an electrocardiogram (ECG) recording - the electrical activity of the heart [3]. The most critical part of beat detection is identifying R peaks in the QRS complex, as they represent ventricular contractions and help to measure heart rate variability, arrhythmias, and other cardiac conditions. Accurate detection of beats helps diagnose life-threatening conditions such as heart block. In addition, detecting R-peaks allows for the precise calculation of the heart rate, leading to the detection of arrhythmias such as bradycardia and tachycardia [4].

Automating this process enhances efficiency, accuracy, and accessibility, benefiting both doctors and patients by reducing workload, being cost-effective and scalable, and enabling

continuous heart monitoring by leveraging wearble devices [5], [6]. Various methods have been developed to detect beat automatically. Early methods involved analyzing the ECG signal using classical signal processing techniques like the Hilbert Transform and Filter Bank methods [7]. The Pan and Tompkins method is widely used in this field by analyzing the slope, amplitude and width of the ECG signal [8]. Machine learning algorithms are used to learn detection rules from a training dataset. Several machine learning models have been used to classify heart failure subtypes, including Support Vector Machines (SVM), K-Nearest Neighbors (KNN), ensemble tree, and logistic regression in Yang et.al's work [9].

Deep learning has significantly advanced ECG beat detection by enabling models to learn complex patterns within ECG signals, leading to improved accuracy in identifying arrhythmias and other cardiac conditions. For instance, a CNN model incorporating an adaptive windowing algorithm attained an R-peak detection accuracy of 98.94% and an F1-score of 99.35% on the MIT-BIH Arrhythmia dataset [1]. However, more complex models typically require more energy to train and have longer inference times, which can limit their use in wearable devices and real-time applications [10]. In this paper, we aim to develop a lightweight model architecture to handle complexity constraints. We are supposed to implement an efficient QRS complex detection algorithm on the resource-constrained STM32F4 microcontroller for real-time ECG signal analysis.

The main contributions of this paper are summarized as follows:

- By combining the strengths of ResNet-style skip connections with a modified Fire module, we propose a lightweight AI model that can be embedded on a microcontroller device.
- We focus on enhancing the overall accuracy and computational efficiency of QRS complex detection.

This paper consists of six sections. Section II discusses the relevant literature in this area. Section III outlines the characteristics of our dataset. Section IV elaborates on the paper's approach to identifying R-peaks in ECG signals. Sections V presents and discusses the result of our proposed method. Finally, section VI summarizes the objectives of this

ITR VN Corporation

979-8-3315-1550-8/25 $31.00 © 2025 IEEE

paper and outlines future research directions.

II. RELATED LITERATURE

This section discusses the need for automatic beat detection, current methods, and the reliability of AI models on devices.

A. Beat detection in healthcare

Accurate ECG beat detection is crucial for diagnosing and monitoring cardiovascular conditions, including arrhythmias and heart rate variability [11], [12]. Traditional manual analysis is time-consuming and error-prone, highlighting the need for automated systems enabling to analyze the vast amount of monitoring data to enhance efficiency, accuracy, and scalability in clinical and offering high-quality and affordable home health monitoring [11].

Current approaches to automatic ECG beat detection leverage signal processing and machine learning techniques to identify QRS complexes, which are key indicators of heartbeats. Traditional techniques, such as the Pan and Tompkins algorithm, enhance QRS complexes by reducing noise and using adaptive thresholding to detect R peaks [1], [13], [14]. Despite its extensive application, the Pan and Tompkins algorithm has certain limitations, especially when dealing with low-quality or noisy signal [1]. More recent advancements leverage deep learning models, including convolutional neural networks (CNNs) and recurrent neural networks (RNNs), which can capture intricate patterns in ECG signals, leading to improved detection accuracy. Šarlija et al. proposed a 1D-CNN-based approach for ECG beat detection, achieving over 99% recall and precision [15]. Furthermore, in the paper of Niroshana et. al., an adaptive window algorithm and CNN model are used to detect heartbeat with 99.08% accuracy and 99.08% F1-score [2].

B. The reliability of AI beat detection model on device

Deploying AI-based ECG beat detection models on edge devices, such as mobile phones, smartwatches, and wearable health monitors, presents both opportunities and challenges in real-world clinical applications. While deep learning models, such as CNNs and RNNs, have demonstrated high accuracy in controlled datasets, their usage of them on edge devices may be limited since they usually take longer to infer and demand more energy to train [16]. Lightweight deep learning models can address the challenges of model rigidity, complexity, and inference speed, offering cardiologist-level accuracy with faster processing times on CPUs, making them suitable for wearable devices [17]. To optimize AI models for deployment, techniques such as quantization, pruning, and knowledge distillation are used to reduce model size while maintaining accuracy. An et.al. introduced a framework using knowledge distillation (KD) to train a lightweight student model from a multi-lead ECG teacher model obtaining a classification accuracy of 96.32% [18]. A lightweight model with fused RNN proposed by Jeon et.al. achieved a cardiologist-level accuracy of 99.80% and conducted ECG beat predictions on a CPU five times faster than a baseline model [17].

III. DATASET

We use a private ECG dataset for training and evaluation, then our model is tested by the MIT-BIH Arrhythmia Database, a widely used benchmark dataset for ECG beat detection and classification.

A. Private ECG Dataset

The present study employed a dataset of more than 90212 three-lead ECG recordings. Each ECG recording lasted for 60 seconds and was digitized at 250 Hz with a 16-bit resolution. The data is split into a training set and a validation set, with 80% for the training set and 20% for the validation set. We adopted a subject-based split to ensure that there is no overlap of subjects between the training and validation sets. This approach helps to prevent data leakage and provides a more realistic assessment of the model's performance.

B. MIT-BIH Arrhythmia Database

This dataset, developed by the Massachusetts Institute of Technology and Beth Israel Hospital, consists of 48 half-minute ECG recordings collected from 48 patients, including both inpatients and outpatients. The ECG signals were sampled at 360 Hz, capturing lead II and V1 or V2 waveforms [19], [20]. Both regular and cardiac arrhythmia-related heartbeats are captured on each 30-minute recording. The dataset provides expert-annotated beat labels, including normal and various arrhythmic beats, making it a valuable resource for training and evaluating automatic beat detection algorithms.

IV. METHODOLOGY

In this paper, we propose a lightweight model for detecting R-peaks in ECG signals. To enhance accuracy, the signal undergoes preprocessing to eliminate noise before being processed by our method. As shown in Fig. 1, the framework comprises two key components: preprocessing and the lightweight model, with each module detailed in the following sections

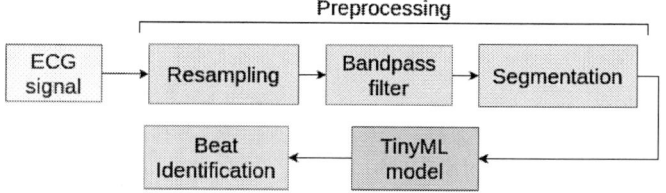

Fig. 1. The overview of our proposed method.

A. Data Preprocessing

Preprocessing ECG data is essential to ensure consistency and improve model performance. First, we resample the signal at 250Hz. A Butterworth band-pass filter is then applied to remove high-frequency noise (above 30Hz) and low-frequency interference (below 0.5Hz). Because our goal is to identify beat, all normal beat annotations (N, L, R, e, j), supraventricular ectopic beats (A, a, J, S), ventricular ectopic beats (V, e), fusion beats (F) are considered as a single "N" category.

979-8-3315-1550-8/25 $31.00 © 2025 IEEE

Additionally, instances labeled as "f" and "Q" are discarded to exclude uncertain or unclassified beats, ensuring a cleaner dataset for model training. We divide the ECG signal into 5-second segments with 20% overlap using the sliding window technique.

B. Model architecture

Our model is inspired by the Fire module from Iandola et al.'s work [21], but we modify its structure to suit our architecture better. Instead of using both squeeze and expand layers, we adopt only the expand layer, which combines 1×1 and 3×3 convolution filters. This adjustment maintains parameter efficiency while preserving the model's ability to capture multi-scale features. Additionally, an inception stage with two parallel convolutions compensates for the reduced depth, each producing half the number of output channels of the modified Fire module [22].

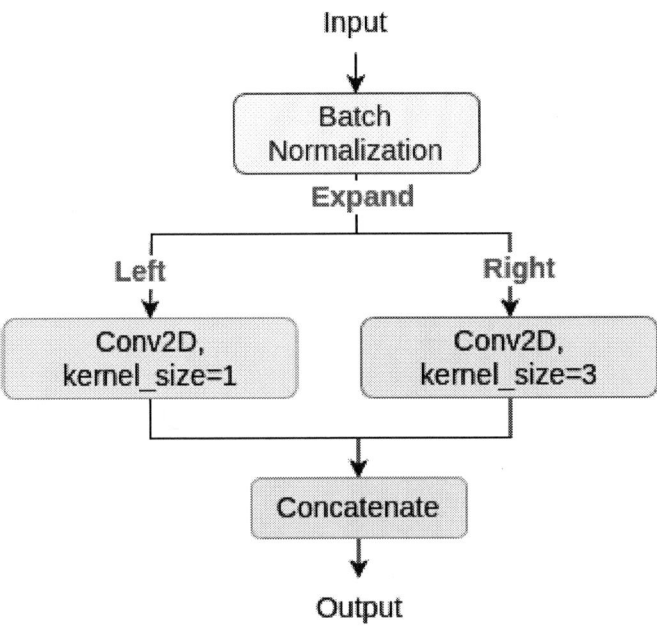

Fig. 2. The architecture of the Fire module used in our model.

This research utilized a 1250-sample one-dimensional ECG segment as input to the model. Fig. 3 shows the overall architecture of our model. The model consists of multiple Conv2D blocks, a Block1D loop, and a classification layer at the end. Our model incorporates a dual-branch design inspired by the ResNet architecture using "Add" layers and squeeze operations of the Fire module. In ResNet, identity shortcut connections enable the model to bypass one or more layers, facilitating deeper architectures and preserving feature representations. We extend this principle while also focusing on parameter efficiency [23]. In our design, one branch consists of a standard Conv2D block with kernel size = 1, working as ResNet-style skip connection. Instead of combining two standard Conv2D blocks in traditional ResNet architecture, the second branch incorporates a Conv2D block and our modified Fire module (labeled "Conv2D squeeze") to reduce the number of parameters while maintaining performance. The predefined filter configurations (f_1, f_2) control the number of filters in each convolutional block, influencing the model's capacity and computational complexity. The values of (f_1, f_2) are chosen from the set:

$$(f_1, f_2) \in \{(8, 8), (8, 24), (24, 8)\}$$

Conv2D block extracts features containing a batch normalization, a ReLU activation, and a dropout before transferring into a convolutional 2D layer.

Block1D loop applies multiple Conv2D squeeze operations and an Add layer, which further enhances feature extraction through depthwise convolutions and residual learning.

Classification layers map the learned features into class probabilities. A dense layer represents the final fully connected layer, indicating that the model performs binary classification. A Softmax activation function at the end is used to generate class probabilities.

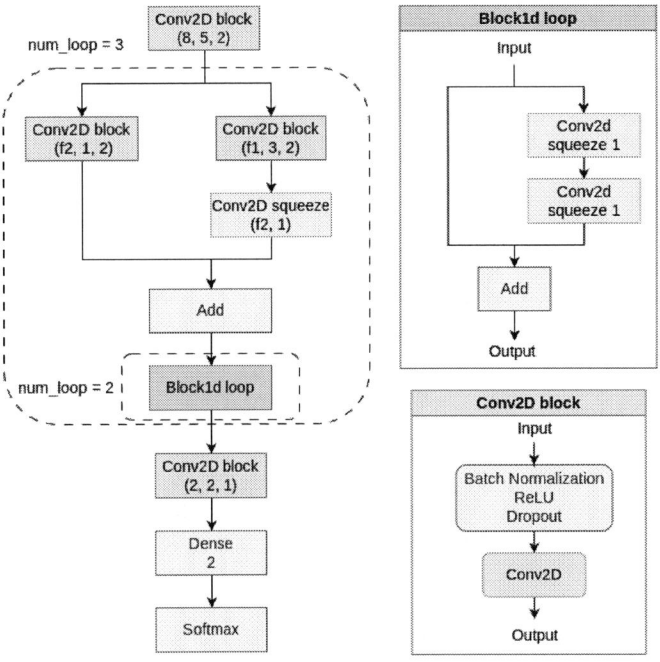

Fig. 3. The architecture of our TinyML.

C. Evaluation metrics

The ANSI/AAMI EC57 standard providing a standardized method for evaluating and reporting the performance of heart beat detection algorithms are used in this work [24]. According to ANSI/AAMI EC57:2012, the algorithm should be evaluated using two parameters:

Sensitivity in ECG beat detection refers to the true positive rate, which is the proportion of correctly detected R-peaks out of all actual R-peaks in the ECG signal. It measures how well a model identifies true heartbeats without missing them.

$$\text{Sensitivity} = \frac{\text{True Positives}}{\text{True Positives} + \text{False Negatives}} \quad (1)$$

In ECG beat identification, positive predictive value (PPV) represents the proportion of detected R-peaks that are actual R-peaks in the ECG signal. It indicates how often a detected R-peak is a true positive rather than a false detection. A high PPV means that most identified R-peaks are correctly located, ensuring the model reliably distinguishes real heartbeats from noise or artifacts.

$$\text{PPV} = \frac{\text{True Positives}}{\text{True Positives} + \text{False Positives}} \quad (2)$$

V. Results and Discussion

To evaluate the performance of our lightweight deep learning model, we conduct a comparative analysis against two representative methods from different domains: deep learning-based and traditional algorithm-based approaches. Specifically, we compare our results with (1) the deep learning-based method proposed by Niroshana et al. [2], which utilizes a convolutional neural network (CNN) combined with an adaptive windowing algorithm, and (2) the traditional algorithm-based method by Khan et al. [1], which introduces an updated version of the Pan-Tompkins algorithm.

We select these two studies for comparison to ensure a comprehensive evaluation of our model's effectiveness across different methodological paradigms. The CNN-based approach from [2] represents a state-of-the-art deep learning model that segments ECG signals into individual beats with high accuracy, while the Pan-Tompkins++ algorithm [1] is a refined version of a widely used traditional signal processing technique. By including both a neural network-based and a rule-based approach, we aim to assess whether our TinyML model can achieve a favorable balance between computational efficiency and detection accuracy.

For a fair and standardized comparison, we evaluate all methods on the MIT-BIH Arrhythmia Database, a widely recognized benchmark dataset for ECG analysis. The results are assessed based on key performance metrics, including sensitivity (Se) and positive predictive value (+P).

Table II presents a quantitative comparison of our proposed method against existing approaches in ECG beat detection. Our model achieves a Se of 99.68% and a +P of 99.72%, outperforming both traditional and deep learning-based methods. Compared to the method in [2], which reported a +P of 99.35%, and [1], which achieved a Se of 99.47% and a +P of 99.6%, our approach demonstrates superior performance in both metrics. While the numerical improvement in performance may seem modest, such gains are meaningful in high-accuracy domains like biomedical signal analysis, where even small margins can translate into a significant reduction in false positives or missed detections in real-world deployments [25].

The results indicate that our TinyML-based model is not only capable of accurately detecting R-peaks but also reduces

TABLE I
EVALUATION OF THE PROPOSED QRS DETECTION METHOD ON 48 MIT-BIH ARRHYTHMIA RECORDINGS.

Recording	TP	FN	FP	Se	+P
100	2272	1	0	99.96	100.00
101	1865	0	2	100.00	99.89
102	2186	1	0	99.95	100.00
103	2082	2	0	99.90	100.00
104	2213	16	5	99.28	99.77
105	2547	25	26	99.03	98.99
106	2020	7	2	99.65	99.90
107	2121	16	77	99.25	96.50
108	1744	19	65	98.92	96.41
109	2531	1	0	99.96	100.00
111	2122	2	11	99.91	99.48
112	2538	1	1	99.96	99.96
113	1794	1	35	99.94	98.09
114	1878	1	0	99.95	100.00
115	1952	1	0	99.95	100.00
116	2386	26	2	98.92	99.92
117	1534	1	15	99.93	99.03
118	2278	0	1	100.00	99.96
119	1987	0	2	100.00	99.90
121	1861	2	1	99.89	99.95
122	2475	1	0	99.96	100.00
123	1518	0	0	100.00	100.00
124	1618	1	0	99.94	100.00
200	2598	3	3	99.88	99.88
201	1956	7	3	99.64	99.85
202	2125	11	2	99.49	99.91
203	2898	82	13	97.25	99.55
205	2652	4	0	99.85	100.00
207	1843	17	1	99.09	99.95
208	2932	23	3	99.22	99.90
209	3002	3	2	99.90	99.93
210	2632	18	2	99.32	99.92
212	2747	1	0	99.96	100.00
213	3248	3	1	99.91	99.97
214	2257	5	1	99.78	99.96
215	3361	2	1	99.94	99.97
217	2206	2	6	99.91	99.73
219	2154	0	1	100.00	99.95
220	2047	1	0	99.95	100.00
221	2426	1	1	99.96	99.96
222	2472	11	6	99.56	99.76
223	2601	4	1	99.85	99.96
228	2040	13	21	99.37	98.98
230	2255	1	1	99.96	99.96
231	1570	1	5	99.94	99.68
232	1780	0	16	100.00	99.11
233	3074	5	0	99.84	100.00
234	2753	0	0	100.00	100.00
Total	**109151**	**343**	**335**	**99.69**	**99.69**

false positives and false negatives more effectively than previous methods. The improvement in positive predictive value suggests that our model generates fewer false detections, making it more reliable for real-world applications. Furthermore, the slight increase in sensitivity implies that our model is better at identifying true R-peaks, ensuring robustness across different ECG recordings.

While the method in [1] follows an updated Pan-Tompkins algorithm for R-peak detection, which processes the ECG signal sequentially, our approach leverages deep learning to extract high-level features directly from raw ECG signals. This eliminates the need for handcrafted feature extraction and

TABLE II
COMPARISON OF OUR PROPOSED METHOD'S PERFORMANCE WITH BASELINE.

Model	TP	FN	FP	Se (%)	+P (%)
Khan et al. [1]	–	0.51%	0.38%	99.47	99.6
Niroshana et al. [2]	108310	–	704	98.92	99.35
Our proposed method	109151	**343**	**335**	**99.69**	**99.69**
textbf0.313%	**0.306%**				

* For our method, both absolute counts and percentage values of TP, FN, and FP are provided. For the other referenced works, only the available values are shown, as the original publications did not include sufficient details to compute or convert them into percentages.

filtering steps, making our model more adaptable to diverse ECG signal characteristics.

Another key advantage of our approach is its efficient inference strategy. Unlike traditional methods that analyze each beat independently or require adaptive windowing for segmentation, our model processes ECG signals in overlapping 5-second windows with a 50% stride. This significantly reduces the number of forward passes required during inference compared to models that operate on a per-beat basis, such as [2] having a maximum 1.4-second input. As a result, our method achieves faster inference while maintaining high detection accuracy, making it well-suited for real-time applications on edge devices.

Furthermore, our advanced model for noise detection has been proven to possess a lightweight design with 20K parameters in STM32F4 micro-controller.

While we successfully deployed our model on a microcontroller, we did not focus on optimizing its execution time or measuring real-time performance. Additionally, our evaluation was conducted on the MIT-BIH dataset, which, despite being widely used, may not fully capture the variability of real-world ECG signals. To enhance the robustness of our approach, future work should explore testing on additional datasets or real-time ECG signals from wearable devices. Furthermore, integrating our model into a real-time wearable ECG system and evaluating its performance under real-world conditions, including motion artifacts and sensor noise, would provide valuable insights into its practical applicability.

VI. CONCLUSION

We proposed a lightweight deep learning model for real-time ECG beat detection, optimized for deployment on resource-constrained devices. Our method integrates preprocessing techniques, including noise removal and label refinement, to enhance signal quality before feeding it into the model. In our model, we adapt the Fire module by omitting the squeeze layer and retaining only the expand layer, which uses both 1×1 and 3×3 convolutions to efficiently capture multi-scale features. Additionally, by embedding these modules within a ResNet-inspired residual framework, our model benefits from improved feature preservation and training stability while maintaining a lightweight design. By combining traditional signal processing with deep learning, our approach achieves accurate and efficient R-peak detection.

Experimental results demonstrate that our model effectively detects R-peaks with high accuracy while maintaining low computational complexity, making it suitable for real-time applications such as wearable and embedded healthcare systems. Compared to baseline methods, our approach improves detection performance and enhances system reliability in diverse ECG signal conditions.

Future work will focus on optimizing the model for further computational efficiency, expanding the dataset to include more diverse ECG patterns. Additionally, we will focus on improving beat detection accuracy in noisy conditions, as noise significantly impacts the performance of existing models. Finally, real-world testing on wearable devices will be conducted to evaluate performance in practical scenarios.

ACKNOWLEDGE

This research is supported by ITR VN Corporation.

CONFLICTS OF INTEREST

The authors have no conflicts of interest to declare.

REFERENCES

[1] N. Khan and M. N. Imtiaz, "Pan-tompkins++: A robust approach to detect r-peaks in ecg signals," 2024. [Online]. Available: https://arxiv.org/abs/2211.03171

[2] S. M. I. Niroshana, S. Kuroda, K. Tanaka, and W. Chen, "Beat-wise segmentation of electrocardiogram using adaptive windowing and deep neural network," *Scientific Reports*, vol. 13, no. 1, p. 11039, 2023. [Online]. Available: https://doi.org/10.1038/s41598-023-37773-y

[3] Y. Liu, Q. Li, R. He, K. Wang, J. Liu, Y. Yuan, Y. Xia, and H. Zhang, "Generalizable beat-by-beat arrhythmia detection by using weakly supervised deep learning," *Frontiers in Physiology*, vol. 13, 2022. [Online]. Available: https://www.frontiersin.org/journals/physiology/articles/10.3389/fphys.2022.850951

[4] A. R. Pérez-Riera, L. C. de Abreu, R. Barbosa-Barros, K. C. Nikus, and A. Baranchuk, "R-peak time: An electrocardiographic parameter with multiple clinical applications," *Annals of Noninvasive Electrocardiology*, vol. 21, no. 1, pp. 10–9, January 2016. [Online]. Available: https://doi.org/10.1111/anec.12323

[5] Q. Qin, J. Li, Y. Yue, and C. Liu, "An adaptive and time-efficient ecg r-peak detection algorithm," *Journal of Healthcare Engineering*, vol. 2017, p. 5980541, September 6 2017.

[6] M. Martínez-Sellés and M. Marina-Breysse, "Current and future use of artificial intelligence in electrocardiography," *Journal of Cardiovascular Development and Disease*, vol. 10, no. 4, p. 175, April 17 2023. [Online]. Available: https://doi.org/10.3390/jcdd10040175

[7] P. Thulasi and V. Sourirajan, "Ecg signal analysis: Different approaches," *International Journal of Engineering Trends and Technology*, vol. 7, pp. 212–216, 01 2014.

[8] M. D'Aloia, A. Longo, and M. Rizzi, "Noisy ecg signal analysis for automatic peak detection," *Information*, vol. 10, no. 2, 2019. [Online]. Available: https://www.mdpi.com/2078-2489/10/2/35

[9] Y. Liu, Q. Li, R. He, K. Wang, J. Liu, Y. Yuan, Y. Xia, and H. Zhang, "Generalizable beat-by-beat arrhythmia detection by using weakly supervised deep learning," *Frontiers in Physiology*, vol. 13, 2022. [Online]. Available: https://www.frontiersin.org/journals/physiology/articles/10.3389/fphys.2022.850951

[10] Y. He, Y. Zhou, Y. Qian, J. Liu, J. Zhang, D. Liu, and Q. Wu, "Cardioattentionnet: advancing ecg beat characterization with a high-accuracy and portable deep learning model," *Frontiers in Cardiovascular Medicine*, vol. 11, 2025. [Online]. Available: https://www.frontiersin.org/journals/cardiovascular-medicine/articles/10.3389/fcvm.2024.1473482

[11] Y. Liu, Q. Li, R. He, K. Wang, J. Liu, Y. Yuan, Y. Xia, and H. Zhang, "Generalizable beat-by-beat arrhythmia detection by using weakly supervised deep learning," *Frontiers in Physiology*, vol. 13, 2022. [Online]. Available: https://www.frontiersin.org/journals/physiology/articles/10.3389/fphys.2022.850951

[12] Y. Ansari, O. Mourad, K. Qaraqe, and E. Serpedin, "Deep learning for ecg arrhythmia detection and classification: an overview of progress for period 2017-2023," *Frontiers in Physiology*, vol. 14, p. 1246746, September 2023, eCollection 2023.

[13] J. Pan and W. J. Tompkins, "A real-time qrs detection algorithm," *IEEE Transactions on Biomedical Engineering*, vol. BME-32, no. 3, pp. 230–236, 1985.

[14] L. Sathyapriya, L. Murali, and T. Manigandan, "Analysis and detection r-peak detection using modified pan-tompkins algorithm," in *2014 IEEE International Conference on Advanced Communications, Control and Computing Technologies*, 2014, pp. 483–487.

[15] M. Šarlija, F. Jurišić, and S. Popovic, "A convolutional neural network based approach to qrs detection," 09 2017, pp. 121–125.

[16] R. Kher, "Signal processing techniques for removing noise from ecg signals," 2019. [Online]. Available: https://api.semanticscholar.org/CorpusID:212573348

[17] E. Jeon, K. Oh, S. Kwon, H. Son, Y. Yun, E.-S. Jung, and M. S. Kim, "A lightweight deep learning model for fast electrocardiographic beats classification with a wearable cardiac monitor: Development and validation study," *JMIR Medical Informatics*, vol. 8, no. 3, p. e17037, March 2020, eCollection 2020.

[18] X. An, S. Shi, Q. Wang, Y. Yu, and Q. Liu, "Research on a lightweight arrhythmia classification model based on knowledge distillation for wearable single-lead ecg monitoring systems," *Sensors*, vol. 24, no. 24, 2024. [Online]. Available: https://www.mdpi.com/1424-8220/24/24/7896

[19] G. B. Moody and R. G. Mark, "The impact of the mit-bih arrhythmia database," *IEEE Engineering in Medicine and Biology Magazine*, vol. 20, no. 3, pp. 45–50, May-June 2001.

[20] A. Goldberger, L. Amaral, L. Glass, J. Hausdorff, P. C. Ivanov, R. Mark, and H. E. Stanley, "Physiobank, physiotoolkit, and physionet: Components of a new research resource for complex physiologic signals," *Circulation*, vol. 101, no. 23, pp. e215–e220, 2000, [Online].

[21] F. Iandola, S. Han, M. Moskewicz, K. Ashraf, W. Dally, and K. Keutzer, "Squeezenet: Alexnet-level accuracy with 50x fewer parameters and ¡0.5mb model size," 02 2016.

[22] N. Beheshti and L. Johnsson, "Squeeze u-net: A memory and energy efficient image segmentation network," in *2020 IEEE/CVF Conference on Computer Vision and Pattern Recognition Workshops (CVPRW)*, 2020, pp. 1495–1504.

[23] K. He, X. Zhang, S. Ren, and J. Sun, "Deep residual learning for image recognition," 2015. [Online]. Available: https://arxiv.org/abs/1512.03385

[24] *Testing and reporting performance results of cardiac rhythm and ST segment measurement algorithms*, Association for the Advancement of Medical Instrumentation (AAMI) Std., 2012. [Online]. Available: https://webstore.ansi.org/standards/aami/ansiaamiec572012r2020

[25] M. H. Murad, L. Lin, H. Chu, B. Hasan, R. A. Alsibai, A. S. Abbas, R. A. Mustafa, and Z. Wang, "The association of sensitivity and specificity with disease prevalence: analysis of 6909 studies of diagnostic test accuracy," *CMAJ*, vol. 195, no. 27, pp. E925–E931, 2023. [Online]. Available: https://www.cmaj.ca/content/195/27/E925

2025 10th IEEE International Conference on Integrated Circuits, Design, and Verification (ICDV)

FPGA-based Design and Implementation of Processing Element Array for Convolutional Neural Networks

Luu Nguyen-Van*, Linh Nguyen-Thi-Thuy*
Chi Hoang-Phuong*†, Minh Nguyen-Duc*,
*School of Electrical and Electronic Engineering, Hanoi University of Science and Technology, Vietnam
†Corresponding Author: chi.hoangphuong@hust.edu.vn

Abstract—**Convolutional Neural Networks (CNNs) are essential in AI for image tasks like facial recognition and object detection. While they improve accuracy, deploying them on hardware is challenging due to high computational and memory demands, requiring efficient accelerators to balance speed, power, and resources. In this study, we designed and implemented a computational unit for CNNs, including components such as a convolutional accelerator, pooling, fully connected layers, and an activation function. This study utilizes *weight stationary* (WS) to minimize data movement and reuse partial sums based on spatial architecture with an array of processing elements (PEs). To the end, the proposed accelerator achieves an inference speed of 33.53 frames per second at 250 MHz, consumes power of 6.34 W and this accelerator achieves a throughput of 54.90 GOPS. The model ontain an accuracy of 98% in software and 95% after hardware simulation with the customized AlexNet model on MNIST dataset.**

Index Terms—**Convolutional neural networks, FPGA, weight stationay, spatial architecture, AlexNet.**

I. INTRODUCTION

CONVOLUTIONAL Neural Networks (CNNs) [1], a specialized form of Deep Neural Networks (DNNs) [2], have transformed Artificial Intelligence (AI) by improving computers' ability to analyze visual data. CNNs learn spatial hierarchies through convolutional layers, making them effective for tasks like facial recognition, object detection, and image classification [3], [4]. With state-of-the-art CNN models [5], [6] such as LeNet (~60K parameters), AlexNet (~60M), and VGG-16 (~138M), inference computation requires handling billions of operations per pass. This leads to significant data movement between on-chip and off-chip memory, which can consume more energy than the computation itself [7], [8]. Therefore, optimizing CNN accelerators involves not only high parallelism for increased throughput but also efficient memory management to reduce data transfer overhead.

Several previous studies have explored the use of field-programmable gate arrays (FPGAs) as a platform for implementing hardware accelerators for CNNs. FPGA-based CNN implementations have been developed by various research groups [12], [15], [22], [23], [24], [25], [26], [27]. Among these, [25], [27], [13], [16] primarily focus on general matrix

multiplication (GEMM) acceleration to enhance computational efficiency. Meanwhile, fusion architectures [26], [12] emphasize reusing intermediate data, significantly reducing off-chip memory transfers. While the work in [22], [23], [24] focuses on optimizing dataflow architectures to minimize data movement and memory accesses and an accelerator in [15], [14] to balance computational and memory demands, improving resource utilization and external memory bandwidth. In addition, systolic array architectures [27], [21], [10] have been widely explored for CNN acceleration on FPGAs, leveraging their deep pipelining, efficient local communication, and reduced global data transfer to achieve high throughput and high frequencies. While these implementations effectively utilize FPGA resources to maximize performance, the continuous increase in FPGA hardware capacity poses new challenges.

This paper proposes an FPGA-based CNN accelerator architecture that optimizes dataflow for efficient data reuse, minimizes off-chip memory access and significantly reduces bandwidth requirements—this is particularly beneficial for deep networks with a high number of weight accesses, such as AlexNet, VGG, and DenseNet, and balances resource utilization while maintaining flexibility across different CNN models. Additionally, our design ensures high processing speed, meeting the performance requirements of real-time applications. Some key contributions of this work are:

(1) *Dataflow-weight stationary*: A spatial architecture where weights remain fixed in PEs, while activations are broadcast for parallel MAC operations. This approach exploits weight (wgt), input feature map (ifmap) and partial sum (psum) reuse, reduces memory accesses, and improves computational throughput.

(2) *PE array configurable for three computation modes:* To increase the flexibility of IP, specifically IP can perform three functions of three layers such as Convolution (CONV), Fully Connected (FC) and Pooling (POOL) layers. Furthermore, parameterization is implemented to facilitate usage across different CNN configurations.

(3) *Utilization of a hierarchical memory structure for data reuse and pipelining:* Three memory levels: register, global buffers and off-chip RAM to reuse partial sum,

979-8-3315-1550-8/25 $31.00 © 2025 IEEE

ifmap, wgt data and perform pipelining on partial sums.

(4) *Implementation and Verification:* This proposed accelerator is implemented and tested on a customized AlexNet network with fixed-point quantization and achieves speed of 33.53 frames per second (FPS) at 250 MHz and consumes power of 6.34 W.

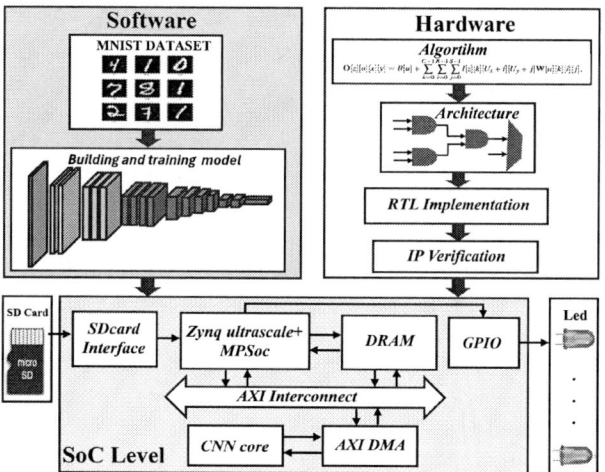

Fig. 1. Overall Design Flow of CNN Accelerator on ZCU104 Platform.

Fig. 1 shows our proposed CNN accelerator flow from model training to hardware deployment on Zynq UltraScale+ MPSoC on ZCU104 board. This paper is organized as follows. Section II provides an overview of CNNs and 3-D convolution. Section III covers system design. Section IV describes the experimental setup, results, and comparisons with previous works. Section V concludes the paper.

II. BACKGROUND OF CNNs

CNNs are constructed from multiple computational layers organized as a directed acyclic graph (DAG) [20]. Each layer extracts an abstraction of data provided by the previous layer, which is referred to as a feature map (FMAP). The most common layers in CNNs are CONV, POOL, and FC layers. In CONV layers, two-dimensional (2-D) filters slide over the ifmap, performing convolution operations to extract feature characteristics from local regions and generating output images or feature maps (ofm). In the case of three-dimensional (3-D) convolution, a batch of 3-D ifmaps is processed by a group of 3-D filters in a layer. In addition, there is a 1-D bias that is added to the filtering result. Given the shape parameters shown in Table I, the computation of a layer is defined as

$$O[z][u][x][y] = ReLU\Bigg(B[u)$$

$$+ \sum_{k=0}^{C-1} \sum_{i=0}^{R-1} \sum_{j=0}^{S-1} I[z][k][U_x + i][U_y + j]\mathbf{W}[u][k][i][j]\Bigg), \quad (1)$$

$$0 < z < N,\ 0 < u < M,\ 0 < y < E,\ 0 < x < F,$$
$$E = \frac{H - R + U}{U}, \quad F = \frac{W - S + U}{U}$$

TABLE I
SHAPE PARAMETERS OF A CNN LAYER

Shape parameter	Description
N	Batch size
M	Number of filters / output feature map channels
C	Number of input feature map channels
H/W	Input feature map height/width
R/S	Filter height/width
E/F	Output feature map height/width

where $O, I, W,$ and B are the matrices of ofm, ifmap, filters, and biases, respectively. U is a given stride size. After the convolutions some activation functions, such as the rectified linear unit (ReLU), are applied to introduce nonlinearity.

III. SYSTEM DESIGN

A. Architecture Overview

Fig. 2 illustrates the block diagram of the architecture and memory hierarchy of the CNN accelerator, which includes a PE array, global buffer, controller block, and ReLU activation function. This block is responsible for performing convolution operations, max pooling, and fully connected layers. The weights, biases, and input feature maps (ifmaps) are stored in off-chip DRAM and read into the accelerator via buffers to reduce latency when accessing off-chip memory. The memory hierarchy consists of three levels: off-chip DRAM, global buffer, and registers for each PE. Each PE in the PE array is accountable support three modes for computing a CONV, POOL, FC operation and accumulating the result through the internal PE register and a global buffer. The accelerator is controlled by a finite state machine (FSM) in controller block.

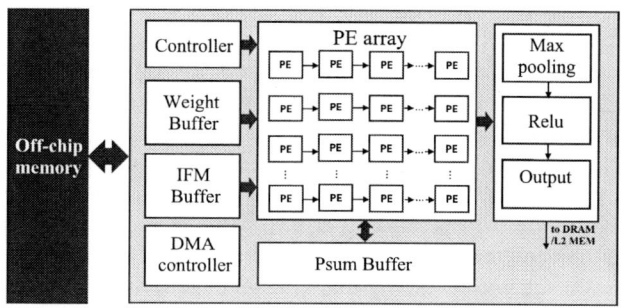

Fig. 2. System architecture overview.

B. Dataflow

Dataflow plays a crucial role in convolutional layers, as it directly impacts computation efficiency, memory access, and accelerator performance, including throughput and on-chip memory storage. Various dataflow strategies, such as weight stationary, input stationary, and output stationary, have been

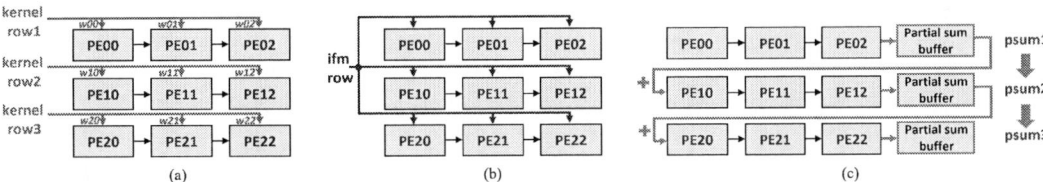

Fig. 3. Dataflow in a PE set for processing a 2-D convolution, suppose in the case where the size of R and S is 3 × 3: (a) Distribution of filter values to the PE array. (b) Streaming ifmap data into the PE array for parallel convolution execution. (c) Row-wise accumulation of partial sums.

studied to optimize these factors [17]. In deep neural networks with a large number of output channels, each filter must be accessed multiple times across different channel computations (e.g., layer CONV3 of VGG-16 contains 256 filters, each of which is applied to all 128 input channels). To mitigate this, weight reuse is maximized by keeping filter weights fixed within the PEs until computations for all corresponding input channels are completed. The WS dataflow minimizes data movement for all data types (ifmap, wgt, psum), simultane-ously and accounts for the energy costs at different levels of the memory hierarchy into account. Data retrieval from high-cost DRAM and the GLB is minimized by maximizing data reuse within low-cost RF memories. As shown in Fig. 3(b), each pixel in the ifmap from the same channel is sequentially streamed to the same set of $R \times S$ PEs. This allows the same pixel to be shared across multiple PEs over time, reducing the need for frequent memory accesses and improving data reuse efficiency. The psum generated in a PE at each cycle is either passed to its neighbor PE or stored back to the global buffer if psum has accumulated enough between one row of the filter weight matrix and S ifmap values. To further optimize data reuse, an asynchronous First-In First-Out (FIFO) buffer—referred to as the psum buffer in Fig. 3(c)—stores and reuses intermediate results for subsequent computations. The number of buffers equals the number of rows in the filter weight matrix, and the FIFO size is equal to the ofm width size, which corresponds to F as defined in (1) in Section II. This architecture minimizes the energy required for weight reads, maximizes convolutional operations, and enables efficient reuse of the filter: 1) *filter reuse*—each filter weight is reused $E \times F$ times within one ifmap channel; 2) *ifms reuse*—each ifm is reused $R \times S$ times within one filter weight matrix.

1-D Convolution Primitive PE array: The WS dataflow divides computation into 1-D convolution primitives, which run in parallel. Each primitive operates on one filter weight and one row of ifmap, generating a single psum per cycle. The psums from different primitives are accumulated to form row partial sums before being stored in the *Partial sum buffer*. By mapping each primitive to a PE, computations for each weight-value and pixel pair remain stationary within the PE. Due to the sliding window mechanism, each PE can utilizes local registers for both wgt reuse and psum accumulation. Since each PE processes only one weight and its corresponding ifmap value at a time, each PE requires two registers—one for

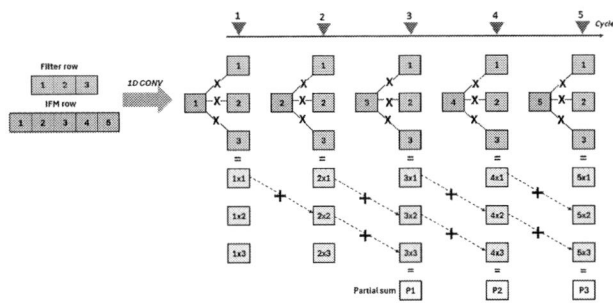

Fig. 4. Processing sequence of a 1-D convolution primitive in a PE. In this example, the filter row size (S) and the ifmap row size (W) are 3 and 5, respectively.

storing the filter weight and one for accumulating the psum. As shown in Fig. 4, the 1-D convolution of a single row of the ifmap and a corresponding row of filter weights is executed. Combined with the PE array in Fig. 3(a). At cycle 1, the ifm = 1 value is multiplied by three filter weights in PE_{00}, PE_{01}, and PE_{02}, resulting in the following states: $PE_{00}(1 \times 1)$, $PE_{01}(1 \times 2)$, $PE_{02}(1 \times 3)$. In cycle 2, the partial sums from the left PEs are passed to the neighboring PEs for accumulation, updating their states to $PE_{00}(1 \times 2)$, $PE_{01}(1 \times 1 + 2 \times 2)$, $PE_{02}(1 \times 2 + 2 \times 3)$. This accumulation process continues: at cycle 3, PE_{02} holds $PE_{02}(1 \times 1 + 2 \times 2 + 3 \times 3)$; at cycle 4, it updates to $PE_{02}(2 \times 1 + 3 \times 2 + 4 \times 3)$; and by cycle 5, the final result is computed as $PE_{02}(3 \times 1 + 4 \times 2 + 5 \times 3)$. This value represents the final partial sum for the weight row with a sliding 1-D window over ifmaps.

2-D Convolution Primitive PE array: While 1-D convo-lution applies the filter row-wise to individual rows of the ifmap, 2-D convolution extends this operation by incorporating multiple rows simultaneously. In this architecture, each PE performs a single MAC operation per cycle, and the compu-tation is organized into PE Sets of size $R \times S$, corresponding to the dimensions of the filter matrix (e.g., in the CONV1 of AlexNet, where the kernel size is 11 × 11, a total of 121 PEs operate in parallel). As shown in Fig. 3(a), the filter weight values are distributed across the PE array, where each PE_{ij} is assigned a unique filter weight value w_{ij}. For example, in the first row, $PE_{00}, PE_{01}, PE_{02}$ correspond to filter weight values w_{00}, w_{01}, w_{02}, respectively. As described in Fig. 3(b) and referenced in the 1-D convolution section, the

979-8-3315-1550-8/25 $31.00 © 2025 IEEE   21

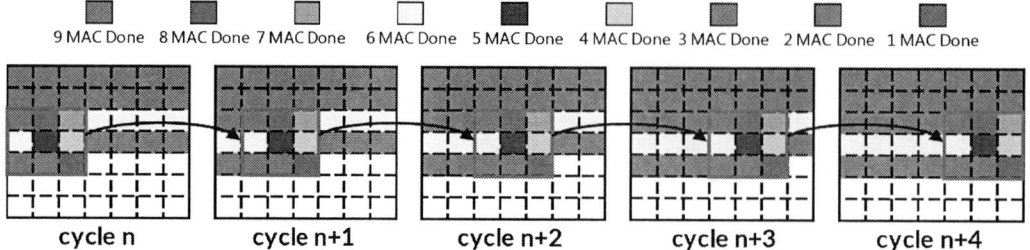

Fig. 5. 2-D convolutional operation with parallel execution, suppose in the case where the size of R and S is 3×3: Each partial sum is progressively computed per cycle. The colored cells represent partial sums, where those of the same color indicate identical completion levels.

ifmap row is broadcast downward, enabling each PE to process its filter weight value and ifmap value per cycle. Each PE row performs 1-D convolution, storing the psum in a *partial sum buffer*. Since the sliding window spans multiple rows, the psum from the upper row is propagated downward to the first PE in the row below, where it is accumulated with the corresponding psum, as illustrated in Fig. 3(c). Specifically, the psum generated from row 1 is stored in the first psum buffer as $psum1$. When processing row 2, its psum $psum2$ is accumulated with the previously stored value $psum1$, and the result is stored in the second psum buffer as $psum1 + psum2$. Similarly, when processing row 3, its psum $psum3$ is added to the accumulated value from the second buffer, resulting in $psum1 + psum2 + psum3$, which is stored in the third psum buffer. Once all rows within the sliding window are processed, the final accumulated value from the psum buffer storing $psum3$ is written to the *OFM buffer* as shown in Fig. 2.

C. Data Reutilization Strategy

Fig. 5 provides a more comprehensive view of maximizing data reutilization. Across cycles, the completion level of data progresses within an $R \times S$ partial sum array. According to (1), the maximum number of times a MAC operation in 1-D CONV can be reused is $R \times S$. This WS dataflow strategy effectively achieves the maximum possible data exploitation. For example, in Fig. 5, each cell represents an ofm element single channel with a corresponding filter weight. Different colors indicate varying levels of completion on a scale of 9 (assuming each ofm requires 9 MAC operations to complete). At cycle n, the psums are at different completion levels, distinguished by color. Moving to cycle $n+1$, the 9 cells within the computation window shift their colors up by one level (e.g., from *yellow* to *orange*), indicating that each has accumulated one additional MAC operation. As cycles progress, the psums gradually reach completion, while the computation window slides downward and repeats the process.

D. Support for Different Layer Types

Fully connected layers [1] follow the same matrix multiplication principles as CONV layers but with a kernel size of $R \times S = 1 \times 1$ and input channels $C = 1$. Unlike CONV layers, where weights remain stationary, FC layers require

both weights and ifm values to be streamed into the PE array, as each weight is used only once per computation.

Max-pooling layers [1] operate by selecting the maximum value within a $K \times K$ pooling window (typically 2×2 in AlexNet), preserving key features while discarding less relevant data. By replacing MAC operations with MAX comparisons in the ALU, the WS dataflow efficiently supports POOL layers. A multiplexer selects the maximum value per PE operation, as shown in Fig. 6, ensuring efficient processing as the pooling window slides with a stride equal to its size, preventing overlap.

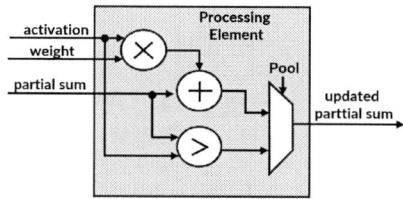

Fig. 6. Proposed PE structure supporting MAC operations and max-pooling mode.

E. Fix-Point representation

For efficient hardware deployment, all weights must be converted into fixed-point representations. To determine the optimal bit-width for fixed-point representation, the output range of each layer in our customized AlexNet needs to be identified. This model achieves a recognition accuracy of 98.62% on the MNIST test set. From Fig. 7, it can be observed that most of the weights in all layers of the network fall within the range of $[-1, 1]$. Therefore, only one bit is needed to represent the sign, with no additional bits are required for the integer part. After quantization, the fixed-point representation is selected to balance precision, range, and efficiency. Weights use a (1,0,12) format to minimize bit-width while preserving precision, whereas activations and images input adopt a (1,9,12) format to handle a larger dynamic range.

IV. EXPERIMENTAL SETUP AND RESULTS

A. Experimental Setup

The proposed accelerator architecture is implemented and tested on customized AlexNet-based model with the architec-

979-8-3315-1550-8/25 $31.00 © 2025 IEEE

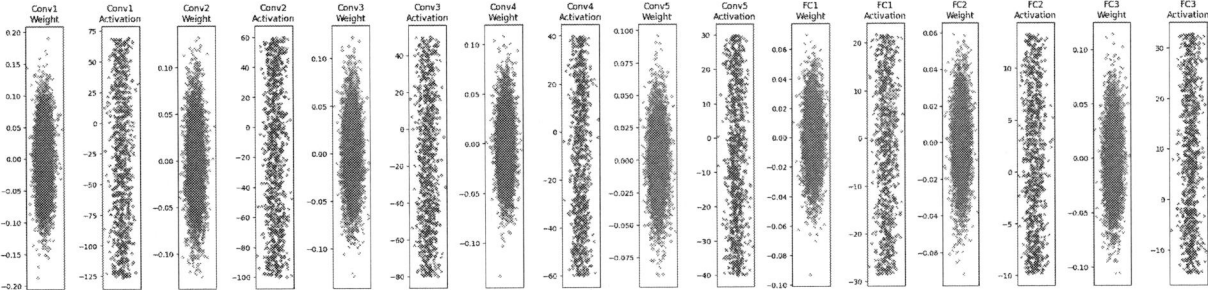

Fig. 7. Range of weight values across hidden layers: From left to right, the histograms represent the distribution of weights for the following layers in order: CONV1, CONV2, CONV3, CONV4, CONV5, and FC1, FC2, FC3.

ture shown in Table II. The training and extraction of post-training parameters for the network were conducted on Google Colab with GPU support (Tesla T4) using the PyTorch library. All network weights are represented in floating-point format. The model was trained using the Stochastic Gradient Descent (SGD) method with the following configuration parameters: Image dataset: MNIST (60,000 training images, 10,000 test images), Number of training samples: 60,000 images, Learning rate: 0.005, Momentum: 0.8, Batch size: 32, Epochs: 20.

TABLE II
CUSTOM ALEXNET NETWORK ARCHITECTURE

Layer	Type	Filter	Input	Output	MAC
0	Conv	11*11*32	227*227*3	55*55*32	11.71M
1	Max	3*3	55*55*32	27*27*32	
2	Conv	5*5*64	27*27*32	27*27*64	37.32M
3	Max	3*3	27*27*64	13*13*64	
4	Conv	3*3*128	13*13*64	13*13*128	12.46M
5	Conv	3*3*128	13*13*128	13*13*128	24.92M
6	Conv	3*3*64	13*13*128	13*13*64	12.46M
7	Max	3*3	13*13*64	6*6*64	
8	FC		6*6*64	4096	4.71M
9	FC		4096	512	1.05M
10	FC		512	10	5.12K
11	Softmax		10	10	
Total					104.65M
Conv vs FC					Conv: 94.5% FC: 5.5%

For deployment, the model runs at 250 MHz on the Zynq UltraScale+ MPSoC on ZCU104 board as shown in Fig. 1, achieving 33.53 FPS with $4.87W$ power consumption using a fixed-point representation. Weights are represented in the (1,0,12) format, while activations and input feature maps use the (1,9,12) format, as described in Section III-E. Table III provides a detailed summary of the resource utilization results for each convolution and fully connected layer of the CNN. As the sliding window size of the convolutional layers increases, so does the number of LUTs, FFs, and DSPs. Moreover, although CONV3, CONV4, and CONV5 layers have different numbers of input and output channels, they consume a similar amount of resources. This is because each channel computation is stored in a fixed buffer, enabling data reuse across channel passes, thereby optimizing partial sum accumulation.

TABLE III
SUMMARY TABLE OF METRICS FOR CUSTOM ALEXNET NETWORK

Layer	Power (mW)	Num.of MAC	Num.of PE	LUT	FF	DSP	BRAM
CONV1	333	11.71M	121	34303	73793	121	-
CONV2	84	37.32M	25	8166	18143	25	-
CONV3	26	12.46M	9	2301	4551	9	-
CONV4	26	24.92M	9	2301	4551	9	-
CONV5	26	12.46M	9	2301	4549	9	-
Sum (all)	6.34W	104.65M	183	93727	125021	183	162.5

B. Comparison with another work

The experimental results demonstrate that the proposed accelerator operates at 250 MHz with a fixed-point 8–16 bit representation, ensuring computational efficiency and reduced hardware complexity. Our design is implemented on the Zynq UltraScale+ ZCU104 FPGA and evaluated using the AlexNet model, in comparison with prior implementations, as shown in Table V. Our accelerator achieves DSP utilization of 183 (11%) and BRAM utilization of 162 (52%), both significantly lower than those of all previous implementations. Despite the reduction in resource utilization, the throughput of our accelerator is 54.90 GOPS, substantially higher than [30] (14.11 GOPS) and [28] (12.09 GOPS), while consuming only 6.34 W of power. This results in a power efficiency of 11.27 GOPS/W, outperforming [10] (0.80 GOPS/W) and [6] (7.48 GOPS/W). Notably, the design yields a DSP efficiency of 0.30 GOPS/DSP, the highest among all compared implementations. The accelerator also obtains a latency of 29.82 ms per image, which is considerably lower than those reported in [30] (102.76 ms), [18] (174 ms), and [28] (123.87 ms), in spite of using fewer resources. Correspondingly, the inference speed reaches 33.53 fps, which surpasses [30] (9.73 fps), [18] (5.75 fps), and [28] (8.07 fps). Competitive performance while maintaining low power consumption and resource utilization makes our design a practical choice for embedded and edge computing applications.

V. CONCLUSION

This study presents a hardware architecture optimized for CNNs, emphasizing efficient computation, reduced data movement, and resource management. By leveraging a weight-

TABLE IV
COMPARISON TO STATE-OF-THE-ART IMPLEMENTATIONS

	[30]	[11]	[18]	[19]	[28]	**Ours**
Year	2019	2020	2022	2022	2024	2024
FPGA	ZCU104	Arria 10 GX1150	Ultra96	Virtex-7 VX485T	7A100T	ZCU104
Frequency (MHz)	300	199	169	100	200	250
CNN	AlexNet	AlexNet	AlexNet	AlexNet	AlexNet	AlexNet
Precision	Int 8 bit	Fixed 8 bit	Fixed 16 bit	Float 32 bit	Float 32 bit	Fixed 8–16 bit
Logic Utilization	230K (33%)	129K (30%)	39.1K (18%)	445.5K (49%)	101.4K	218K (32%)
DSP Utilization	696 (40%)	300 (20%)	334 (92%)	2.5K (88%)	240	183 (11%)
BRAM Utilization	198.5 (64%)	1.1K (40%)	332 (76%)	644 (31%)	N/A	162 (52%)
Throughput (GOPS)	14.11	80.04	51	86.91	12.09	54.90
Power Consumption (W)	17.67	N/A	4.70	7.19	1.61	6.34
Power Efficiency (GOPS/W)	0.80	N/A	10.85	12.08	7.48	8.65
DSP Efficiency (GOPS/DSP)	0.02	0.27	0.15	0.03	0.05	**0.30**
Latency/Image (ms)	102.76	18.24	174	15.32	123.87	**29.82**
Inference Speed (fps)	9.73	54.82	5.75	65.27	8.07	**33.53**

stationary dataflow on a spatially arranged PE array, our design minimizes data transfers and maximizes filter reuse, lowering energy consumption and efficiently utilizing limited hardware resources. Experimental results demonstrate its suitability for low-power embedded systems. Future work will explore extending the PE array from 2D to 3D, as proposed in prior research [29], [9], to further enhance computational throughput, reduce latency, and improve scalability for design.

REFERENCES

[1] S. Albawi, T. A. Mohammed, and S. Al-Zawi, "Understanding of a Convolutional Neural Network," in ICET, 2017, Antalya, Turkey.

[2] The Swiss AI Lab IDSIA "Deep learning in neural networks: An overview" Jan. 2015.

[3] Genevieve Sapijaszko, Wasfy B. Mikhael, "An Overview of Recent Convolutional Neural Network Algorithms for Image Recognition" 2018.

[4] Yifan Chen; Shuang Wang; Yunpeng Ge, "A Survey on the Applications of Image Classification Based on Convolution Neural Network" 2022.

[5] H. Yu, L. T. Yang, Q. Zhang, D. Armstrong, and M. J. Deen, "Convolutional neural networks for medical image analysis: State-of-the-art, comparisons, improvement and perspectives," 2021.

[6] M. Sarigül, B. M. Ozyildirim, and M. Avci, "Differential convolutional neural network", Aug. 2019.

[7] R. Hameed, W. Qadeer, M. Wachs, O. Azizi, A. Solomatnikov, B. C. Lee, S. Richardson, C. Kozyrakis, and M. Horowitz, "Understanding sources of inefficiency in general-purpose chips", 2010.

[8] M. Horowitz, "Computing's energy problem (and what we can do about it)", Feb. 2014.

[9] Yu-Hsin Chen, Tushar Krishna, Joel S. Emer, Fellow, "Eyeriss: An Energy-Efficient Reconfigurable Accelerator for Deep Convolutional Neural Networks", 2017.

[10] S. Das, A. Roy, K. K. Chandrasekharan, A. Deshwal, and S. Lee, "A systolic dataflow based accelerator for CNNs", Oct. 2020.

[11] A. Ghaffari and Y. Savaria, "CNN2Gate: Toward Designing a General Framework for Implementation of Convolutional Neural Networks on FPGA," Apr. 2020.

[12] Q. Xiao, Y. Liang, L. Lu, S. Yan, and Y.W. Tai, "Exploring heterogeneous algorithms for accelerating deep convolutional neural networks on fpgas" 2017.

[13] N. Suda, V. Chandra, G. Dasika, A. Mohanty, Y. Ma, S. Vrudhula, J. Seo, and Y. Cao, "Throughput-optimized OpenCL-based FPGA accelerator for large-scale convolutional neural networks," 2016.

[14] Zhang, Chen, et al. "Optimizing FPGA-based accelerator design for deep convolutional neural networks" 2015.

[15] J. Qiu, J. Wang, S. Yao, K. Guo, B. Li, E. Zhou, J. Yu, T. Tang, N.Xu, S. Song et al., "Going deeper with embedded fpga platform for convolutional neural network," 2016.

[16] Guo, Kaiyuan, et al. "Angel-eye: A complete design flow for mapping CNN onto embedded FPGA" (2017).

[17] Q. Nie and S. Malik, "CNNFlow: Memory-driven Data Flow Optimization for Convolutional Neural Networks," Mar. 2023.

[18] A. Alhussain and M. Lin, "Hardware-Efficient Template-Based Deep CNNs Accelerator Design," FL, USA, 2022.

[19] S. M. Sait, A. El-Maleh, M. Altakrouri, and A. Shawahna, "Optimization of FPGA-based CNN accelerators using metaheuristics," Sep. 2022.

[20] K. T. Chitty-Venkata and A. K. Somani, "Neural architecture search survey: A hardware perspective", 2022.

[21] C. Zhang et al., "Caffeine: Towards Uniformed Representation and Acceleration for Deep Convolutional Neural Networks" 2016.

[22] Y. Ma et al., "Optimizing Loop Operation and Dataflow in FPGA Acceleration of Deep Convolutional Neural Networks," in FPGA, 2017.

[23] U. Aydonat et al., "An OpenCL Deep Learning Accelerator on Arria 10," in FPGA, 2017.

[24] J. Zhang et al., "Improving the Performance of OpenCL-based FPGA Accelerator for Convolutional Neural Network," in FPGA, 2017.

[25] L. Bai, Y. Zhao, and X. Huang, "A CNN accelerator on FPGA using depthwise separable convolution" Oct. 2018.

[26] Q. Xiao, Y. Liang, L. Lu, S. Yan, and Y.-W. Tai, "Exploring heterogeneous algorithms for accelerating deep convolutional neural networks on FPGAs" Jun. 2017.

[27] X. Wei, C. H. Yu, P. Zhang, Y. Chen, Y. Wang, and H. Hu, "Automated systolic array architecture synthesis for high throughput CNN inference on FPGAs" Jun. 2017.

[28] Y. Xu, J. Luo, and W. Sun, "Flare: An FPGA-Based Full Precision Low Power CNN Accelerator with Reconfigurable Structure," *Sensors*, vol. 24, no. 7, p. 2239, Mar. 2024.

[29] J. M. Joseph, A. Samajdar, L. Zhu, R. Leupers, S. K. Lim, and T. Pionteck, "Architecture, dataflow and physical design implications of 3D-ICs for DNN-accelerators" Apr. 2021.

[30] M. Zhang, L. Li, H. Wang, Y. Liu, H. Qin, and W. Zhao, "Optimized Compression for Implementing Convolutional Neural Networks on FPGA," Electronics, vol. 8, no. 3, p. 295, Mar. 2019.

979-8-3315-1550-8/25 $31.00 © 2025 IEEE

Efficient AI Model and Hardware Architecture Based on CNN for Arrhythmia Prediction

Duc-Huy Pham[1,2], Thi-Minh-Tuyen Huynh[1,2], Tuan-Kiet Tran[1,2], Thanh-Dat Bui[1,2],
Cong-Kha Pham[3], and Huu-Thuan Huynh[1,2]

[1] University of Science, Ho Chi Minh City, Vietnam
[2] Vietnam National University, Ho Chi Minh City, Vietnam
Emails: {htmtuyen, trtkiet, hhthuan}@hcmus.edu.vn
[3] University of Electro-Communications (UEC), Tokyo, Japan
Email: phamck@uec.ac.jp

Abstract—The development of Edge Artificial intelligence (AI) is increasingly prevalent in real-time recognition, classification, and prediction applications, driven by advances in embedded hardware and lightweight, optimized AI models with low computational requirements. However, deploying lightweight models also faces challenges on hardware with limited resources, including ensuring computational efficiency and utilization, as well as controlling latency and accuracy. In this study, we propose an effective algorithm for the Convolutional Neural Network (CNN) model in arrhythmia prediction applications. Additionally, a hardware architecture is designed to be compatible with this model. The model is optimized through compression and quantization techniques to integer format, making it suitable for edge computing devices. The hardware architecture is implemented using the proposed methods, including the Weight Stationary architecture of the Systolic Array (SA), efficiently organizing data flow, and integrating 8-bit integer computation algorithms for full Processing Element (PE) utilization and resource optimization. Our experimental results can reach over 80% of PE utilization and throughput achieved 60.6 GOP/s.

Index Terms—Edge AI, CNNs, ECG classification, hardware accelerators, Systolic Array, PE utilization.

I. INTRODUCTION

Recent research has focused on developing lightweight AI models to optimize performance in edge computing devices. By refining network architectures or applying model compression techniques, such as reducing the number of parameters and layers, AI models can operate efficiently on resource-constrained hardware. Field-programmable gate arrays (FPGAs) offer the ability to integrate multiple hardware functions and provide flexibility in programming, allowing them to meet the complex computational requirements of neural network models [1]. In applications such as healthcare, disease diagnosis systems require high precision and the ability to provide the earliest possible predictions. Effective optimization of AI models requires a co-design approach that integrates both hardware acceleration and software-based algorithmic improvements [2]. Widely used compressed models such as MobileNets, SqueezeNets, ShuffleNets, and so on are characterized by a reduction of parameters. Still, they typically have a large number of layers and varying

kernel sizes, affecting hardware deployment efficiency. Additionally, model compression techniques such as running can degrade parallel computational capabilities, while compressed sparse models may lead to suboptimal hardware utilization [2]. Therefore, selecting an appropriate model is crucial for designing a hardware accelerator with minimal trade-offs. In CNNs model, convolutional layers typically account for approximately 85% of the total layers [3]. These layers perform convolution operations between the input image and the kernel sets, resulting in frequent data reuse. Systolic Array is a parallel matrix computation architecture proposed by H.T. Kung in 1978 [4]. Due to its efficient handling of matrix operations, this architecture has gained significant interest in the research community and has been widely applied in CNN models to accelerate convolutional and fully connected layers [5]. Several studies have focused on leveraging its advantages, including applying it to DNN (Deep Neural Network) models for matrix computation, reducing latency while achieving high throughput and low cost [6], [7], minimizing hardware design complexity and high flexibility [8], [9], and efficiency compared to other designs [8]. However, most designs involve trade-offs to balance performance and resource utilization [10]. To enhance computational efficiency, Shi et al. [11] proposed a parallel computation design for convolution kernels using an addition tree architecture for 5×5 kernel size. In study [12], a SA design was employed to reuse both data and weights, combined with a multi-level storage technique to improve computational efficiency. Implementing lightweight CNN models on hardware can maximize resource utilization, reduce external memory reliance, and improve computational performance [13]. By combining this approach with a fully pipelined operation structure, the design achieves 53.3 GOP/s on LeNet and 41.9 GOP/s on MobileNet. In CNN models, as the network depth increases, input data size decreases, and the use of varying kernel sizes may result in inactive PEs in certain layers. This leads to resource inefficiency and a reduction in processing throughput density [13]. Furthermore, this paper presents a hardware architecture for predicting arrhythmia symptoms, leveraging a Systolic Array, aiming to achieve high computational efficiency and optimal resource utilization. The main contributions of this study include:

- Implementing quantization for 8-bit integer convolution while maintaining high accuracy.
- Mapping convolution accelerators to CNN model sizes to maximize PE utilization.
- Efficiently managing intermediate data streams to achieve high performance.

II. PROPOSED METHODOLOGY

A. CNN Model

The CNN model is implemented to classify arrhythmia symptoms and assist in predicting cardiovascular diseases. The network architecture is based on traditional VGGNet and AlexNet, which are characterized by a large number of layers and parameters, posing challenges in computation and storage. To mitigate the parameter overhead, the Knowledge Distillation method [14] is applied. Specifically, the teacher model consists of 14 layers with a total of 8,890,824 parameters, achieving an accuracy of 99.9%. Meanwhile, the student model is optimized by reducing the architecture to 12 layers, while also decreasing the number of kernels by more than 50%, resulting in a total of 1,070,104 parameters, while still maintaining an accuracy of 99%.

TABLE I: Summary of the CNN Model Architecture.

Layer	Output Size	Kernel Size
Input	(128, 128, 1)	-
Conv2D_0	(126, 126, 16)	(3, 3, 1, 16)
Conv2D_1	(124, 124, 16)	(3, 3, 16, 16)
Conv2D_2	(60, 60, 32)	(3, 3, 16, 32)
Conv2D_3	(28, 28, 64)	(3, 3, 32, 64)
Conv2D_4	(12, 12, 128)	(3, 3, 64, 128)
Conv2D_5	(10, 10, 128)	(3, 3, 128, 128)

Table I presents the statistics of convolutional layers in the model, detailing the output size after each convolution operation and the corresponding number of kernels. All kernels have a fixed size of 3×3, and the number of kernels increases following a power-of-two progression. However, the model parameters used during inference are typically represented as 32-bit floating-point numbers (FP32), which poses significant challenges for hardware implementation. Due to the wide numerical representation range of floating-point values, models with a large number of parameters require extensive computational resources, increasing the complexity of hardware design. Moreover, utilizing FPGA resources, such as digital signal processing (DSP) blocks, for FP32 computations leads to substantial resource consumption, potentially affecting overall system efficiency. To address this issue, numerous studies [15], [16] have proposed alternative methods, such as mapping model parameters to integer representations, which facilitate computation while maintaining model accuracy. Notable techniques in this domain include Quantization Aware Training (QAT) and Post-Training Quantization (PTQ). In this study, we adopt the Post-Training Quantization approach, utilizing 8-bit integer representation to optimize computational efficiency while preserving model

performance. The general formula for convolution operations is given as follows:

$$Y = b + \sum XW \qquad (1)$$

In the convolution operation, the key components include X as the input image, W as the convolutional layer weights, b as the bias parameter, and Y as the output of the convolution. When implemented on hardware, representing these parameters in the floating-point format can lead to high computational costs. Therefore, to reduce computational complexity and optimize hardware resources, model parameters are often mapped from a floating-point to an 8-bit integer domain through a quantization process. This process is performed using two key factors: scale (s) and zero-point (z), which help preserve model accuracy while minimizing the need for floating-point computations. The quantization formula is defined as follows:

$$X_q = round(\frac{x}{s} + z) \qquad (2)$$

Where x is the original floating-point value, s is the scale factor that maps the range of floating-point values to the smaller range of 8-bit integers, effectively determining the precision, z is zero-point, an integer used to shift the quantized values so that zero in float maps to an integer (often in the center of the integer range), and X_q is the quantized value as an integer. Based on equations (1) and (2), and following the method presented in [17], all components involved in the convolution operation are quantized. Specifically, the bias, weights, and activations are represented in their quantized forms: bias as B_q with scale s_b and zero-point z_b, weights as W_q with scale s_w and zero-point z_w, and inputs as X_q with scale s_x and zero-point z_x. The output is then represented in quantized form as Y_q, with scale s_y and zero-point z_y. By incorporating these quantized representations into the convolution operation and rearranging terms, the final quantized inference formula can be expressed as formula (3):

$$Y_q = z_y + \frac{s_b}{s_y}(B_q - z_b) + \frac{s_x s_w}{s_y}\left[\left(\sum X_q W_q\right) \\ - \left(z_w \sum X_q\right) - \left(z_x \sum W_q\right) + z_x z_w\right] \qquad (3)$$

In an integer quantized model, the ReLU activation function is typically fused with the Convolution layer using layer fusion to optimize computation. The output y' of the ReLU operation contains only non-negative values, as expressed in equation (4):

$$y' = \text{ReLU}(y, 0) = \begin{cases} 0 & \text{if } y < 0 \\ y & \text{if } y \geq 0 \end{cases} \\ = s_{y'}(Y'_q - z_{y'}) \qquad (4)$$

According to the results after training the model, both variables z_b and z_W are equal to zero. Based on the properties of the quantization method, we have $s_b = s_x s_w$, allowing the expression to be simplified when applying ReLU after

979-8-3315-1550-8/25 $31.00 © 2025 IEEE

transformation of activation output quantization Y_q' as shown in equation (5):

$$Y_q' = \begin{cases} z_{y'} & \text{if } M < N \\ \frac{s_b}{s_{y'}}[B_q + (\sum X_q W_q - z_x \sum W_q)] + z_{y'} & \text{if } M \geq N \end{cases}$$
(5)

With $M = \sum X_q W_q$; $N = \sum z_x W_q - B_q$. In the transformed expression, the components N and the shift factor of $\frac{s_b}{s_y}$ as defined in [17], can be precomputed offline before inference. This reduces the computational cost of the scale-down process to 8-bit during convolution operations. The 8-bit integer representation model has been implemented on low-performance edge computing embedded devices, maintaining accuracy while achieving impressive inference times. The evaluation results demonstrate the hardware-friendly nature of this approach, optimizing computational efficiency and resource utilization. Based on the previously presented integer computation transformation formula, a C-based model was implemented to simulate the hardware architecture for performing inference. The output of each layer and the overall accuracy of the C model were compared with the reference results from the Python model. Using the same algorithm and post-training weight set, the accuracy of the C model was used to verify that the hardware accelerator's outputs match those of the C model at each layer as well as the entire model.

B. Hardware Architecture Design

1) Overview Architecture: Figure 1 presents the overall architecture of the proposed CNN hardware accelerator, designed to optimize the computational efficiency of the convolutional layer. The central component is a 9×16 processing element (PE) array, which undertakes all convolution-related computations. The weights and input data are stored in external off-chip memory and are retrieved using an input loader and a weight fetcher. The output from the Systolic Array undergoes processing through a ReLU activation module and a scale-down module before being written back to external memory. Additionally, an output buffer is employed to cache intermediate values during the convolution computations, thereby enhancing processing throughput.

2) Sytolic Array Architecture: In this study, we proposed utilizing the Weight Stationary architecture, a variant of the Systolic Array (SA) [18], as illustrated in Figure 2b. The key feature of this architecture is that weights are preloaded into each column of the PE array. A key characteristic of this architecture is that the weights remain stationary within each PE and do not change until a new set of weights is fetched into the PE array for the next channel. During computation, each input channel sequentially slides through the corresponding channels of all kernels in that layer. As shown in Figure 2, the weight k_0 of each filter is loaded into individual PE columns, while the input feature map (ifm) moves across all these columns containing k_0. The CNN architecture is designed with an initial filter count of 16, which increases exponentially by powers of 2. Consequently, the number of

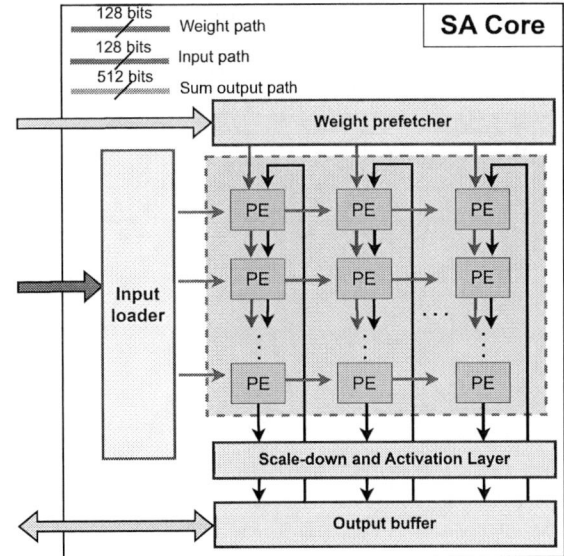

Fig. 1: Systolic Array Core.

columns in the PE array is set to 16. Furthermore, since all kernels have a fixed size of 3×3, the PE array consists of 9 rows to accommodate convolution operations efficiently. This array configuration is chosen to maximize weight reuse, minimize data transfer overhead, and enhance computational performance. Our CNN model workload significantly exceeds the capacity of the Systolic Array, the array is insufficient for handling complete loop unrolling. Consequently, the workload is partitioned into smaller tiles, which are sequentially processed by the 9×16 PE array.

Figure 4 shows the architecture of our processing element. It incorporates an 8-bit multiplier to perform feature and weight value multiplications, alongside a 32-bit accumulator to aggregate the multiplier results and the partial sum (psum) of the above PE into the current psum. To mitigate the latency associated with external memory access, each PE is equipped with a small buffer to store the next weight required for computation. This buffer has a capacity of 16 bytes, which is optimized for the input dataflow described in Input datapath part.

3) Data path: The computation array is structured from PEs interconnected through data streams, enabling flexible data flow control during computations. This design consists of three main data paths, including the Input datapath, the Weight datapath, and the Output datapath. Each is responsible for a specific function in the convolutional processing pipeline.

- Input datapath: As illustrated in the overall architecture in Figure 1, the data path of the input is transferred from external memory via a 128-bit Avalon bus interface. This data is then reformatted into columns before computation within the PE array. The input loader plays a crucial role in buffering input data and converting it into columns (Image to Column). However, a significant challenge arises from the unformatted data of post-convolutional computation, which can impede subsequent processing

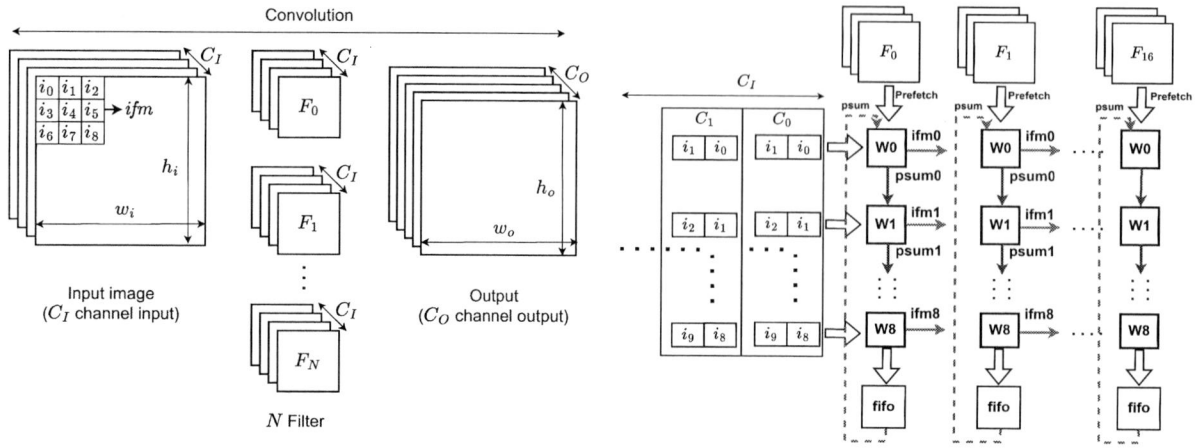

(a) Computation of Convolutional Layer.

(b) Weight Stationary and Data Flow of PE Array Architecture.

Fig. 2: Applying a given CNN layer to Systolic Array: (a) Computation of Convolutional layer, (b) Weight Stationary and Data Flow of PE Array Architecture.

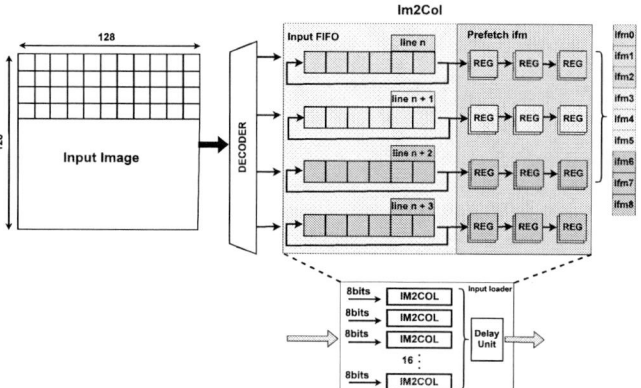

Fig. 3: Input Loader Architecture.

stages. To address this issue, 16 Im2Col modules have been implemented within the input loader. According to each calculating input feature map channel, the corresponding Im2Col block is selected. This implementation ensures that the workload is partitioned into multiple partitions, each comprising 16 kernels, thereby facilitating smoother processing. Each Im2Col block is dedicated to processing one line of the current input feature map, as shown in Figure 3. When the kernel slides to the end of the input feature map line, the input loader switches to the next indexed Im2Col block and loops until the 16th Im2Col block is reached, yielding the final value of the 3D convolution for each line. Simultaneously, a new line is loaded into all input loaders. This design allows the proposed architecture to hide the latency of external memory access while effectively handling unformatted data and ensuring the PEs remain consistently busy. This method enables the architecture to consume a minimal size of intermediate buffer while maintaining throughput close to theoretical values. In order to reduce the latency of external memory access, we propose using a buffer to store four rows of the input feature map (three

processing rows and one prefetching row). The system preloads one additional row while waiting for Im2Col to complete. With a stride of 1, the system then processes the next three rows. This data-loading approach helps reduce internal memory usage while maintaining high computational efficiency for the SA core. For each row, we prefetch three pixels to hide the latency of processing. Post-Im2Col data is passed through a delay unit and then arranged into nine rows, each delayed by one clock cycle and propagating in a pipeline manner within the SA architecture. The input columns sequentially move across the PE array columns, corresponding to the input feature map from the left side being equal to the input feature map from the right side ($ifmap\_f\_left = ifmap\_f\_right$). This structured data flow optimizes weight reuse and significantly reduces input data transfer overhead.

- Weight Datapath: The Weight Stationary architecture is a method that keeps weights fixed within each PE. According to the model parameter summary of Table I, the number of weights increases in deeper layers, particularly in fully connected layers. Therefore, storing weights directly in external memory is necessary. However, since external memory stores not only the weights but also the output feature maps, frequent memory accesses result in significant power consumption and increased latency during read/write operations. This is primarily due to the high reuse of input feature maps for convolution operations. The CNN model is designed with the number of kernels increasing exponentially (powers of two), starting from 16 kernels, facilitating the efficient reuse of the PE array. To reduce weight fetch time from external memory and avoid processing delays in PE operations, we incorporate a small local memory in each PE, capable of storing 16 weight values, as shown in Figure 4, each represented in an 8-bit format. After processing each channel of the input feature map, new weights stored in the buffer will be loaded for calculation.

979-8-3315-1550-8/25 $31.00 © 2025 IEEE

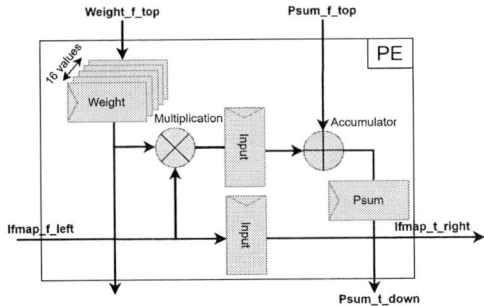

Fig. 4: Processing Element Design.

TABLE II: FPGA Resource Utilization.

Resource	BRAMs	DSP	Logic
Utilization	228	176	24878
Available	11721	5760	933120
Utilization (%)	2	3	2.6

- Output Datapath: Following the convolutional computation, the output must undergo scale-down prior to being stored in external memory. Each column is allocated for the computation of a specific filter. Intermediate output values are temporarily stored in a buffer and subsequently reloaded into the partial sum (psum) of the first row of the PE array to compute the next channel of the current processing line. The red arrow in Figure 2b illustrates the data flow of the intermediate output convolution. This data path is active only from the second channel to the 16th channel during the computation process of each line. Therefore, only a FIFO with a depth of 128 is required for each column of the PE array, ensuring that the PE remains consistently active. In instances where the number of filters exceeds 16, the computation must be divided into multiple partitions. After completing 16 iterations of calculation, the intermediate data is stored back into external memory. Subsequently, new input data and weights are loaded and reloaded upon the completion of calculations for each line.

III. EXPERIMENTS AND RESULTS

The CNN model is applied with a specific objective to achieve optimal results and is evaluated on edge computing devices. Based on this, we develop a dedicated hardware accelerator for a specific application. Furthermore, using the same algorithm, dataset, and trained weights, the outputs generated by the hardware implementation were validated against those from the C model, which simulates the hardware behavior. The C model outputs had previously been verified against those produced by the standard Python implementation. Therefore, the correctness and numerical accuracy of the hardware inference are preserved. We assess computational performance on the Intel Stratix 10 GX using Verilog. This board provides a total of 933K Adaptive Logic Modules (ALMs), 5760 Digital Signal Processing (DSP) slices, and

TABLE III: Comparison With Different Hardware Accelerators.

Approaches	[13]	[11]	[12]	Ours
BRAMs	16	530	54.5	228
DSP	301	449	204	176
FF	64747	95523	66569	33142
Logic	45904	125350	25276	24878
Frequency (MHz)	100	150	150	200
Throughput (GOP/s)	53.3	115.2	28.8	60.6
Throughput Density(GOP/s/DSP)	0.122	0.26	0.14	0.344
Power (W)	0.49	N/A	3.8	8.7
Precision (bits)	16-bit fixed	16-bit fixed	16-bit fixed	8 bits integer
Platform	XAZU3EG	Zynq ZC706	ZCU102	Intel Stratix 10

11.7K Block BRAMs (M20K). The design is synthesized and simulated to analyze resource utilization and computational efficiency. Table II summarizes the resource utilization, indicating that the proposed design consumes only a small fraction of the available hardware resources. The maximum operating frequency, as determined by Quartus, is reported in Table III, where the architecture achieves 200 MHz. The total throughput of all convolution layers can reach 60.6 GOP/s, with a throughput density of 0.344 GOP/s. The compact model proposed in this study achieved high computational density while maintaining a balanced resource utilization by employing 8-bit integer arithmetic. A key metric for evaluating hardware efficiency in Systolic Array architectures is the PE utilization rate, particularly when processing computational workloads for a CNN model. High PE utilization is indicative of optimized hardware performance, minimizing idle resources, and maximizing throughput. Figure 5 shows the PE utilization rate of our architecture when performing six convolution layers of our model with about 246,672 pa-

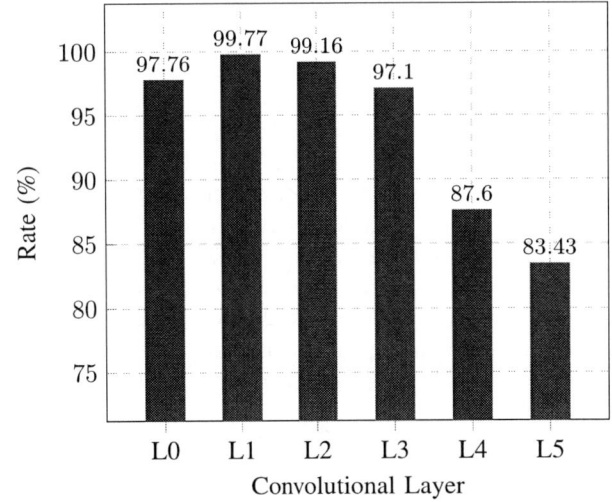

Fig. 5: PE Utilization Rate Across Convolutional Layers.

rameters. The results indicate that the experimental outcomes closely approximate the peak utilization rate. The highest value is observed at layer 1 and decreases for subsequent layers. According to the study [19], the PE utilization rate is calculated using the formula given in equation (6).

$$Util_{cycle} = \frac{Num.activatedPEs}{Num.totalPEs} \quad (6)$$

The elevated utilization in early layers (L0, L1, and L2) can be attributed to the compatibility between the kernel dimensions and the PE array size, which minimizes the overhead associated with loading new weights. In contrast, layers L4 and L5 experience slightly lower utilization due to the increased number of weight sets, necessitating additional data transfers from external memory. Overall, our proposed architecture achieves high efficiency with the convolution layer because the data in the output buffer retain their original format, eliminating the need for reshaping and allowing the input loader block to directly use it in the next layer. Additionally, the data preparation process (pre-Im2Col) occupies a small proportion compared to the convolution computation process.

IV. CONCLUSION

A hardware-oriented compression architecture was designed based on a simplified and optimized VGGNet CNN model, utilizing Knowledge Distillation and Quantization techniques. These methods achieved an 8.3× reduction in model weights, significantly lowering resource consumption, computational complexity, and memory requirements, while maintaining minimal accuracy loss. In the case study, the architecture follows the classical Weight Stationary structure of the Systolic Array (SA), characterized by uniform kernel sizes (3×3) and an exponentially increasing number of filters. This design enables efficient reuse of the PE array, improves PE utilization, and balances hardware resources. The implementation achieves an operating frequency of 200 MHz with 144 PEs and demonstrates higher throughput density compared to related studies. However, some limitations remain, particularly the latency introduced by the Im2Col process, which could be mitigated using techniques such as pipelining and prefetching. In addition, the architecture remains limited in flexibility, making it less suitable for lightweight models such as those with small-scale convolutions. Furthermore, the training data were derived from the MIT-BIH golden dataset, which may not fully reflect real-world scenarios. However, the applied weight compression and network depth reduction techniques effectively reduce hardware demands while preserving competitive performance, indicating a strong potential for deployment in high-efficiency edge AI systems, particularly in medical applications. In the future, based on the current architecture, further development will be undertaken to ensure the hardware can accommodate various kernel sizes while maintaining high performance. Additionally, integration into SoC will be implemented to enable real-time recognition.

ACKNOWLEDGMENT

This research is funded by University of Science, VNU HCM under grant number: DTVT 2024-02.

REFERENCES

[1] T. Mohaidat and K. Khalil, "A survey on neural network hardware accelerators," *IEEE Transactions on Artificial Intelligence*, 2024.

[2] T. Yuan, W. Liu, J. Han, and F. Lombardi, "High performance cnn accelerators based on hardware and algorithm co-optimization," *IEEE Transactions on Circuits and Systems I: Regular Papers*, vol. 68, no. 1, pp. 250–263, 2020.

[3] Y. Zhao, X. Chen, R. Xu, J. Wei, J. Lu, and B. Yuan, "A research and design of lightweight convolutional neural networks accelerator based on systolic array structure," in *Proceedings of the 6th International Conference on Robotics and Artificial Intelligence*, 2020, pp. 241–247.

[4] H.-T. Kung, *Why systolic architecture?* Design Research Center, Carnegie-Mellon University, 1982.

[5] R. Xu, S. Ma, Y. Guo, and D. Li, "A survey of design and optimization for systolic array-based dnn accelerators," *ACM Computing Surveys*, vol. 56, no. 1, pp. 1–37, 2023.

[6] W. Huang, H. Wu, Q. Chen, C. Luo, S. Zeng, T. Li, and Y. Huang, "Fpga-based high-throughput cnn hardware accelerator with high computing resource utilization ratio," *IEEE Transactions on Neural Networks and Learning Systems*, vol. 33, no. 8, pp. 4069–4083, 2021.

[7] Y. Xu, S. Wang, N. Li, and H. Xiao, "Design and implementation of an efficient cnn accelerator for low-cost fpgas," *IEICE Electronics Express*, vol. 19, no. 19, pp. 20 220 370–20 220 370, 2022.

[8] Y.-H. Chen, T.-J. Yang, J. Emer, and V. Sze, "Eyeriss v2: A flexible accelerator for emerging deep neural networks on mobile devices," *IEEE Journal on Emerging and Selected Topics in Circuits and Systems*, vol. 9, no. 2, pp. 292–308, 2019.

[9] R. Xu, S. Ma, Y. Wang, X. Chen, and Y. Guo, "Configurable multi-directional systolic array architecture for convolutional neural networks," *ACM Transactions on Architecture and Code Optimization (TACO)*, vol. 18, no. 4, pp. 1–24, 2021.

[10] T. Mohaidat and K. Khalil, "A survey on neural network hardware accelerators," *IEEE Transactions on Artificial Intelligence*, 2024.

[11] Y. Shi, T. Gan, and S. Jiang, "Design of parallel acceleration method of convolutional neural network based on fpga," in *2020 IEEE 5th International Conference on Cloud Computing and Big Data Analytics (ICCCBDA)*. IEEE, 2020, pp. 133–137.

[12] C. Zhang, X. Wang, S. Yong, Y. Zhang, Q. Li, and C. Wang, "An energy-efficient convolutional neural network processor architecture based on a systolic array," *Applied Sciences*, vol. 12, no. 24, p. 12633, 2022.

[13] R. Wu, B. Liu, P. Fu, and H. Chen, "An efficient lightweight cnn acceleration architecture for edge computing based-on fpga," *Applied Intelligence*, vol. 53, no. 11, pp. 13 867–13 881, 2023.

[14] G. Hinton, "*Distilling the Knowledge in a Neural Network*," *arXiv preprint arXiv:1503.02531*, 2015.

[15] T. Sledevič and A. Serackis, "mnet2fpga: A design flow for mapping a fixed-point cnn to zynq soc fpga," *Electronics*, vol. 9, no. 11, p. 1823, 2020.

[16] M. Cho and Y. Kim, "Fpga-based convolutional neural network accelerator with resource-optimized approximate multiply-accumulate unit," *Electronics*, vol. 10, no. 22, p. 2859, 2021.

[17] B. Jacob, S. Kligys, B. Chen, M. Zhu, M. Tang, A. Howard, H. Adam, and D. Kalenichenko, "Quantization and training of neural networks for efficient integer-arithmetic-only inference," in *Proceedings of the IEEE conference on computer vision and pattern recognition*, 2018, pp. 2704–2713.

[18] A. Samajdar, J. M. Joseph, Y. Zhu, P. Whatmough, M. Mattina, and T. Krishna, "A systematic methodology for characterizing scalability of dnn accelerators using scale-sim," in *2020 IEEE International Symposium on Performance Analysis of Systems and Software (ISPASS)*. IEEE, 2020, pp. 58–68.

[19] R. Xu, S. Ma, Y. Wang, X. Chen, and Y. Guo, "Configurable multi-directional systolic array architecture for convolutional neural networks," *ACM Transactions on Architecture and Code Optimization (TACO)*, vol. 18, no. 4, pp. 1–24, 2021.

2025 10th IEEE International Conference on Integrated Circuits, Design, and Verification (ICDV)

High-PSR Capacitor-Less LDO with Enhanced Bandgap Reference in 65nm CMOS Technology

Viet Ngo[1], Cuong Huynh[1]

[1]*Faculty of Electrical and Electronics Engineering*
Ho Chi Minh City University of Technology, Ho Chi Minh City, Viet Nam
Email: viet.ngo208k21@hcmut.edu.vn, hpmcuong@hcmut.edu.vn

Abstract—This paper presents a capacitor-less low-dropout regulator (LDO) designed and simulated in a 65nm CMOS process, achieving high power supply rejection (PSR) across the low-to-mid frequency range while ensuring enhanced loop stability. The proposed LDO integrates a bandgap reference (BGR) with two key techniques to improve PSR performance. Stability is further enhanced through Miller compensation and feedforward compensation. Overall results indicate that the LDO achieves a PSR ranging from 70 dB to 96 dB at 1 kHz under process, voltage, and temperature (PVT) variations. The measured line regulation is 0.04 mV/V, with a quiescent current of 140 µA. The regulator supports a maximum load current of 300 mA while operating over a supply voltage range of 1.8V to 2.5V.

Index Terms—Low dropout regulator (LDO), LDO regulator, Capacitorless LDO, Bandgap reference (BGR), frequency compensation, power-supply rejection (PSR).

I. INTRODUCTION

Modern Power Management Integrated Circuits (PMICs) commonly incorporate both switching DC-DC converters and low dropout regulators (LDOs) to deliver stable and noise-free voltage supplies, as illustrated in Fig. 1. These circuits are vital for noise-sensitive applications, including analog-to-digital converters (ADCs) and voltage-controlled oscillators (VCOs) [2, 3]. To meet these requirements, an LDO must ensure high power supply rejection (PSR), minimal dropout voltage, and low power consumption while maintaining efficient operation across various load conditions. In System-on-Chip (SoC) applications,

LDO design must also minimize silicon area while maintaining high performance to optimize overall cost and power efficiency. Traditional LDOs rely on large off-chip capacitors for stability [1], [2], which are unsuitable for SoCs. To address this, capacitor-less LDOs (CL-LDOs) address this by using small on-chip capacitors and advanced frequency compensation to ensure stability under all conditions.

PSR is a critical LDO performance metric and has been extensively studied [3], [4]. Ripple feedforward techniques have shown promise in enhancing PSR [5], but most designs assume an ideal voltage reference, which is unrealistic. In reality, the PSR of the bandgap reference (BGR) significantly impacts the overall PSR of the LDO, particularly at low and moderate frequencies. A common approach to mitigate this issue is inserting an RC low-pass filter at the BGR's reference voltage (V_{REF}) input. However, large RC components are area-inefficient for SoCs.

This paper proposes a capacitor-less LDO regulator that achieves high PSR in the low-to-mid frequency range (100 Hz to 100 kHz) by incorporating a high-PSR bandgap reference design. Additionally, the proposed design maintains full stability without large external capacitors using optimized frequency compensation.

The remainder of this paper is organized as follows: Section II explores the relationship between LDO and BGR in PSR performance. Section III presents the proposed high-PSR BGR design. Section IV details the capacitor-less LDO architecture. Section V discusses simulation results, and Section VI concludes the paper.

II. CORRELATION BETWEEN THE PSR OF LDO AND BANDGAP REFERENCE

In the conventional LDO architecture shown in Fig. 2 of [7], the PSR performance is determined by three

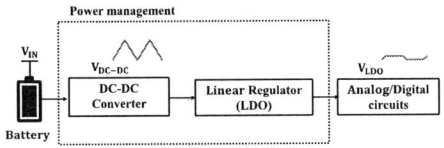

Fig. 1. PMIC application

979-8-3315-1550-8/25 $31.00 © 2025 IEEE

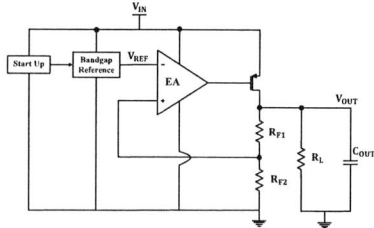

Fig. 2. Conventional LDO Architecture

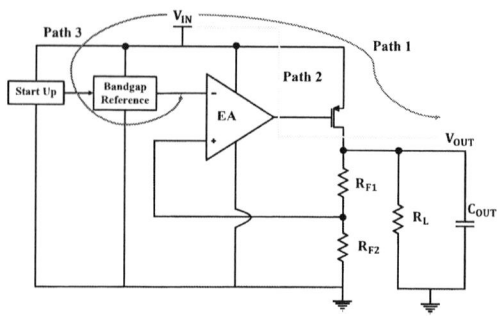

Fig. 3. PSR path of LDO

primary paths in Fig. 3. Path 1 is governed by the feedback loop of the LDO and the output impedance of the pass device. Path 2 arises from the PSR characteristics of the error amplifier. Finally, path 3 arises from the finite PSR of the bandgap reference.

Furthermore, PSR of LDO can be analyzed across three frequency regions. The first region corresponds to DC and low frequencies (up to approximately 100 Hz), denoted as PSR_{dc}. The second region, covering moderate frequencies (1 kHz – 100 kHz), is referred to as PSR_{MF}. Lastly, the high-frequency region, where PSR is evaluated beyond 100 kHz, is designated as PSR_{HF}.

The first thing to analysis is the output impedance of LDO at node V_{OUT} is given by

$$Z_{OUT} = [(R_{F1} + R_{F2}) \| Z_L] \quad (1)$$

where Z_L is the total load impedance (without feedback resistances R_{F1} and R_{F2}) that appear at node V_{OUT}, $Z_L = R_L \| \left(R_{ESR} + \frac{1}{sC_0} \right)$

From Fig. 4, applying Mason's rule, the PSR from Path 1 is

$$\left. \frac{V_{OUT}}{V_{IN}} \right|_{pass} = \frac{\left(g_{mp} + \frac{1}{r_{dsp}} \right) \cdot Z_{OUT}}{1 + g_{mp} Z_{OUT} \cdot \beta \cdot H_{EA0}(s) + \frac{1}{r_{dsp}} \cdot Z_{OUT}} \quad (2)$$

where g_{mp} and r_{dsp} are the transconductance and channel resistance of the pass transistor, M_p, $H_{EA}(s) = \frac{A_{EA}}{1 + \frac{s}{\omega_{EA}}}$, A_{EA} is the DC gain, ω_{EA} is the dominant pole of the error amplifier, and β is the ratio of the feedback network, given by $\beta = \frac{R_{F2}}{R_{F1} + R_{F2}}$.

At low frequencies, the PSR of Path 1 is primarily dominated by the gain of the error amplifier A_{EA}, which dominates equation (2). As frequency increases, the gain of the error amplifier $H_{EA}(s)$ drops due to its dominant pole, causing PSR degradation at high frequencies. The low- and high-frequency PSR expressions for Path 1 are:

When low frequency, s = 0:

$$\left. \frac{V_{OUT}}{V_{IN}}(s=0) \right|_{pass} = \left. PSR_{DC} \right|_{pass} \approx \frac{1}{\beta \cdot A_{EA}} \quad (3)$$

When high frequency, s = ∞:

$$\left. \frac{V_{OUT}}{V_{IN}}(s=\infty) \right|_{pass} = \left. PSR_{HF} \right|_{pass} \approx \frac{\left(g_{mp} + \frac{1}{r_{dsp}} \right) Z_{OUT}}{1 + \frac{1}{r_{dsp}} \cdot Z_{OUT}} \quad (4)$$

For Paths 2 and 3, the corresponding transfer functions are expressed as follows:

$$\left. \frac{V_{OUT}}{V_{IN}} \right|_{EA} = \frac{PSR_{EA} \cdot H_{EA}(s) \cdot g_{mp} \cdot Z_{OUT}}{1 + g_{mp} Z_{OUT} \cdot \beta \cdot H_{EA}(s) + \frac{1}{r_{dsp}} Z_{OUT}} \quad (5)$$

$$\left. \frac{V_{OUT}}{V_{IN}} \right|_{REF} = \frac{PSR_{REF} \cdot H_{EA}(s) \cdot g_{mp} \cdot Z_{OUT}}{1 + g_{mp} Z_{OUT} \cdot \beta \cdot H_{EAO}(s) + \frac{1}{r_{ds \cdot p}} Z_{OUT}} \quad (6)$$

where PSR_{EA} is the PSR of the error amplifier, and PSR_{REF} is the PSR of the bandgap reference circuit. From equation (5) and (6), it can be observed that the PSR of Paths 2 and 3 is influenced by PSR_{EA} and PSR_{REF}, which amplify the total PSR contribution of these paths. As the frequency increases, the dominant pole of the error amplifier shifts downward, causing the PSR of Paths 2 and 3 to approach zero, effectively blocking V_{IN} ripple from reaching V_{OUT}

In summary, all three paths affect PSR at low frequencies, as shown in Fig. 5. This work focuses on enhancing PSR in the low-to-mid frequency range, which is critical for applications like post-DC-DC converter LDOs. A key strategy is improving PSR_{BG}, boosting rejection performance where it matters most.

III. PROPOSED HIGH PSR BANDGAP REFERENCE

The proposed bandgap reference (BGR) in Fig. 6 includes a conventional BGR core, output impedance boosting, a Miller capacitor (C_C) and a bypass capacitor to V_{IN} (C_V). It provides a stable 1.1V output over

979-8-3315-1550-8/25 $31.00 © 2025 IEEE

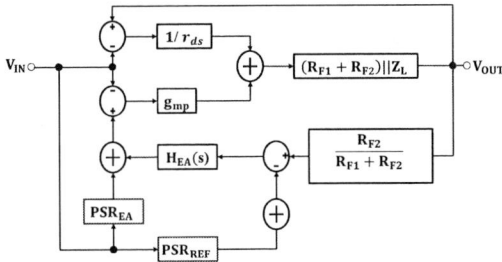

Fig. 4. Signal flow of LDO

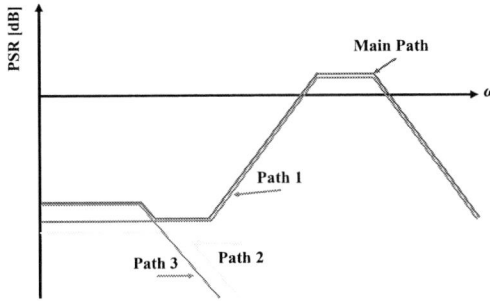

Fig. 5. Total PSR of LDO and PSR due to each path in frequency domain

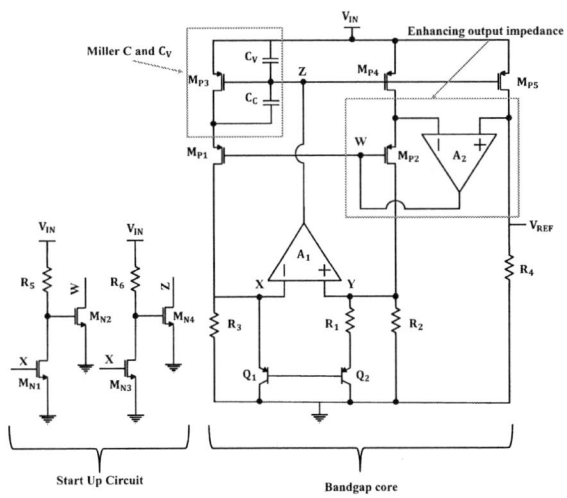

Fig. 6. Proposed BGR Schemitic

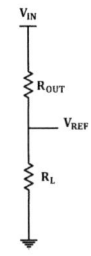

Fig. 7. Simple analsyis PSR of BGR

a supply range of 1.8V to 2.5V. The output voltage is given by:

$$= \frac{R_4}{R_2}\left(V_{CTAT} + \frac{R_2}{R_1}V_{PTAT}\right) \quad (7)$$

which V_{CTAT} is complementary to absolute temperature, and V_{PTAT} is proportional to it. Proper resistor selection balances both terms for temperature-independent output.

To improve DC PSR, the design increases the output impedance of V_{REF}. Capacitors C_C and C_V enhance stability and extend PSR bandwidth.

A. Enhancing Output Impedance

To boost the output impedance of the BGR, a feedback loop using an operational amplifier (opamp) is employed. This feedback mechanism increases the output impedance R_{OUT} by a factor of $1 + A_{v,A2}$. A simplified model is shown in Fig. 7, where a voltage divider represents $\frac{V_{REF}}{V_{IN}}$, corresponding to the BGR's PSR:

$$\frac{V_{REF}}{V_{IN}} = \frac{R_{OUT}}{R_L + R_{OUT}} = \frac{1}{1 + \frac{R_{OUT}}{R_L}} \quad (8)$$

where $R_{OUT} = (r_{o5} \| R_4)(1 + A_{v,A2})$. As equation (8) shows, increasing R_{OUT} improves PSR, especially R_L is fixed by surrounding circuits.

Additionally, the operational amplifier used to boost the output impedance must exhibit both low power consumption and high gain.

B. C_C and C_V frequency compensation

The second technique uses C_V to stabilize the feedback loop of the operational amplifier A_1. Additionally, C_C extends the PSR bandwidth of the BGR.

A simplified small-signal BGR model from [10] is used to derive the PSR in equation (9).

$$\frac{V_{REF}}{V_{VIN}} = (g_m R_4 \| R_0) \cdot \frac{1}{1 + g_{mA_1} r_{oa_1} g_m (R_A - R_B)} \cdot \frac{(1 + s/z_1)}{(1 + s/p_1)(1 + s/p_2)} \quad (9)$$

$$z_1 = \frac{1}{r_{oa_1} C_Z} \quad p_1 = \frac{1}{(R_4 \| R_O) C_{OUT}} \quad p_2 = \frac{g_{ma} g_m (R_A - R_B)}{C_Z}$$

With enhanced output impedance and added C_C and C_V, the PSR is given in equation (10):

$$\frac{V_{REF}}{V_{VIN}} = \frac{(g_m R_4 \| R_O)}{1 + g_{mA_2} r_{oA_2}} \cdot \frac{1}{1 + g_{mA_1} r_{oA_1} g_m (R_A - R_B)} \cdot \frac{(1 + s/z_1)}{(1 + s/p_1)(1 + s/p_2)}$$

$$z_1 = \frac{1}{r_{oa_1} C_C \left(1 + \frac{R_B}{R_{\infty o1}} + g_{mA1} R_B\right)} \quad (10)$$

$$p_1 = \frac{1}{(R_4 \| R_O) C_{OUT}} \quad p_2 = \frac{g_{mA1} g_m (R_A - R_B)}{C_V + C_C \left(1 + \frac{R_B}{R_{a_1}} + g_{mA1} g_m R_A R_B + g_m R_B + g_{mA1} R_B\right)}$$

From equation (10), the z_1 depends on the newly introduced C_C. A smaller C_C improves the PSR bandwidth, while p_1 and p_2 experience slight variations.

979-8-3315-1550-8/25 $31.00 © 2025 IEEE

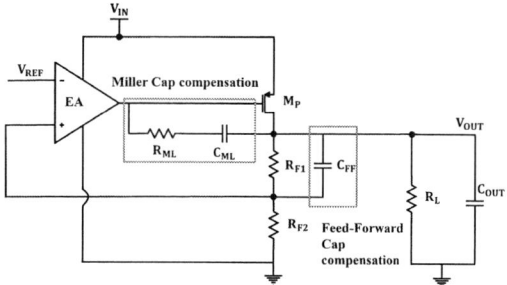

Fig. 8. Proposed LDO

C_V mainly improves the loop gain of A_1, enhancing feedback stability.

IV. PROPOSED LOW DROPOUT VOLTAGE

LDOs rely on feedback, making stability essential to prevent oscillations. Industrial standards require a phase margin over 60°. While traditional designs use large external capacitors to set the dominant pole, modern CL-LDOs avoid this.

In this design, internal compensation using Miller capacitance and feedforward capacitors ensures stability with only a 1 pF load, as shown in Fig. 8.

A. Miller Capacitance compensation

To improve stability, the Miller effect is used for pole splitting, separating poles to increase phase margin. Additionally, introducing a resistor shift the zero from a positive to a negative value or enables pole-zero cancellation. thereby increasing the phase margin. Based on the circuit, the compensated LDO's poles and zeros are given by:

$$z_{1,\text{LDO}} = \frac{1}{C_{\text{ML}}\left(\frac{1}{g_{mp}} - R_{\text{ML}}\right)}$$

$$p_{1,\text{LDO}} = \frac{1}{R_{\text{OUT,EA}}\, g_{mp} R_{\text{OUT,LDO}} C_{\text{ML}}} \quad (11)$$

$$p_{2,\text{LDO}} = \frac{g_{mp}}{C_{\text{OUT,LDO}}}$$

From equation (11), when $R_{ML} = \frac{1}{g_{mp}}$, the zero is canceled. If $R_{ML} > \frac{1}{g_{mp}}$, negative zero is introduced. Alternatively, matching z_1 with p_2, achieving pole-zero cancellation. Carefully optimizing R_{ML} and C_{ML} ensures the best stability performance.

B. Compensation of Feedforward

Another commonly used compensation technique in feedback loops involves the pole-zero pair formed by the feedback resistor. The corresponding pole and zero are given by:

$$z_{1,\text{LDO}} = \frac{1}{C_{\text{FF}} R_{F1}}$$

$$p_{1,\text{LDO}} = \frac{1}{C_{\text{FF}}\left(R_{F1} \| R_{F2}\right)} \quad (12)$$

This feed-forward compensation technique is highly effective when using a large R_{F1}, as it increases the separation between the poles and zeros, allowing for better compensation. Additionally, placing it near the unity-gain frequency (UGF) enhances stability. In the proposed LDO design, achieving a high β close to one improves the PSR. As a result, C_{FF} does not need to be excessively large, minimizing its overall impact. The value of C_{FF} is not to much. So the value for C_{FF} depends on R_{F1} and R_{F2}, and while its effect is limited, it provides a slight increase in phase margin, enhancing stability.

V. SIMULATION RESULTS

The proposed LDO circuit, shown in Fig. 8, was simulated in Cadence Virtuoso using a 65nm CMOS technology. The total layout area of the LDO, including the bandgap reference (BGR) circuitry, is 0.25 mm^2, with the BGR occupying approximately 50 of the total area. The LDO provides a regulated output voltage of 1.2V and supports a load current I_{LOAD} ranging from 500 μA to 300 mA, while operating with an input supply voltage varying from 1.8V to 2.5V.

As discussed in Section III, the proposed BGR enhances PSR from 38 dB to nearly 90 dB using an impedance-enhancement technique but reduces PSR bandwidth due to increased capacitance shifting the zero lower, as shown in Fig. 9. Fig. 10 shows that without C_C, the PSR holds until a zero appears at approximately 3 kHz. Introducing a small C_C shifts the zero forward, extending bandwidth to around 10 kHz. However, using a larger C_C (e.g., 30 fF or 100 fF) diminishes this effect due to the behavior of the system's transfer function, as analyzed in Section III.

Table 1 summarizes the BGR's performance over process-voltage-temperature (PVT) variations in the layout implementation. The temperature coefficient (Tempco) remains within an acceptable range, and in the worst-case scenario, the PSR is maintained at 73 dB. These specifications ensure robust performance within the LDO design. The line transient response of the proposed LDO is shown in Fig. 11. In this simulation, the input voltage V_{IN} varies from 1.8V to

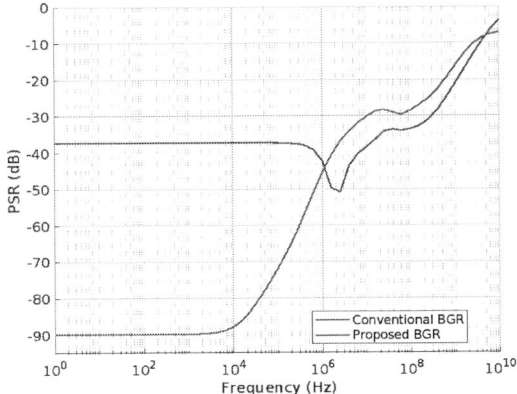

Fig. 9. PSR between Conventional BGR and Proposed BGR

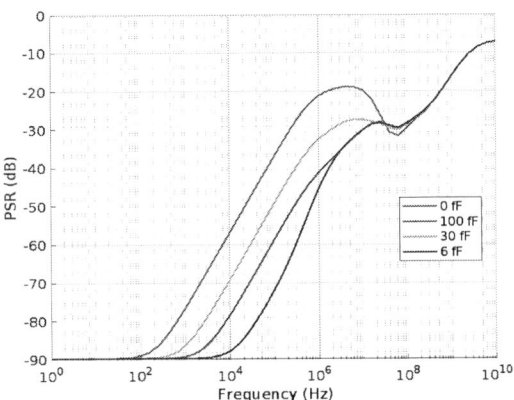

Fig. 10. PSR of Proposed BGR with sweeping C_C

2.5V with a rise and fall time of 2 µs. Due to the high PSR, the maximum overshoot is limited to 8 mV, while the maximum undershoot is 4 mV.

Additionally, Fig. 12 illustrates the full-scale load transient response, where the load current I_{load} transitions from its minimum (500 µA) to maximum (300 mA) with a rise and fall time of 2 µs. The results demonstrate robust performance, with both overshoot and undershoot reaching approximately 40 mV across

TABLE I
PERFORMANCE BGR SUMMARY

Technology	65nm
Area $[mm^2]$ x10^{-3}	77.44
Input voltage range V_{IN} (V)	1.8 - 2.5
V_{REF} (V)	1.1
Temperature Range [mA]	-40 to 125
TC (ppm/^oC) [V]	10 - 18
PSR (dB@DC)	73 - 116
PSR (dB@10KHz)	66 - 91

all test cases. These findings confirm that the capacitorless LDO operates stably with minimal oscillations.

Fig. 13 presents the measured PSR of the LDO under both minimum and maximum load conditions with a load capacitance C_{Load} = 1pF. The results highlight a trade-off between achieving high PSR at DC and reduced bandwidth. The worst-case PSR is approximately 70 dB, while the highest recorded PSR reaches nearly 97 dB. The overall performance of the proposed LDO, along with a comparison to other works, is summarized in Table II.

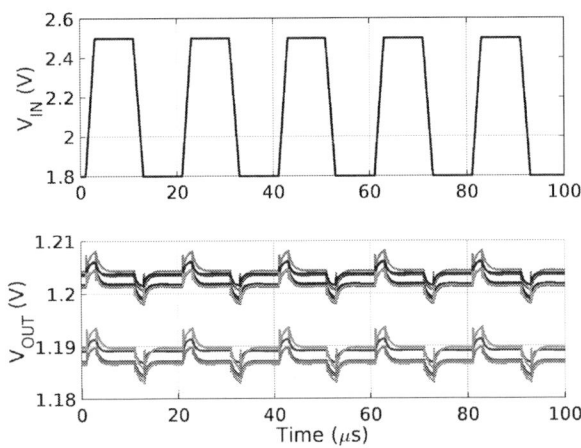

Fig. 11. Line transient response of LDO

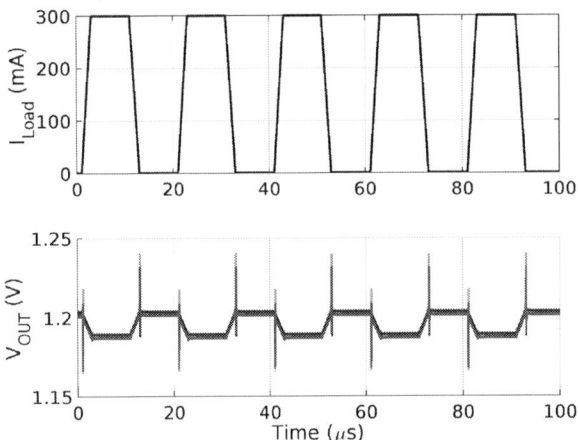

Fig. 12. Load transient response of LDO

VI. CONCLUSION

The proposed LDO, built in 65nm CMOS, shows better overall performance than previous works [8]

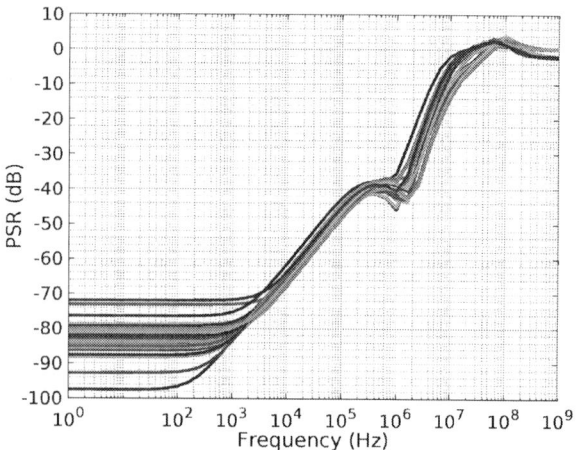

Fig. 13. Proposed LDO

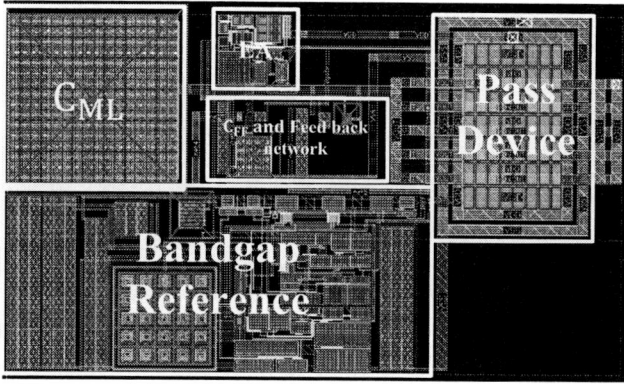

Fig. 14. Layout of LDO

TABLE II
PERFORMANCE SUMMARY AND COMPARISON

	[8]	[9]	This Work
Technology	CMOS 180nm	CMOS 65nm	CMOS 65nm
Area [mm^2] x10^{-3}	-	87	235
V_{IN} (V)	2.5	1.2	1.8 - 2.5
Imax [mA]	200	25	300
V_{OUT} [V]	1.8	1	1.2
V_{DROP} [mV]	200	200 - 400	200
I_Q [uA]	44.75	300	140
PSR (dB@1KHz)	59	70	70 - 96
ΔV_{OUT} (Full load transient)	< 140mV	-	< 80mV
Line Regulation (mV/V)	-	3.8	0.04

and [9]. It uses an improved bandgap reference and internal compensation to reach a PSR of 70–96dB at low frequencies and keeps 60dB at 10kHz. This is higher than the 59dB in [8] and similar to [9], but with lower power use. The design supports up to 300mA of load current with a quiescent current of just 140µA, including the BGR, which is much lower than [9]. It also has better transient response, keeping voltage swings under 80mV, compared to 140mV in [8]. The line regulation is improved to 0.04mV/V, much better than 3.8mV/V in [9].

Overall, this capacitor-less LDO suits SoC use well, offering high PSR, low power, and minimal external parts in a compact design.

REFERENCES

[1] M. Ho, K. N. Leung and K. -L. Mak, "A Low-Power Fast-Transient 90-nm Low-Dropout Regulator With Multiple Small-Gain Stages," in IEEE Journal of Solid-State Circuits, vol. 45, no. 11, pp. 2466-2475, Nov. 2010.

[2] K. N. Leung and Y. S. Ng, "A CMOS Low-Dropout Regulator With a Momentarily Current-Boosting Voltage Buffer," in IEEE Transactions on Circuits and Systems I: Regular Papers, vol. 57, no. 9, pp. 2312-2319, Sept. 2010.

[3] G. A. Rincon-Mora and P. E. Allen, "A low-voltage, low quiescent current, low drop-out regulator," in IEEE Journal of Solid-State Circuits, vol. 33, no. 1, pp. 36-44, Jan. 1998.

[4] F. Lavalle-Aviles, J. Torres and E. Sánchez-Sinencio, "A High Power Supply Rejection and Fast Settling Time Capacitor-Less LDO," in IEEE Transactions on Power Electronics, vol. 34, no. 1, pp. 474-484, Jan. 2019.

[5] L. Chen, Q. Cheng, J. Guo and M. Chen, "High-PSR CMOS LDO with embedded ripple feedforward and energy-efficient bandwidth extension," 2015 28th IEEE International System-on-Chip Conference (SOCC), Beijing, China, 2015.

[6] M. El-Nozahi, A. Amer, J. Torres, K. Entesari and E. Sanchez-Sinencio, "High PSR Low Drop-Out Regulator With Feed-Forward Ripple Cancellation Technique," in IEEE Journal of Solid-State Circuits, vol. 45, no. 3, pp. 565-577, March 2010.

[7] J. Torres et al.,"Low Drop-Out Voltage Regulators: Capacitor-less Architecture Comparison," in IEEE Circuits and Systems Magazine, vol. 14, no. 2, pp. 6-26, Secondquarter 2014

[8] C. Yang and X. Hong, "A capacitor-less LDO with sub-bandgap voltage reference and DFC compensation strategy," 2017 IEEE 2nd Advanced Information Technology, Electronic and Automation Control Conference (IAEAC), Chongqing, China, 2017.

[9] Y. Lim, J. Lee, S. Park, and J. Choi, "An External-Capacitor-less Low- Dropout Regulator with Less than 36dB PSRR at All Frequencies from 10kHz to 1GHz Using an Adaptive Supply-Ripple Cancellation Technique to the Body-Gate," in Proc. IEEE Custom Integrated Circuits Conf. (CICC), pp. 1-4, May 2017.

[10] L. Wang et al., "Design of high-PSRR current-mode bandgap reference with improved frequency compensation," 2016 IEEE International Conference on Electron Devices and Solid-State Circuits (EDSSC), Hong Kong, China, 2016.

2025 10th IEEE International Conference on Integrated Circuits, Design, and Verification (ICDV)

Inductorless 5.405 GHz Fractional-N PLL for RF Synthesis with 5.6 mW Power Consumption

Thi Viet Ha Nguyen, Cong-Kha Pham
Dept. of Computer and Network Engineering
The University of Electro-Communications
Tokyo, Japan
vietha@vlsilab.ee.uec.ac.jp, phamck@uec.ac.jp

Xuan Thanh Pham, Manh Kha Hoang
School of Electrical & Electronic Engineering
Hanoi University of Industry
Hanoi, Vietnam
thanhpx@haui.edu.vn, khahoang@haui.edu.vn

Abstract—For energy-efficient RF synthesis in IoT and mobile applications, this work introduces an enhanced Fractional-N Phase-Locked Loop (PLL) architecture. The design achieves a power consumption of less than 5.6 mW while delivering superior performance in terms of phase noise and loop bandwidth. To mitigate $\Delta\Sigma$ quantization noise, a novel adaptive digital noise filter is implemented, replacing the conventional synchronous delay line. A multi-loop PLL structure is incorporated to improve frequency accuracy and minimize jitter. Furthermore, a high-resolution digital phase detector (DPD) is utilized to reduce phase variation. Implemented in a 180 nm CMOS process, the PLL exhibits an in-band phase noise of less than -105 dBc/Hz, an integrated jitter of approximately 1 ps$_{rms}$, and a loop bandwidth exceeding 15 MHz. These results demonstrate the proposed PLL's suitability for high-precision, low-power wireless communication systems.

Index Terms—Phase Locked-Loop, fractional-N PLL, $\Delta\Sigma$ noise, multi-loop PLL, digital phase detector, RF synthesis.

I. INTRODUCTION

Phase-locked loops (PLLs) [1]-[3] are essential components in modern RF and communication systems, providing critical frequency synthesis for applications ranging from wireless connectivity (Wi-Fi, Bluetooth, 5G) to high-speed wireline communication and precision timing. Fractional-N PLLs [4], [5], in particular, have gained prominence due to their fine frequency resolution, spectral efficiency, and dynamic agility, crucial for multi-standard transceivers and advanced clocking circuits [6]-[8].

However, fractional-N PLLs face significant challenges in high-speed, low-power applications. A primary limitation is the $\Delta\Sigma$ modulation, which, while enabling fine resolution, introduces quantization noise that degrades phase noise performance [7], leading to peaking, increased jitter, and spectral impurities. Balancing loop bandwidth and phase noise suppression is another critical concern. While wider bandwidths suppress VCO noise [9], they also allow more quantization noise to pass, impacting performance. Traditional narrow-band filtering techniques address this but compromise frequency settling time and responsiveness [10]-[11]. Furthermore, charge-pump-based architectures introduce nonlinearities, exacerbating noise folding and spur generation, which are detrimental to applications demanding low phase noise and high dynamic range. The increasing demand for low-power, high-precision

frequency synthesis in IoT, 5G, and satellite communication necessitates innovative PLL architectures.

This work presents a novel cascaded PLL architecture designed to overcome these challenges. Our multi-pronged approach integrates an adaptive digital noise filter to dynamically suppress $\Delta\Sigma$ quantization noise while maintaining low power consumption. We also employ a multi-loop frequency synthesis technique that optimizes the bandwidth-phase noise trade-off, significantly reducing jitter. A digital phase detector (DPD) replaces conventional charge-pump-based detectors, enhancing phase tracking accuracy and reducing noise folding. Implemented in 180 nm CMOS, the proposed architecture achieves sub-5.6 mW power consumption, an in-band phase noise of -105 dBc/Hz, and integrated jitter below 1 ps$_{rms}$, demonstrating superior performance compared to existing ring-based and inductor-less PLLs.

II. PRIOR ART

Several strategies have been explored to mitigate the challenges of Fractional-N PLLs. One common approach employs a feedforward digital-to-analog converter (DAC) to cancel $\Delta\Sigma$ quantization noise [12]. While effective in reducing noise peaking, this method suffers from stringent component matching requirements, leading to residual noise and spurious artifacts, and increased power consumption due to complex calibration procedures, hindering its use in ultra-low-power designs.

Another strategy focuses on loop bandwidth control, typically using a narrow bandwidth (typically $f_{REF}/200$ to $f_{REF}/1000$) to suppress high-frequency $\Delta\Sigma$ quantization noise [13]-[14]. However, this results in slow frequency acquisition, limited VCO noise suppression, and increased sensitivity to process, voltage, and temperature (PVT) variations, making it unsuitable for frequency-agile systems.

Multi-loop cascaded PLLs, using a pre-scaling integer-N PLL to boost reference frequency, have emerged as a promising solution. While pushing quantization noise to higher frequencies and enabling wider bandwidth operation, these architectures often rely on synchronous delay-line-based noise filtering, which incurs excessive power consumption and implementation complexity.

979-8-3315-1550-8/25 $31.00 © 2025 IEEE

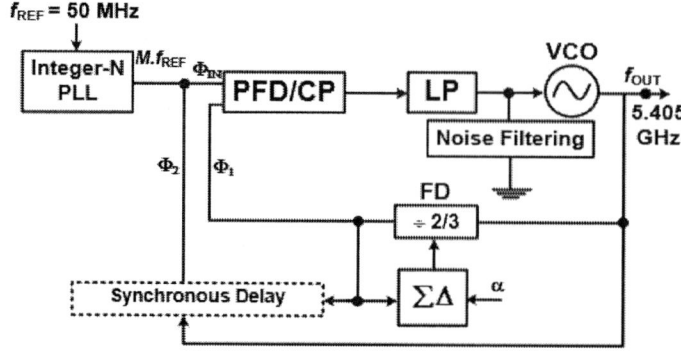

Fig. 1. The proposed Fractional-N PLL architecture .

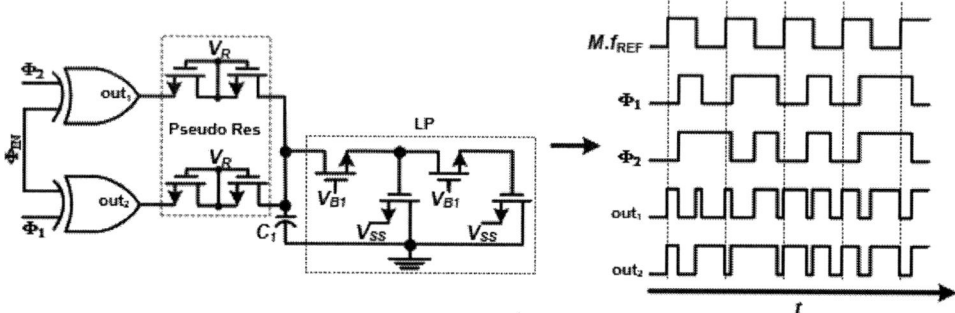

Fig. 2. The waveforms of PFD.

Advanced PLL architectures have also been proposed. Ring-oscillator-based PLLs reduce area and power by eliminating inductors, but their low-Q nature results in poor phase noise performance. Reference injection and phase interpolation techniques enhance spectral purity but add design complexity due to high-speed interpolators and precise timing control. Sub-sampling phase detectors (SSPDs) improve noise performance but require high-Q reference signals, challenging compact, low-power implementations. Given these limitations, a novel approach is needed one that combines the power efficiency of inductor-less architectures with significantly enhanced phase noise suppression and jitter performance.

III. PROPOSED ARCHITECTURE

A. Overall Architecture and Noise Filtering

Phase-locked loops (PLLs) are closed-loop control systems designed to generate an output signal (f_{OUT}) that is synchronized in both frequency and phase with a reference signal (f_{REF}). As illustrated in Fig. 1, a Fractional-N PLL typically consists of a phase/frequency detector (PFD), charge pump (CP), loop filter (LF), voltage-controlled oscillator (VCO), and frequency divider (FD).

The PFD compares the phase and frequency of the reference signal with the feedback signal from the VCO, generating an error signal. The CP converts this error signal into a proportional control voltage. The LF then smooths this control voltage, eliminating high-frequency noise and stabilizing the

system. The VCO generates the output signal, with its frequency controlled by the LF. Finally, the FD scales the VCO output frequency to match the reference frequency.

In the proposed Fractional-N PLL, noise filtering is incorporated to mitigate quantization noise introduced by $\Delta\Sigma$ modulation. This noise can degrade phase noise, increase jitter, and cause spectral impurities. Traditional noise cancellation methods are power-intensive and complex. However, the adaptive digital noise filter implemented in this design efficiently suppresses quantization noise with minimal power overhead, resulting in improved phase noise and jitter performance. The key advantage of this approach is its ability to minimize power overhead while achieving significant phase noise reduction and jitter suppression. Through precise digital control, the filter operates within the feedback loop of the PLL, attenuating unwanted quantization noise before it reaches the voltage-controlled oscillator (VCO), where it would otherwise degrade the output signal quality. The result is a cleaner output signal with reduced spurious components, better spectral purity, and improved synchronization with the reference signal. In addition, the adaptive nature of the filter ensures robust performance across different process, voltage, and temperature (PVT) variations, maintaining stability and efficiency under various operating conditions. This enables the PLL to perform reliably in a range of practical applications, from high-speed data converters to low-noise clock generators.

This design employs a Digital Phase Detector (DPD) utiliz-

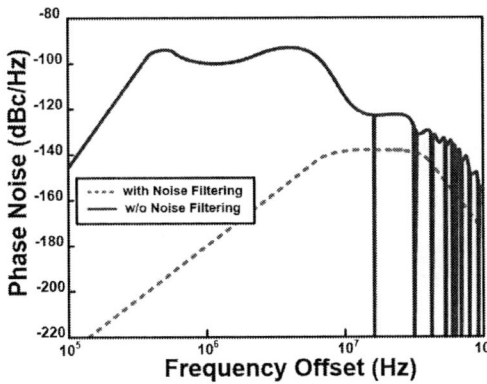

Fig. 3. Fractional-N PLL $\Delta\Sigma$ phase noise contribution with or without noise filtering.

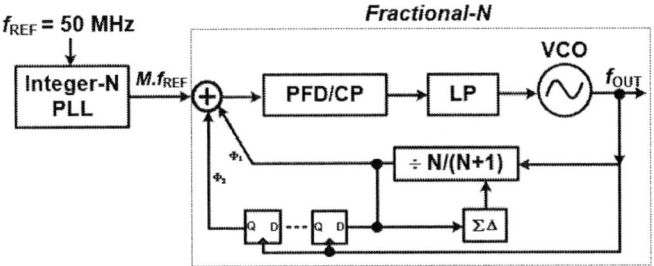

Fig. 4. Cascaded in PLL structure.

ing XOR gates for efficient phase detection in Phase-Locked Loop (PLL) systems. An XOR gate compares the reference clock and feedback clock, generating a pulse-width modulated (PWM) signal proportional to their phase difference. The XOR gate's output is connected to a pseudo-resistor, which functions as a high-value resistive element. This converts the PWM signal into a voltage suitable for subsequent processing. The pseudo-resistor enables a compact, low-power implementation while maintaining a stable voltage response, even at low-frequency phase differences. To generate the phase difference $\Phi_1 + \Phi_2$, where Φ_1 is the output phase of the PLL's frequency divider and Φ_2 is a delayed version of Φ_1, the following steps are performed: First, $\Phi_{IN} - \Phi_1$ and $\Phi_{IN} - \Phi_2$ are computed using separate XOR gates, where Φ_{IN} is the output phase of the Integer-N PLL. Fig. 2 illustrates the DPD's waveforms, demonstrating how the XOR gates calculate the phase errors between the reference phase and the divider's past output phases. Reference multiplication is a critical technique in Phase-Locked Loop (PLL) systems, influencing both the output frequency and noise characteristics. This method involves multiplying the reference signal by a factor to increase its frequency prior to comparison with the feedback signal. Fig. 3 illustrates the impact of reference multiplication on noise contributions within the PLL system. The figure also incorporates the effect of a feedback noise filter, highlighting the inherent trade-offs between noise performance and frequency resolution in PLL design.

B. Cascaded PLL

To minimize flip-flop count in the noise filter while maintaining system performance, a dual-loop PLL architecture is employed. An initial Integer-N PLL multiplies the reference frequency from 50 MHz to approximately 1 GHz, which is then fed into the Fractional-N loop. This elevated reference frequency allows the delta-sigma modulator to operate at a higher rate, effectively shifting quantization noise to higher frequencies and mitigating its impact on phase noise. The cascaded PLL structure is depicted in Fig. 4.

C. Synchronous Delay Block

In the proposed Fractional-N PLL architecture, a Synchronous Delay Block ensures precise timing alignment, particularly when scaling the reference frequency from 50 MHz to 1 GHz using an Integer-N PLL. This block is critical for maintaining the effectiveness of the noise filtering process. As the noise power spectral density ($S_Q(f)$ in Eq. (1)) is inversely proportional to the reference frequency, increasing f_{REF} reduces quantization noise. However, it also necessitates precise synchronization to prevent phase misalignment. By implementing a synchronous delay, the design ensures that the notch filter at 50 MHz and additional traps effectively suppress spurious tones while preventing metastability in the high-frequency flip-flop-based filtering. Here, n represents the order of the noise shaping filter in the Delta-Sigma modulator.

$$S_Q(f) \propto \frac{1}{f^n} \qquad (1)$$

IV. BUILDING BLOCKS

A. VCO Design

A three-stage ring oscillator VCO, controlled by bias currents (V_B, V_{B1}), is designed using inverters and transmission gates for precise frequency tuning, as shown in Fig. 5. Each oscillator stage consists of a CMOS inverter paired with a transmission gate. The control voltage V_{VCT} modulates the transmission gates' resistance, thereby affecting the stage delay. As V_{VCT} varies, the transmission gate resistance adjusts, altering the propagation delay of each stage. Because the total ring oscillator delay is the sum of these stage delays, this mechanism effectively controls the VCO's oscillation frequency. This design achieves a tunable frequency range from 93 Hz to 1.67 GHz for a 0 V to 1.8 V control voltage, and a phase noise of -105 dBc/Hz.

B. Delta-sigma Modulation

To reduce the power consumption of the delta-sigma modulator (DSM) in CMOS processes, this paper proposes an optimized DSM architecture for a Fractional-N PLL operating with a 50 MHz reference and a 5.405 GHz output, as shown in Fig. 6. The primary source of power inefficiency in high-speed DSMs is the need for high bit-width accumulators, which are essential for maintaining fine frequency resolution at higher reference frequencies. Conventional single-loop, high-order

979-8-3315-1550-8/25 $31.00 © 2025 IEEE

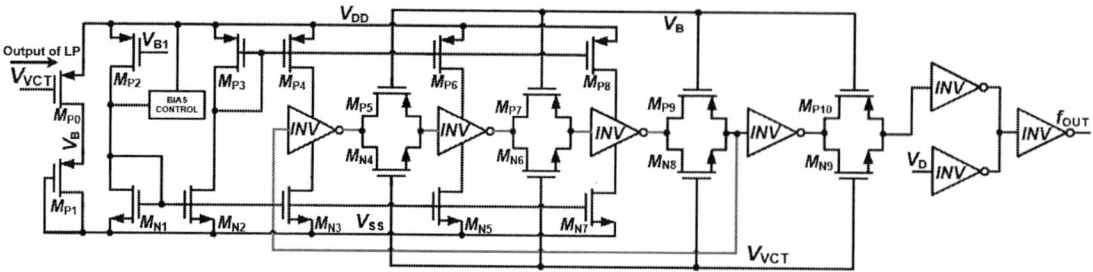

Fig. 5. The schematic of VCO.

DSM designs require large accumulators to ensure stability, particularly due to the limited linear range of the phase-frequency detector (PFD). For example, a third-order DSM with a 16-bit input typically needs accumulators exceeding 20 bits, resulting in significant power overhead. This design addresses the high bit-width issue by introducing a hybrid Delta-Sigma Modulator (DSM) structure that reduces bit-width while preserving noise shaping performance. The proposed architecture employs a two-stage quantization process. The first stage consists of two cascaded 4-bit first-order DSMs, which extract and propagate carry information from the least significant bits (LSBs) of the 16-bit input. These outputs are then processed by a subsequent third-order DSM with an 8-bit input, which performs higher-order noise shaping to suppress quantization artifacts. This partitioning reduces the computational complexity of the main DSM while effectively suppressing shaped quantization noise. Furthermore, the co-efficients of the third-order stage are carefully optimized to minimize circuit overhead, allowing multipliers and subtractors to be implemented using simple logic operations, such as inverters, instead of power-intensive arithmetic units.

$$NTF = \frac{(1 - z^{-1})^3}{1 + z^{-1} - 4z^{-2} + 2.25z^{-3}} \quad (2)$$

The proposed DSM achieves superior noise performance while significantly reducing power and area consumption.

Simulation results, illustrated in Fig. 7, show that while the intermediate outputs of the first-order DSM stages contain periodic spurious tones, these are effectively randomized at the input of the third-order DSM. This creates a dithering effect that minimizes spectral artifacts.

The final output spectrum confirms that residual spurs are positioned at harmonics of the reference frequency, where they are effectively attenuated by the PLL loop dynamics. The noise transfer function is calculated using

C. Noise Filtering

To reduce phase noise and enhance stability, this work incorporates an active noise filtering circuit as shown inf Fig. 8 within the PLL loop filter. This circuit uses transconductance-based filtering to suppress high-frequency noise before it reaches the Voltage-Controlled Oscillator (VCO), thereby re-ducing timing jitter and improving spectral purity. Eq. (2).

The proposed noise filtering architecture comprises multiple transconductance stages and capacitive elements, forming an active low-pass filter with enhanced noise rejection.

Initially, the transconductance amplifier G_{m1} converts the control voltage V_{IN} into a corresponding current, enabling precise control of the loop dynamics. The impedance Z_T (Eq. 3 [14]) and capacitor C_s form the primary noise filtering network, attenuating unwanted high-frequency components.

$$Z_T(s) = \frac{1 + (C_2/G_{m2} + G_{m3})s^2}{C_s s[1 + G_{m1}(C_2/G_{m2} + G_{m3})s]} \quad (3)$$

Subsequently, cascaded transconductance amplifiers G_{m2} and G_{m3} provide additional filtering and stabilization, ensuring only the necessary control signal components reach the VCO. The capacitor C_2 at the output further refines the filtering response, optimizing the loop's noise shaping characteristics. By integrating active transconductance elements, this noise filtering approach offers superior phase noise suppression compared to traditional passive filters. It provides enhanced control over loop bandwidth and dynamic response, making it ideal for applications requiring low-jitter, high-purity fre-quency synthesis.

V. POST LAYOUT SIMULATION RESULTS

This work presents a cascaded fractional-N synthesizer designed for 5.405-GHz RF standards, eliminating the need for LC oscillators and charge pumps. A feedback noise filter integrated into the second loop effectively suppresses noise while maintaining a moderate loop bandwidth. The results demonstrate the significant potential of inductor-less synthesizers, offering a promising alternative for modern RF applications.

Fig. 9 shows the post-layout simulated PLL phase noise and the main noise contributions from the two VCOs and the delta-sigma modulator. Figure 10 presents the output spectrum of the PLL with a center frequency of 5.405 GHz and a 10 MHz span, showing a main signal at -10.5 dBm and a noise floor at -100 dBm. Figure 11 displays the measured fractional spur versus frequency offset of the PLL and the simulated spectrum of the $\Delta\Sigma$ modulator. The largest spur is at -72.5 dBc.

The proposed cascaded frequency synthesizer is imple-mented in 180 nm CMOS technology, occupying a compact 185 μm×150 μm die area (Fig. 12). It operates with a 50 MHz

979-8-3315-1550-8/25 $31.00 © 2025 IEEE

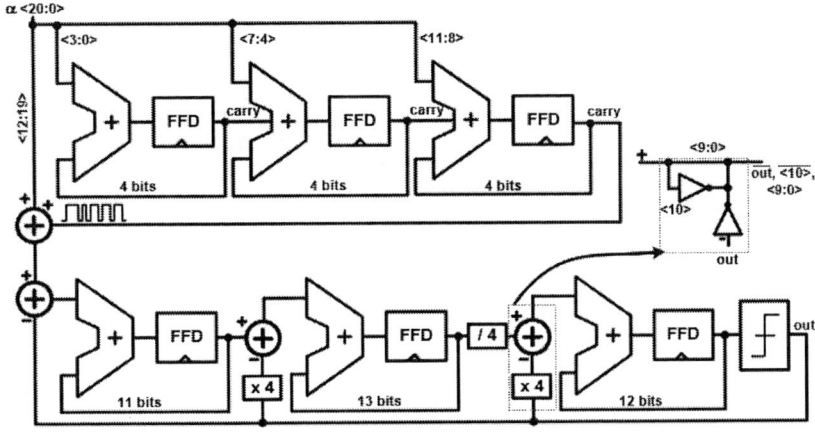

Fig. 6. $\Delta\Sigma$ modulation architecture in Fraction-N PLL.

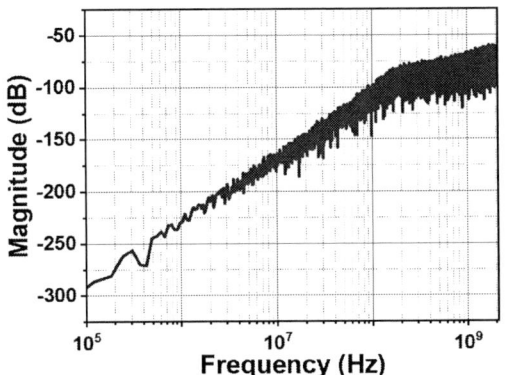

Fig. 7. Spectrum of $\Delta\Sigma$ 3-order in Fraction-N PLL.

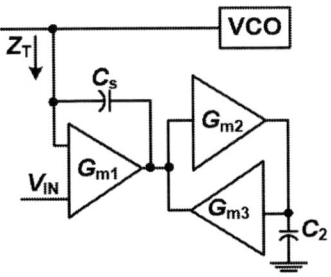

Fig. 8. Noise filtering architecture in PLL.

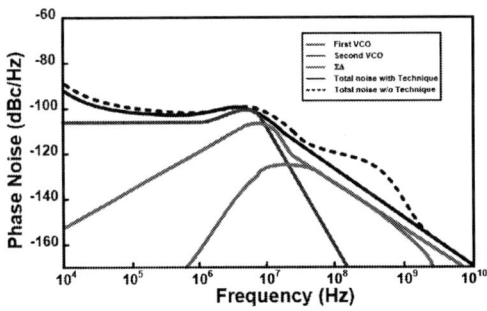

Fig. 9. Overall post-layout simulated PLL phase noise and main contributions.

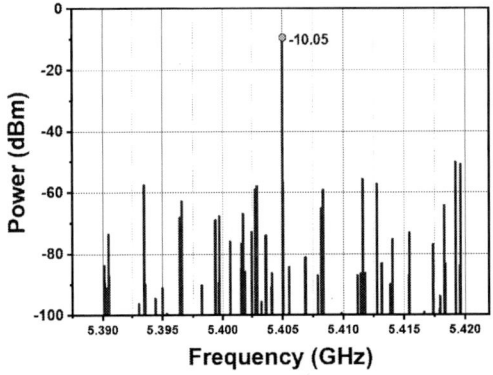

Fig. 10. Phase domain model of PLL.

reference crystal oscillator and generates an output frequency range of 5.3 GHz to 5.5 GHz. Powered by a 1.8 V supply, the total power consumption is 5.6 mW, distributed as follows: 28.85% for the integer-N PLL, 66.23% for the delta-sigma modulator and phase detector, 0.54% for the PFD/CP and loop filter, and 7.38% for the VCO and noise filtering circuit. This distribution ensures stable frequency synthesis with optimized power efficiency. The total power consumption of the PLL is shown in Fig. 13. Finally, Table I compares the specifications of the proposed PLL structure with other existing PLL designs.

VI. CONCLUSION

This paper introduces a cascaded fractional-N synthesizer for 5.405 GHz RF applications, eliminating the need for LC oscillators and charge pumps. A feedback noise filter in the second loop effectively manages noise while maintaining stable loop bandwidth.

The overall performance demonstrates the significant potential of inductorless designs, offering superior noise suppression and high operational efficiency.

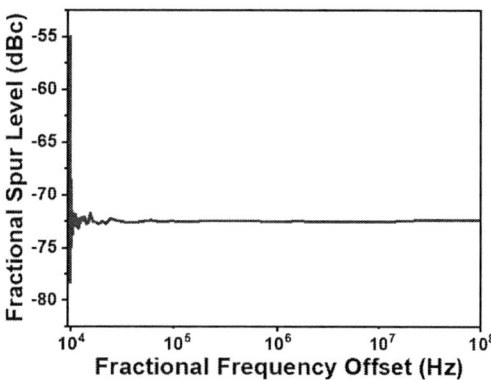

Fig. 11. Fractional spur versus frequency offset.

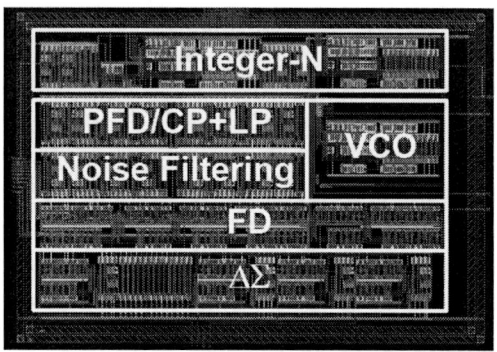

Fig. 12. Layout of this proposed PLL.

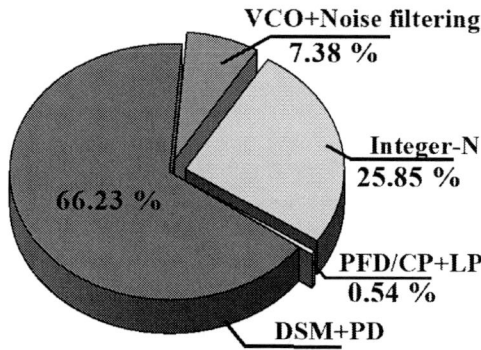

Total power consumption: 5.6 mW

Fig. 13. Total power consumption.

TABLE I
PERFORMANCE SUMMARY

	This work	[15]	[16]	[17]
Oscillator Topology	Ring	LC	Ring	Ring
Reference Freq. (MHz)	50	50	1 G	153.6
Freq. Range (GHz)	5.3-5.5	1.06-1.6	0.05-0.324	1.9-6.1
Phase noise (dBc/Hz)	-105	-90	-98.7	-140
Ref. Spur (dBc)	-72.5	-57	-	-
FOM (dB)	-234.5	-	-157.06	-153
Power (mW)	5.6	11	1.86	10.1
Area (mm^2)	0.028	-	0.008	0.055
Tech. (nm)	180	180	180	28
Data type	Sim	Sim	Sim	Sim

ACKNOWLEDGMENT

The VLSI chip in this study was fabricated in the chip fabrication program of VLSI Design and Education Center (VDEC), the University of Tokyo, with collaboration of Rohm Corporation and Toppan Printing Corporation.

REFERENCES

[1] L. Kong and B. Razavi, "A 2.4 GHz 4 mW integer-N inductorless RF synthesizer," IEEE J. Solid-State Circuits, vol. 51, no. 3, pp. 626–635, Mar. 2016.

[2] W. S. Chang, P. C. Huang, and T. C. Lee, "A fractional-N divider-less phase-locked loop with a sub-sampling phase detector," IEEE J. Solid-State Circuits, vol. 49, no. 12, pp. 2964–2975, Dec. 2014.

[3] B. D. Muer and M. S. J. Steyaert, "A CMOS monolithic-controlled fractional-N frequency synthesizer for DCS-1800," IEEE J. Solid-State Circuits, vol. 37, no. 7, pp. 835–844, Jul. 2002.

[4] Y. P. Fu, L. M. Li, D. M. Wang, "A Fractional-N Divider for Phase Locked Loop with Delta-Sigma Modulator and Phase-Lag Selector," IEEE Int. Symp. on Radio-Freq. Integer. Technol., Nov. 2018, pp. 1-4.

[5] T.-H. Tsai, M.-S. Yuan, C.-H. Chang et al., "14.5 A 1.22ps Integrated Jitter 0.25-to-4GHz Fractional-N ADPLL in 16nm FinFET CMOS," in 2015 IEEE International Solid-State Circuits Conference - (ISSCC) Digest of Technical Papers, 2015, pp. 1–3.

[6] Y. Yorozu, M. Hirano, K. Oka, and Y. Tagawa, "Electron spectroscopy studies on magneto-optical media and plastic substrate interface," IEEE Transl. J. Magn. Japan, vol. 2, pp. 740–741, August 1987 [Digests 9th Annual Conf. Magnetics Japan, p. 301, 1982].

[7] G. Vlachogiannakis et al., "A self-calibrated fractional-N PLL for WiFi 6/802.11ax in 28nm FDSOI CMOS," in Proc. IEEE Eur. Solid State Circuits Conf., Sep. 2019, pp. 105–108.

[8] L. Kong and B. Razavi, "A 2.4-GHz 6.4-mW fractional-N inductorless RF synthesizer," IEEE J. Solid-State Circuits, vol. 52, no. 8, pp. 2117–2127, Aug. 2017.

[9] C.-F. Liang and P.-Y. Wang, "A wideband fractional-N ring PLL using a near-ground pre-distorted switched-capacitor loop filter," in ISSCC Dig. Tech. Papers, Feb. 2015, pp. 1–3.

[10] J. Kim et al., "32.4 A 104fsrms-jitter and –61dBc-fractional spur 15 GHz fractional-N subsampling PLL using a voltage-domainquantization-error cancelation technique," in Proc. IEEE Int. Solid-State Circuits Conf., Feb. 2021, pp. 448–449..

[11] C.-M. Hsu, M. Z. Straayer, and M. H. Perrott, "A low-noise wide BW 3.6-GHz digital fractional-N frequency synthesizer with a noise-shaping time-to-digital converter and quantization noise cancellation," IEEE J. Solid-State Circuits, vol. 43, no. 12, pp. 2776–2786, Dec. 2008.

[12] C. Weltin-Wu, E. Temporiti, D. Baldi, and F. Svelto, "A 3GHz fractional N all-digital PLL with precise time-to-digital converter calibration and mismatch correction," in IEEE Int. Solid-State Circuits Conf. (ISSCC) Dig. Tech. Papers, Feb. 2008, pp. 344–345.

[13] Jang, S.; Chae, M.; Park, H.; Hwang, C.; Choi, J, "A 5.5 µs-Calibration-Time, Low-Jitter, and Compact-Area Fractional-N Digital PLL Using the Recursive-Least-Squares (RLS) Algorithm", In Proceedings of the 2024 IEEE International Solid-State Circuits Conference (ISSCC), San Francisco, CA, USA, 18–22 February 2024; pp. 190–192.

[14] Santiccioli, A.; Mercandelli, M.; Bertulessi, L.; Parisi, A.; Cherniak, D.; Lacaita, A.L.; Samori, C.; Levantino, S, "A 66-fs-rms Jitter 12.8-to-15.2-GHz Fractional-N Bang–Bang PLL with Digital Frequency-Error Recovery for Fast Locking", IEEE J. Solid-State Circuits 2020, 55, 3349–3361.

[15] Kashani, Z.G.; Avanji, S.A.I. Fully integrated fractional-N phase-locked loop for GNSS standards. IET Microw. Antennas Propag. 2019, 13, 2391–2395.

[16] Sun, L.; Luo, Y.; Deng, Z.; Wang, J.; Liu, B, "Novel Power-Efficient Fast-Locking Phase-Locked Loop Based on Adaptive Time-to-Digital Converter-Aided Acceleration Compensation Technology", Electronics 2024, 13, 3586.

[17] Kang, B.; Kim, Y.; Son, H.; Kim, S, "A 0.055 mm2 Total Area Triple-Loop Wideband Fractional-N All-Digital Phase-Locked Loop Architecture for 1.9–6.1 GHz Frequency Tuning", Electronics 2024, 13, 2638.

979-8-3315-1550-8/25 $31.00 © 2025 IEEE

Effect of Temperature on the Stability of SnSe Nanoribbons as a Channel Material for Field-Effect Transistors

1st Nilüfer Ertekin
Electrical and Electronics Engineering Department of Yalova University, Yalova, Türkiye
School of Engineering, Electrical, Electronic and Computer Engineering, The University of Western, Perth, Australia
nilufer.ertekin@yalova.edu.tr

2nd Wen Lei, *Member, IEEE*
School of Engineering, Electrical, Electronic and Computer Engineering, The University of Western Australia
Perth, Australia
wen.lei@uwa.edu.au

Abstract—Tin monoselenide (SnSe) nanoribbons, as one-dimensional Sn-based channel materials, offer potential for the fabrication and development of nanotransistors. However, their fundamental properties, behavior under transistor setups, and response to temperature variations require further investigation to assess their performance. In this regard, the present study uses density functional theory (DFT) to examine the structural, electronic, and thermal properties of SnSe nanolayers as a benchmark and armchair SnSe nanoribbons as channel materials for transistors. The results indicate that SnSe monolayers are stable and possess an indirect bandgap of 1.4 eV, which limits charge transport, which agrees with previous studies. In contrast, armchair SnSe nanoribbons exhibit a direct bandgap of 0.84 eV, confirming their semiconducting nature and suitability as channel materials for nanotransistor applications. Investigation of the temperature effects in the 300 to 850 K range reveals that between 300 and 425 K, the temperature has no significant effect on the drain current, indicating reliable device performance within this range. However, between 425 and 650 K, the drain current increases linearly with rising temperature. Above 650 K, the drain current decreases with increasing temperature due to the effects of lattice vibrations.

Keywords— *Nanotransistor, SnSe nanoribbon, Temperature effects, Drain current, Density functional theory (DFT).*

I. INTRODUCTION

Nanoelectronics, a specialized branch of nanotechnology, has evolved over the past 50 years, driven by the continuous miniaturization of transistors [1]. It has significantly impacted the electronics industry by enabling the development of smaller, faster, and more energy-efficient devices. Key applications include high-density information storage, compact computing systems, logic circuit design, and nanoscale components such as nanowires. Nanoscale field-effect transistors (FETs) are increasingly miniaturized for modern circuits. They consist of source, drain, and gate terminals. They regulate current flow through the channel between the drain and source by applying voltage to the gate terminal. These transistors are widely used in logic circuits and energy storage-based memory devices [2]–[4].

Various materials are employed as channel in nanotransistors, each exhibiting distinct electronic properties [5]. Among these, two-dimensional (2D) semiconductors composed of atomic layers demonstrate superior electronic, thermal, and mechanical characteristics compared to their bulk counterparts [6], include graphene [7], transition metal dichalcogenides such as MoS_2 and WS_2, and black phosphorus [8]–[10]. This material is utilized in the fabrication of transistors, such as in the creation of nanolayers, nanotubes, and nanosheets, and nanoribbons. Among them, nanoribbons, which have narrow widths compared to their length, provide unique electronic and mechanical features as channel materials [11].

Tin monoselenide (SnSe) and diselenide ($SnSe_2$) are particularly prominent in optoelectronics due to their thermal properties and high absorption coefficient [12], [13]. These materials adopt a structure with SnSe functioning as a p-type semiconductor and $SnSe_2$ as an n-type semiconductor [14]. $SnSe_2$ has been extensively explored as a Sn-based channel material for bottom gate transistors [12]–[16]; however, the fundamental properties of SnSe as a channel material remain unclear.

Previous experimental research on SnSe has primarily focused on transistor fabrication [17], [18], while simulations have provided insights into its transport properties [19]. However, the fundamental characteristics of SnSe as a transistor material, as well as the effects of environmental conditions such as temperature, humidity, and dust, have not been clearly reported. In this regard, investigating the response and behavior of SnSe under these conditions can yield accurate data, aiding in the design and engineering of nanotransistors for diverse operational environments.

Current research considers the SnSe nanoribbon (SnSeNR) as channel materials for Field-Effect Transistors(FET) and employs density functional theory (DFT) [20] method for simulations. It probes the stability of the nanoscale FET transistors under wide range of temperature. To accomplish this goal, the study started with SnSe monolayers and phonon spectrum and band gaps calculations were performed for validation of simulation results. In the follow, the SnSeNR with armchair chirality was modeled and similar teste were performed on it. Then, SnSeNR integrated into nanotransistor setup and gate-source, drain-source voltages was applied on it to investigate its metrics in different temperature conditions to improve its performance.

II. SIMULATION METHOD

All DFT simulations in this study were performed using the Quantum ESPRESSO [21] and SIESTA [22] software.

These software programs play an important role in performing geometric optimizations and computing the electronic properties of SnSe nanolayers and nanoribbons in transistor applications. for proposed system and simulation setups, to assess the electron correlation and exchange interactions, the generalized gradient approximation with the Perdew–Burke–Ernzerhof [23] function was applied. moreover, an iterative approach was performed to integrate the Kohn–Sham equations [24], and in the meantime the Ultrasoft Pseudopotentials were utilized to improve accuracy and computational power. To ensure precise energy convergence and stabilize atomic forces, the convergence thresholds and force minimization were set at 10^{-6} Ry and 0.001 Ry/Bohr, respectively. The band gap energy was calculated by computing the energy difference between the valence band maximum (VBM) and the conduction band minimum (CBM) [25]. Phonon calculations were performed to evaluate the system's dynamic stability. For transport property calculations, SIESTA was used with a single-zeta polarized basis set. To ensure convergence, a maximum of 500 self-consistent field iterations was used, and k point sampling of $1\times1\times100$ was defined for Brillouin zone integration using the Monkhorst-Pack scheme. A real-space mesh cutoff of 100 Ry was applied.

The Landauer–Büttiker formalism [26], which characterizes electron transport with quantum effects in mesoscopic systems, was applied to describe the current between the drain and source. By calculating the transmission probability of electrons across the channel, the current was determined. This method allows for the investigation of conductance, resistance, and quantum effects within the channel, where the current is generated. Eq. (1) represents the Landauer–Büttiker formalism [26].

$$I(V_{ds}, V_g) = \frac{2e}{h} \int_{-\infty}^{+\infty} \{T(E, V_{ds}, V_g[f_s(E - \mu_S) - f_s(E - \mu_D)]\} dE \quad (1)$$

In the above equation, V_{ds} is the different voltage between drain and source and V_g signifies the gate voltage. $\mu_{S(D)}$ is the electrochemical potential, $f_{S(D)}$ signifies Fermi-Dirac distribution functions for the source. The average of the k-dependent transmission coefficients at different gate and bias voltages are shown by $T(E, V_{bias}, V_g)$. The k-dependent transmission coefficients ($T_{k\parallel}(E)$), the Green's function ($G_{k\parallel}(E)$), self-energy coupling terms (Γ) can be found from below equations:

$$T_{k\parallel}(E) = Tr[\Gamma^S_{k\parallel}(E)G_{k\parallel}(E)\Gamma^D_{k\parallel}(E)G^{\dagger}_{k\parallel}(E)] \quad (2)$$

$$G_{k\parallel}(E) = \left[(E + i\eta^+)I - H_{k\parallel} - \Sigma^S_{k\parallel} - \Sigma^D_{k\parallel}A\right]^{-1} \quad (3)$$

$$\Gamma^{L(R)}_{k\parallel}(E) = i\left[\Sigma^{L(R)}_{k\parallel} - \left(\Sigma^{L(R)}_{k\parallel}\right)^{\dagger}\right] \quad (4)$$

where $k\parallel$ is a reciprocal lattice vector orthogonal to the transmission direction. $G_{k\parallel}(E)$ and $G_{k\parallel}\dagger(E)$ represent the retarded and advanced Green's functions, respectively. Γ^S and Γ^D are the coupling functions between the channel and the source and drain electrodes, respectively. $H_{k\parallel}$, $\eta+$, and E denote the identity matrix, the Hamiltonian matrix of the channel, an infinitesimally small value greater than zero, and the energy, respectively. The self-energy coupling terms for the source (Γ^S) and drain (Γ^D) are given by Eq. (2), representing the contributions from the left and right

electrodes. The electrostatics between the channel and the source/drain contacts are handled by solving the Poisson equation self-consistently using a real-space solver. The level broadening is represented by the corresponding term [27], [28].

III. RESULT AND DESCUTION

A. SnSe monolayer

Figure 1 depicts the different viewpoint of SnSe monolayer and its unit lattice. The Sn and Se were highlighted in an orthorhombic pattern. The lattice constants of this arrangement are 4.26 Å and 4.44 Å, which align with the results found in previous study [29]. To create this monolayer, the unit cell reported in parallel direction. For the simulation setup, the vacuum region was set to a size of 20 Å along the z:[001] direction to prevent interlayer interactions and create a free surface perpendicular to the x-y plane. The thickness of the layer was assumed to be 2.73 Å. Additionally, periodic boundary conditions were applied in the x:[100] and y:[010] directions.

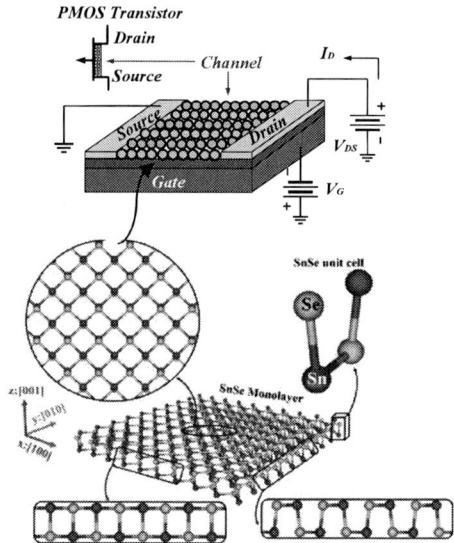

Fig. 1. The initial all-atomic configuration of SnSe monolayer as channel material for bottom gate transistors. The grey and yellow spheres represent Sn and Se atoms, respectively.

After energy minimization of the SnSe monolayer, the phonon spectrum, partial density of states (PDOS), and band structure were computed and are presented in Figure 2. Following structural modeling and energy minimization, the optimized configuration underwent phonon calculations to assess its dynamic stability. The phonon spectrum provides critical insights into the vibrational modes and frequencies of the material, serving as a key indicator of its stability. As shown in Figure 2.a, the phonon dispersion diagram exhibits no negative frequencies, confirming the structural stability of the SnSe monolayer.

The absence of imaginary modes indicates that the material remains dynamically stable under the given simulation conditions. If negative frequencies were observed, it would signify structural instability and errors in the simulation setup. Once the structural stability was established, electronic structure calculations were conducted to analyze the energy band characteristics of the monolayer. As depicted in Figure 2.b, the PDOS emphasizes the separation of s, p, and d orbitals relative to the Fermi level, further confirming the semiconducting nature of the SnSe monolayer. Additionally, Figure 2.c illustrates that the computed bandgap of the SnSe structure is 1.4 eV, which is in agreement with previously reported values [30], [31]. This bandgap suggests that SnSe possesses suitable electronic properties for potential applications in nanoelectronics devices.

Table 1 compares the band gaps of different semiconductor monolayers, including MoS_2, $MoSe_2$, WS_2, and WSe_2, with that of SnSe. Based on this data, it is evident that the SnSe sheet has the lowest band gap among this group, indicating that electron excitation in this system requires less energy compared to the other samples. Lower operational energy for electron transport leads to reduced heat generation and operating temperature, which positively affects the cycle life of the device. In light of these considerations, it can be concluded that the SnSe monolayer is a promising candidate for use as a channel material in nanotransistors, offering advantages such as lower operating heat, reduced temperature, and decreased energy requirements for electron transport.

Table 1 Intercomparison between band gap value and type of different semiconductor monolayer [32], [33].

Monolayer	SnSe	MoS_2	$MoSe_2$	WS_2	WSe_2
Band gap (eV)	1.40	1.76	1.43	1.95	1.62
Band gap type	Indirect	Direct/indirect			

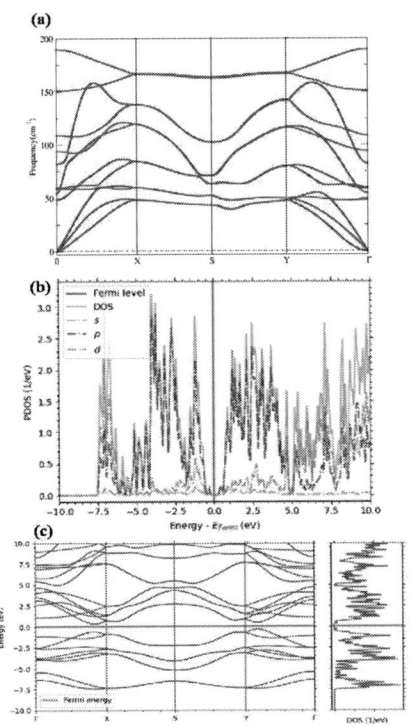

Fig. 2. (a) Phonon spectrum, (b) partial density of states (PDOS), and (c) the band structure of SnSe monolayer.

B. SnSe nanoribbon

Building upon the stabilized monolayer structure, SnSe nanoribbons were designed in two distinct edge configurations: zigzag and armchair. The choice of edge configuration significantly influences the electronic and transport properties of the nanoribbons, making their structural analysis critical for potential transistor applications. The electronic structure of these nanoribbons varies based on their edge morphology, leading to fundamentally different transport behaviors. The zigzag type of SnSe nanoribbon has a zero-band gap and manner as metal [16], [32] because of the metallic manner, it is not proper candidate for FET transistors. On the other hand, armchair type of SnSe nanoribbons, due to band gap in the range of 0.8 to 1.4 eV, provide a semiconductor nature and are suitable for nanotransistors [16]. In the following, the DFT simulation evaluated the electronic and transmission attributes of the armchair SnSe nanoribbon with seven lattice constant lengths and its potential for nanotransistor applications.

Figure 3.a illustrates the SnSe nanoribbon, meticulously simulated with an optimized lattice constant of 4.27 Å and a total length of 15.57 Å along the z:[001] axis. This configuration includes a vacuum layer of 20 Å thickness, introduced perpendicular to the nanoribbon surface along the y:[010] direction, to eliminate interactions between periodic images of atoms. The resulting lattice constant of 4.27 Å, obtained after careful optimization, aligns well with the pervious reported data [32]. Subsequently, the PDOS and band structure of the nanoribbon were computed, and the results were visually presented in Figures 3.b and 3.c, respectively. The PDOS diagram, which depicts the contribution of s, p, and d orbitals to the total density of states (DOS), confirms the semiconducting nature of the SnSe nanoribbon. The electronic structure analysis reveals the presence of a direct band gap of 0.84 eV, indicating that the CBM and VBM occur at the same k-point.

A comparative analysis of the band structures of the SnSe nanolayer (Figure 2.c) and the SnSe nanoribbon (Figure 3.c) clearly demonstrates the transition from an indirect band gap in the monolayer to a direct band gap in the nanoribbon. This transformation occurs due to the quantum confinement effect introduced by reducing the dimensionality from a two-dimensional nanolayer to a one-dimensional nanoribbon and edge effects (chirality effects, typically about armchair and zigzag type effects). This key finding highlights the potential of SnSe nanoribbons with armchair type as channel materials for transistor applications.

C. Transistor set-up

To further validate the feasibility of SnSe nanoribbons in nanoelectronics devices, additional electronic transport simulations were conducted. These simulations aim to evaluate critical transistor parameters like drain-source current, which will be discussed in the following sections.

As explained in the previous sections, the SnSe nanoribbon with an armchair edge structure (ANR-SnSe) exhibits promising characteristics for nano-FET transistor applications and must be evaluated based on transistor performance criteria. To achieve this, a bottom gate transistor configuration is employed as the foundation for the study. the transistor structure for quantum transport calculations is divided into three main regions: electrode 1 (drain), electrode 2 (source), and the intermediate region, which serves as the scattering region or channel.

To implement this setup, the modeled nanoribbon from Figure 3.a is placed within the framework depicted in Figure

4. In this configuration, the electrodes are constructed by repeating two ANR-SnSe, while the channel region consists of five. This results in electrode and channel thicknesses of two- and five- ANR-SnSe, respectively, ensuring a well-defined and stable structure for transport calculations. The chosen configuration prevents direct interaction between the electrodes, thereby enabling accurate modeling of electron transport within the system. Additionally, the transport direction is consistently aligned along the z-axis to maintain computational accuracy and facilitate a precise evaluation of the nanoribbon's suitability as a transistor channel material.

current can be attributed to changes in carrier mobility and carrier scattering, particularly phonon scattering, which affect charge transport in the nanoribbon channel.

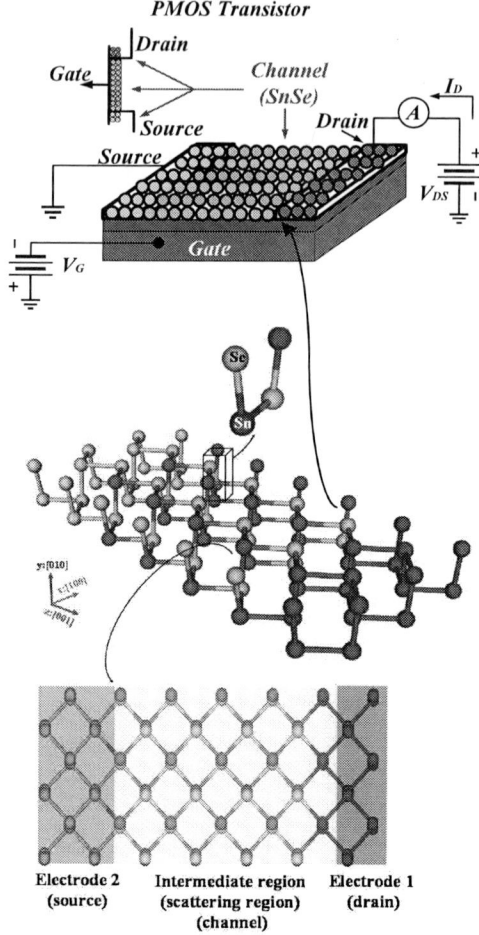

Fig. 4. A bottom gate transistor configuration of SnSe include electrode and channel thicknesses of two- and five ANR-SnSe, respectively. The system structure for quantum transport calculations is divided into three main regions: electrode 1 (drain), electrode 2 (source), and the intermediate region, which serves as the scattering region or channel.

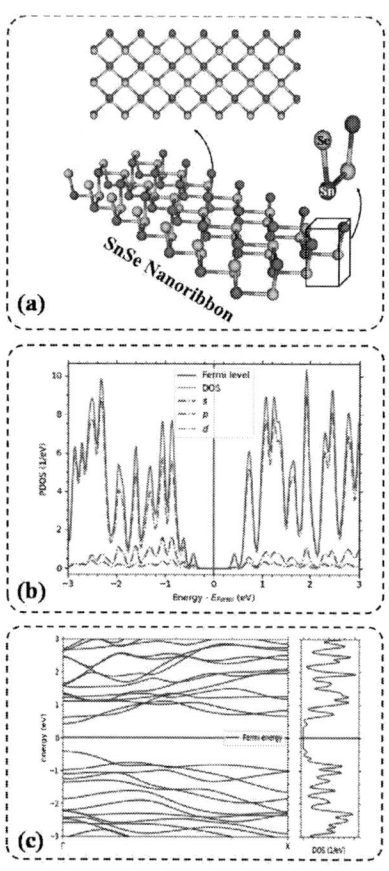

Fig. 3. (a) Structure, (b) partial density of states (PDOS), (c) band structure calculated for SnSe nanoribbon using DFT method in current research.

D. Effects of temperature on drain current of SnSe nanoribbon

This section examines the effects of temperature on the drain-source current of the transistor to assess its thermal performance. To achieve this, the transistor's behavior and current were analyzed under varying temperature conditions. Temperatures above room temperature, assumed to be 300 K, were applied to the device, and the current was calculated and recorded following the transmission run. Subsequently, the temperature was increased to 850 K, and the current was computed at each temperature increment. The gate and drain-source voltages were set to -1.2 V and 2 V, respectively. As shown in Figure 5, temperature variations within the range of 300 K to 425 K did not significantly impact device performance, indicating that the transistor operates reliably within this temperature window. The minimal fluctuations in

Figure 5 illustrates that for temperatures exceeding 425 K, the drain current increases sharply linearly up to 650 K and then begins to decrease. Higher temperatures (from 425 to 650 K) reduce the band gap, allowing more electrons (or holes) to transition across the bandgap and contribute to conduction. Additionally, the increase in trap states due to thermal activation facilitates the generation of more free carriers, further enhancing conduction. However, when the temperature exceeds the threshold of 650 K, the behavior changes again, as additional physical parameters become active beyond this point, significantly influencing the drain current.

After 650 K, a reduction in drain-source current was observed. Above this temperature, phonon scattering and lattice vibrations become dominant, reducing the mobility of charge carriers. Thermal vibrations lead to the formation of an atomic landscape of hills and valleys, commonly referred to as a disordered sheet. This landscape affects charge carrier dynamics and charge transport properties [35], [36]. At higher temperatures, thermally activated recombination increases, leading to a decrease in free charge carriers and a subsequent drop in current. Consequently, carrier concentration dominates at lower temperatures, increasing the current, whereas at higher temperatures, mobility degradation due to phonon scattering leads to a decline in current.

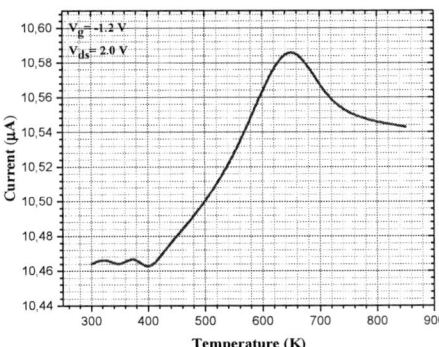

Fig. 5. Drain- source current across temperature variations within the range of 300-850

In current research, the length and width of nanoribbon set to specific values which is common and acceptable for DFT study and revealing transport features. However, it is clear that the different length and width of nanoribbon can change the electronic characteristic of SnSe nanoribbon. Increasing length and decreasing width lead to rising resistivity of nanoribbon. This lead to operating of system with more temperature and produced heat during operation can damage to other part of electronic circuit and reduce the life time of devices. In this concern, choosing appropriate dimension which can lead to independent results to dimension is critical for commercial applications.

In the case of device-level implications, decreasing the size and weight of electronic components leads to more flexibility for electronic packaging devices. It decreases the size of the final product. The size-reduced devices provide a customer-friendly market and low-cost packing services. However, cooling system and interface connection limitations will appear when the size decreases. In the case of cooling, smaller sizes cause poor flexibility when designing cooling systems. As results, produced temperature remain in system and residual heat cause increasing the temperature of electronic components, such as nanotransistors. Increasing the temperature of channel materials in nanotransistors leads to operating limitations by manipulating electron transfer. In the case of SnSe nanoribbon, the results reveal that between 300 K and 425 K, the efficiency of the channel is independent of temperature, which is a positive effect. However, for more than 425 K, the drain current increased linearly with temperature and disturbed channel operation.

IV. Conclusion

This study investigates SnSe nanomaterials as potential channel materials for nanotransistors. Specifically, SnSe nanolayers and armchair nanoribbons were examined to establish benchmarks and develop transistor setups using the density functional theory (DFT) method. In both cases, structural, and electronic properties were analyzed. Validation through phonon spectrum analysis, PDOS, and band structure calculations confirmed the stability of monolayer SnSe, revealing an indirect bandgap of 1.4 eV, which limits charge transport. To enhance performance, dimensional confinement

was explored by transitioning to one-dimensional (1D) nanoribbons. In this context, the SnSe nanoribbon with an armchair configuration was investigated, yielding a direct bandgap energy of 0.84 eV. This result confirms its semiconducting nature and suitability as a channel material for nanotransistor applications.

Regarding the application of nanotransistors under varying conditions and temperatures, the drain-source current was examined at temperatures above room temperature to assess the thermal stability of the device. The results indicate that temperature variations between 300 K and 425 K did not significantly affect device performance, demonstrating reliable transistor operation within this range. The observed thermal stability highlights the potential of the proposed FET for nanoelectronics applications. However, within the temperature range of 425 K to 650 K, the drain current increased linearly with temperature. Beyond the threshold temperature of 650 K, this trend reversed, and the current decreased due to the impact of lattice vibrations.

Acknowledgment

This study is supported by the Scientific and Technological Research Council of Türkiye, TÜBITAK- 2219 - International Postdoctoral Research Fellowship Program for Turkish Citizens. It is also funded by the TÜBITAK- 2224-A Grant Program for Participation in Scientific Meetings Abroad. The authors express their gratitude to TÜBİTAK for its support.

References

[1] P. Gargini, F. Balestra, and Y. Hayashi, "Roadmapping of nanoelectronics for the new electronics industry," *Appl. Sci.*, vol. 12, no. 1, p. 308, 2022, doi: 10.3390/app12010308.

[2] R. Chau, "Benchmarking nanotechnology for high-performance and low-power logic transistor applications," *2004 4th IEEE Conf. Nanotechnol.*, vol. 4, no. 2, pp. 3–6, 2004, doi: 10.1109/nano.2004.1392230.

[3] B. Yu and M. Meyyappan, "Nanotechnology: Role in emerging nanoelectronics," *Solid. State. Electron.*, vol. 50, no. 4, pp. 536–544, 2006, doi: 10.1016/j.sse.2006.03.028.

[4] D. Sadighbayan, M. Hasanzadeh, and E. Ghafar-Zadeh, "Biosensing based on field-effect transistors (FET): Recent progress and challenges," *TrAC - Trends Anal. Chem.*, vol. 133, p. 116067, 2020, doi: 10.1016/j.trac.2020.116067.

[5] P. Razavi, G. Fagas, I. Ferain, R. Yu, S. Das, and J. P. Colinge, "Influence of channel material properties on performance of nanowire transistors," *J. Appl. Phys.*, vol. 111, no. 12, 2012, doi: 10.1063/1.4729777.

[6] J. Jiang *et al.*, "Schottky-barrier quantum well in two-dimensional semiconductor nanotransistors," *Mater. Today Phys.*, vol. 15, p. 100275, 2020, doi: 10.1016/j.mtphys.2020.100275.

[7] L. Xu *et al.*, "Graphene–Silicon Hybrid MOSFET Integrated Circuits for High-Linearity Analog Amplification," *IEEE Electron Device Lett.*, vol. 43, no. 11, pp. 1886–1889, 2022, doi: 10.1109/LED.2022.3204950.

[8] A. Pon, A. Bhattacharyya, and R. Rathinam, "Recent Developments in Black Phosphorous Transistors: A Review," *J. Electron. Mater.*, vol. 50, no. 11, pp. 6020–6036, 2021, doi: 10.1007/s11664-021-09183-1.

[9] X. Zhang, H. Zhao, X. Wei, Y. Zhang, Z. Zhang, and Y. Zhang,

"Two-dimensional transition metal dichalcogenides for post-silicon electronics," *Natl. Sci. Open*, vol. 2, no. 4, p. 20230015, 2023, doi: 10.1360/nso/20230015.

[10] A. S. Kumar, V. Bharath Srinivasulu, C. Ganesh, V. Jukuru, T. Valluru, and D. P. S. S. S. K. Vamsi, "Principle Study of MoS2FET at lower Channel Lengths," in *Journal of Physics: Conference Series*, IOP Publishing, 2024, p. 12080. doi: 10.1088/1742-6596/2837/1/012080.

[11] L. Li *et al.*, "Fast near-infrared photodetectors from p-type SnSe nanoribbons," *Nanotechnology*, vol. 34, no. 24, p. 245202, 2023, doi: 10.1088/1361-6528/acc1eb.

[12] A. M. El-Mahalawy, S. A. Mansour, A. R. Wassel, A. E. Mohamed, and S. E. Ali, "Impact of structural and optical properties tunability of SnSe2 thin films on its optoelectronic properties," *Surfaces and Interfaces*, vol. 33, p. 102251, 2022, doi: 10.1016/j.surfin.2022.102251.

[13] M. Kumar, S. Rani, P. Vashishtha, G. Gupta, X. Wang, and V. N. Singh, "Exploring the optoelectronic properties of SnSe: a new insight," *J. Mater. Chem. C*, vol. 10, no. 44, pp. 16714–16722, 2022, doi: 10.1039/d2tc03799h.

[14] P. A. Fernandes, M. G. Sousa, P. M. P. Salomé, J. P. Leitão, and A. F. Da Cunha, "Thermodynamic pathway for the formation of SnSe and SnSe2 polycrystalline thin films by selenization of metal precursors," *CrystEngComm*, vol. 15, no. 47, pp. 10278–10286, 2013, doi: 10.1039/c3ce41537f.

[15] A. Sirohi and J. Singh, "2D Bi2O2 Se Based Highly Selective and Sensitive Toxic Non-Condensable Gas Sensor," *IEEE Trans. Nanotechnol.*, vol. 21, pp. 794–800, 2022, doi: 10.1109/TNANO.2022.3226507.

[16] Y. Yang, Y. Zhou, Z. Luo, Y. Guo, D. Rao, and X. Yan, "Electronic structures and transport properties of SnS-SnSe nanoribbon lateral heterostructures," *Phys. Chem. Chem. Phys.*, vol. 21, no. 18, pp. 9296–9301, 2019, doi: 10.1039/c9cp00427k.

[17] S. H. Cho, K. Cho, N. W. Park, S. Park, J. H. Koh, and S. K. Lee, "Multi-Layer SnSe Nanoflake Field-Effect Transistors with Low-Resistance Au Ohmic Contacts," *Nanoscale Res. Lett.*, vol. 12, pp. 1–6, 2017, doi: 10.1186/s11671-017-2145-2.

[18] S. Liu, Y. Chen, S. Yang, and C. Jiang, "SnSe field-effect transistors with improved electrical properties," *Nano Res.*, vol. 15, no. 2, pp. 1532–1537, 2022, doi: 10.1007/s12274-021-3698-z.

[19] M. Zhang *et al.*, "The electronic transport properties of zigzag phosphorene-like MX (M = Ge/Sn, X = S/Se) nanostructures," *Phys. Chem. Chem. Phys.*, vol. 19, no. 26, pp. 17210–17215, 2017, doi: 10.1039/c7cp02201h.

[20] E. Engel, *Density functional theory*. Springer, 2011.

[21] P. Giannozzi *et al.*, "QUANTUM ESPRESSO: A modular and open-source software project for quantum simulations of materials," *J. Phys. Condens. Matter*, vol. 21, no. 39, p. 395502, 2009, doi: 10.1088/0953-8984/21/39/395502.

[22] J. M. Soler *et al.*, "The SIESTA method for ab initio order-N materials simulation," *J. Phys. Condens. Matter*, vol. 14, no. 11, pp. 2745–2779, 2002, doi: 10.1088/0953-8984/14/11/302.

[23] J. P. Perdew, K. Burke, and M. Ernzerhof, "Generalized gradient approximation made simple," *Phys. Rev. Lett.*, vol. 77, no. 18, pp. 3865–3868, 1996, doi: 10.1103/PhysRevLett.77.3865.

[24] W. Kohn and L. J. Sham, "Self-consistent equations including exchange and correlation effects," *Phys. Rev.*, vol. 140, no. 4A, p. A1133, 1965, doi: 10.1103/PhysRev.140.A1133.

[25] N. ERTEKİN, "Structural and electronic properties of fluorine-doped lithium oxide as a solid electrolyte interphase for lithium air batteries," *J. Sci. Reports-A*, no. 055, pp. 94–103, 2023, doi: 10.59313/jsr-a.1318117.

[26] S. Datta and H. van Houten, *Electronic Transport in Mesoscopic Systems* , vol. 49, no. 5. Cambridge university press, 1996. doi: 10.1063/1.2807624.

[27] M. Jafarinaeimi and M. Berahman. "Spin transport in ScSe2 nanoribbon field effect transistor." *The Journal of Physical Chemistry C* 127.28 (2023): 13782-13788.

[28] K. Tamersit, "Performance enhancement of an ultra-scaled double-gate graphene nanoribbon tunnel field-effect transistor using channel doping engineering: Quantum simulation study," *AEU - Int. J. Electron. Commun.*, vol. 122, p. 153287, 2020, doi: 10.1016/j.aeue.2020.153287.

[29] D. Liu, B. Qin, and L. D. Zhao, "Lattice plainification advances highly effective SnSe crystalline thermoelectrics," *Kexue Tongbao/Chinese Sci. Bull.*, vol. 68, no. 21, pp. 2716–2718, 2023, doi: 10.1360/TB-2023-0553.

[30] L. Bin Shi, M. Yang, S. Cao, Q. You, Y. Y. Niu, and Y. Z. Wang, "Elastic behavior and intrinsic carrier mobility for monolayer SnS and SnSe: First-principles calculations," *Appl. Surf. Sci.*, vol. 492, pp. 435–448, 2019, doi: 10.1016/j.apsusc.2019.06.211.

[31] A. Batool, Y. Zhu, X. Ma, M. I. Saleem, and C. Cao, "DFT study of the structural, electronic, and optical properties of bulk, monolayer, and bilayer Sn-monochalcogenides," *Appl. Surf. Sci. Adv.*, vol. 11, p. 100275, 2022.

[32] E. S. Kadantsev and P. Hawrylak, "Electronic structure of a single MoS 2 monolayer," *Solid State Commun.*, vol. 152, no. 10, pp. 909–913, 2012, doi: 10.1016/j.ssc.2012.02.005.

[33] S. Deng, L. Li, and M. Li, "Stability of direct band gap under mechanical strains for monolayer MoS2, MoSe2, WS2 and WSe2," *Phys. E Low-Dimensional Syst. Nanostructures*, vol. 101, pp. 44–49, 2018, doi: 10.1016/j.physe.2018.03.016.

[34] K. Tyagi, K. Waters, G. Wang, B. Gahtori, D. Haranath, and R. Pandey, "Thermoelectric properties of SnSe nanoribbons: A theoretical aspect," *Mater. Res. Express*, vol. 3, no. 3, p. 35013, 2016, doi: 10.1088/2053-1591/3/3/035013.

[35] R. S. Sánchez-Carrera, P. Paramonov, G. M. Day, V. Coropceanu, and J. L. Brédas, "Interaction of charge carriers with lattice vibrations in oligoacene crystals from naphthalene to pentacene," *J. Am. Chem. Soc.*, vol. 132, no. 41, pp. 14437–14446, 2010, doi: 10.1021/ja1040732.

[36] D. Kim, A. Aydin, A. Daza, K. N. Avanaki, J. Keski-Rahkonen, and E. J. Heller, "Coherent charge carrier dynamics in the presence of thermal lattice vibrations," *Phys. Rev. B*, vol. 106, no. 5, p. 54311, 2022, doi: 10.1103/PhysRevB.106.054311

2025 10th IEEE International Conference on Integrated Circuits, Design, and Verification (ICDV)

A 12-bit 100MS/s SAR ADC with Sub-Radix and Optimize Digital Delay Path

Pham-Hoang-Long Ho[1,2], Thien-Lam Van[1,2] and Cuong Huynh[1,2]

[1]*Faculty of Electrical and Electronics Engineering, Ho Chi Minh University of Technology,*
268 Ly Thuong Kiet Street, District 10, Ho Chi Minh City, Vietnam
[2]*Vietnam National University Ho Chi Minh City,*
Linh Trung Ward, Thu Duc District, Ho Chi Minh City, Vietnam
Email: long.hophg@hcmut.edu.vn, lam.vanthien@hcmut.edu.vn, hpmcuong@hcmut.edu.vn

Abstract—**This paper introduces a 12-bit sub-radix successive approximation register (SAR) analog-to-digital converter (ADC) operating at 100 MS/s. The design employs an optimized digital control logic with a reduced delay path to enhance conversion speed; meanwhile, shortening the decision time in the SAR algorithm increases susceptibility to comparator instability or settling of capacitor digital-to-analog (CDAC). To mitigate this, a redundancy bit is implemented to correct erroneous decisions. The proposed architecture achieves a signal-to-noise-and-distortion ratio (SNDR) over 70 dB, corresponding to an effective number of bits (ENOB) of 11 at the Nyquist rate, and consumes 2.85mW with a 1.2 V power supply.**

Index Terms—**Analog-to-digital converter (ADC), SAR ADC, sub-radix SAR ADC, redundancy bit, CMOS.**

I. INTRODUCTION

High-speed analog-to-digital converters (ADCs) are essential for modern communication systems, data acquisition, and instrumentation. However, achieving high sampling rates and medium resolution (10-16 bits) while maintaining low power consumption remains a significant challenge. Existing ADC designs frequently employ oversampling techniques to achieve high resolution at the cost of limited input bandwidth [1], or compromise efficiency by consuming significant power to maintain performance [2]. Recently, successive approximation register (SAR) ADCs have gained attention because of their efficient power consumption and simple architecture.

The SAR ADC utilizes a successive approximation approach to digitize an analog input signal. An asynchronous SAR ADC is illustrated in Fig. 1, its core components include a CDAC to generate the reference voltage, SAR logic to control the CDAC, and generate the clock signal for the comparator to operate. These elements operate asynchronously with clock signals, leading

to dynamic power consumption. However, a fundamental limitation of this architecture is its bit-by-bit conversion process. Specifically, for an N-bit resolution, the ADC requires at least N+1 clock cycles per sampling period to complete the conversion. Consequently, an effective strategy to enhance the overall ADC's speed is to accelerate the operation of each individual component. Recent research has introduced a split-CDAC architecture that reduces the number of unit capacitors by multiples, thereby decreasing the capacitive load on bit branches. Additionally, a double-tail comparator is used in place of a single Strong-ARM latch to enhance comparison speed. As a result, the primary performance bottleneck is shifted to the digital delay path in the SAR logic and CDAC settling. This research proposes the SAR ADC with the optimize delay path in SAR logic.

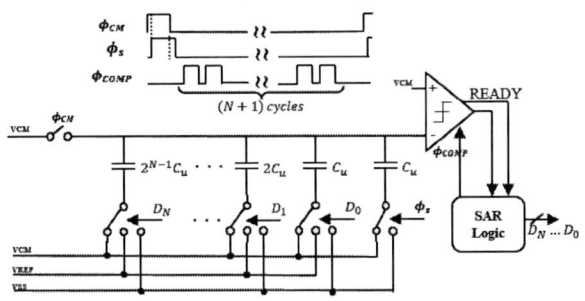

Fig. 1. Basic diagram of asynchronous SAR ADC.

II. PROPOSED CIRCUIT

Figure 2 illustrates the architecture of the proposed SAR ADC. To achieve medium resolution and high sampling rates, the design incorporates an alternative bootstrapped switch to mitigate the impact of charge sharing due to parasitic capacitance of large sampling

979-8-3315-1550-8/25 $31.00 © 2025 IEEE

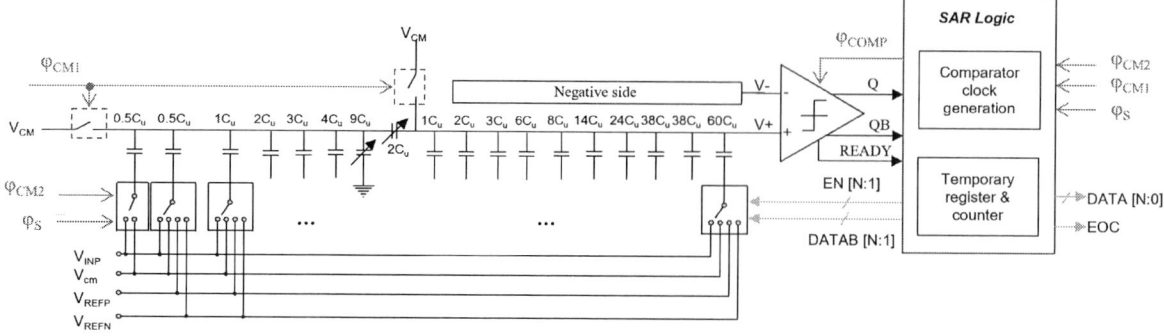

Fig. 2. Schematic of proposed SAR ADC.

switches. A triple-tail comparator is employed to enhance speed for 12-bit conversions by leveraging two initial amplification stages for amplifying small residue supporting reduce comparator decision time. Additionally, the split-CDAC utilizing a tri-reference (V_{REFP}, V_{CM}, V_{REFN}) switching scheme significantly reduces the capacitor array size, allowing faster DAC settling. To maintain high accuracy with high bit-conversion speed, four redundancy bits are integrated to correct errors caused by DAC imperfections and reduce per-bit decision times without compromising accuracy.

The operation of the SAR ADC within a single sampling cycle is illustrated in Fig. 3, implementing a tri-reference switching scheme. In this scheme, ϕ_{CM1} and ϕ_S serve as non-overlapping clocks for bottom-plate sampling, while ϕ_{CM2} facilitates common-mode swapping. Following the sampling phase, the conversion phase commences under the control of the SAR logic and proceeds sequentially until the final, LSB, comparison is completed.

A. Asynchronous SAR logic

The asynchronous SAR logic has three functions: generating the comparator clock, ϕ_{COMP} and $\bar{\phi}_{COMP}$; storing the comparator's result whenever the READY signal is HIGH; and sending a control signal to switch decoders. Typical asynchronous clock generation is shown in fig.4a, where the comparator resets and introduces a delay between successive comparisons. This delay, implemented using a chain of inverters, allows the CDAC to settle to the correct level. However, a major drawback of this approach is that inverter delay is highly sensitive to process variations, requiring additional programmable bits for compensation. Moreover, applying the same delay to every conversion wastes timing margins. To address this, [6] proposed using two different delay

paths depending on the conversion order. However, in designs with long digital paths, parasitic capacitance can degrade the maximum speed of the SAR logic. The proposed approach is shown in fig.4b. In the proposed

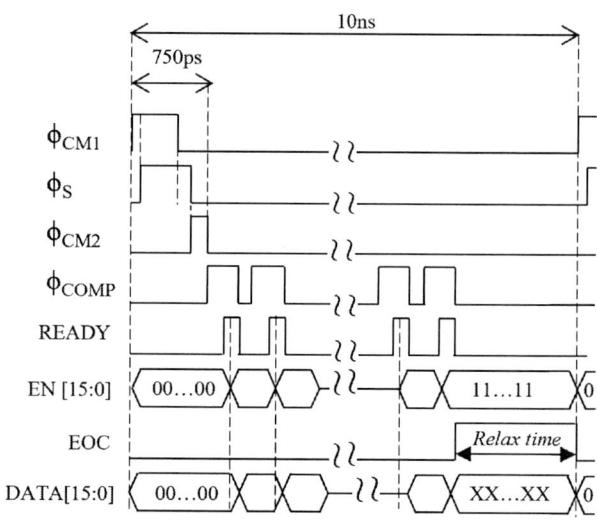

READY is a signal from comparator when the result is ready

Fig. 3. Waveform of proposed SAR ADC.

design, the comparator performs its first comparison after the sampling phase and the common-mode voltage swap from the top plate to the bottom plate. Once the comparison result is ready, the comparator resets, and the next comparison proceeds after a short delay. This delay, determined by the total propagation delay of the NAND gate and inverter, allows the SAR logic to store new data while ensuring CDAC settling. Consequently, this design prioritizes speed over accuracy; however, any resulting errors can be corrected by incorporating a redundancy bit. Fig. 5 illustrate the implementation of the comparator

979-8-3315-1550-8/25 $31.00 © 2025 IEEE   50

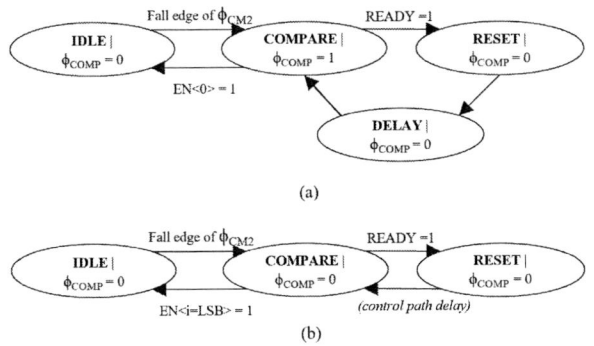

Fig. 4. (a) Typical and (b) proposed FSM of ϕ_{COMP} generation.

clock generator, which includes a temporary register file and shift registers. The temporary register utilizes both comparator outputs to maintain symmetrical capacitor loading, ensuring low offset.

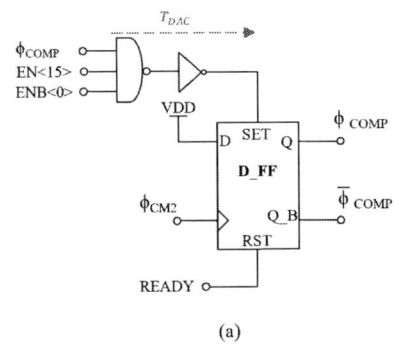

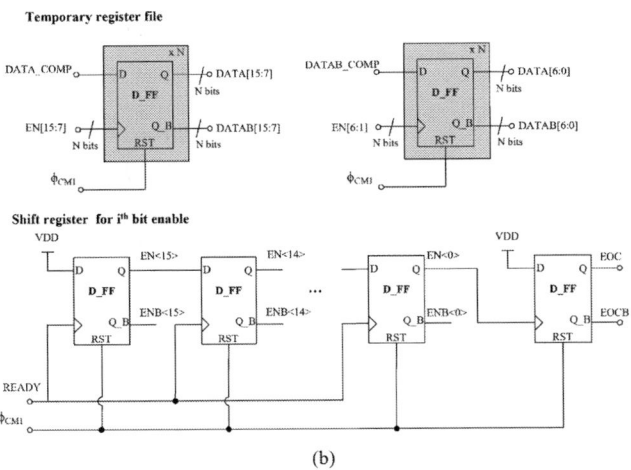

Fig. 5. (a) Comparator clock generator, (b) Memory of SAR logic

B. Modified split CDAC with redundancy bit

Fig. 2 shows the top-level architecture of a 12-bit SAR ADC with four redundancy bits, resulting in a

total output of 16 bits. The LSB and MSB branchs contribute 11 and 2037 steps, respectively, fulfilling the 2048 steps requirement for a 12-bit ADC using tri-reference (V_{REFP}, V_{CM}, V_{REFN}). The bridge capacitor is used to split the CDAC into two segments, MSBs and LSBs, making the MSB capacitor smaller and lowering the total capacitance; the MSB branch now uses 192 unit capacitors instead of the original 2037 unit capacitors. This greatly reduces the capacitor load at each bit branch, allowing for a faster conversion time compared to a full binary system shown in Fig. 1. This reduces the capacitor load at each bit branch significantly; thus, a faster conversion time compared to a full binary one shown in Fig. 1 is achieved. Meanwhile, split capacitor topology has its disadvantages: over-rail swing of the LSB branch and mismatch in the bridge capacitor. These problems can be solved by adding dummies to the cap in the LSB branch [6] [7].

The value of dummy capacitor is determined by selecting an integer multiple of the unit capacitor for the bridge capacitor (1). Choosing C_{bridge} as $2C_u$, C_{dummy} is approximately equal total capacitor in LSB branch which effectively reducing the LSB branch swing by half as well.

$$(C_{dummy} + C_{totalLSB})//C_{bridge} = C_u \qquad (1)$$

Intuitively, a 12-bit split CDAC offers faster operation compared to a 16-bit implementation, it requires no error decision during bit conversion, which is either impractical or imposes stringent constraints on other circuit blocks. Therefore, trading some speed for improved accuracy is a reasonable and practical design choice.

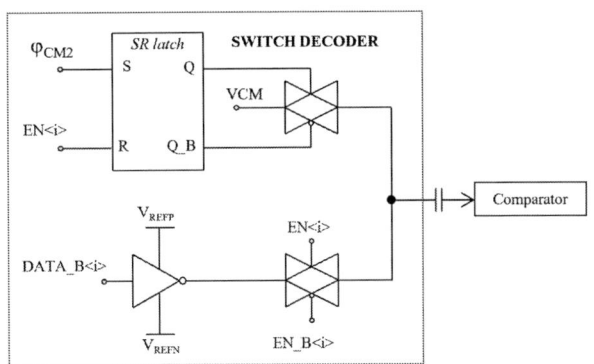

Fig. 6. Tri-reference decoder for CDAC.

The bottom-plate sampling is preferred for its linearity merits [8] and insensitivity to parasitic capacitance [6]. The insensitivity to parasitic capacitance of bottom-plate

979-8-3315-1550-8/25 $31.00 © 2025 IEEE   51

sampling is only applied for the MSB branch; the bridge capacitor and dummy capacitor need to be calibrated. This is another advantage of this topology in terms of limited analog calibration parts.

Motivated by [9], this work employs inverters combined with transmission gates to implement the decoder for the ith bit conversion, as illustrated in Fig. 6. This prevent floating node during bit-conversion. Moreover, The switch decoder is uniformly sized across all bit branches, with its drive strength optimized for the most significant bit (MSB) branch. For the dummy capacitor, no voltage decoder is required; instead, the reset terminal of the SR latch is directly connected to ϕ_{CM1}.

C. Comparator

The triple-tail comparator is preferred for increasing speed and still consuming power dynamically. The decision time of regenerative comparators is highly sensitive to the amplitude of the input signal, with smaller input differences resulting in prolonged decision times. In high-resolution applications, such as 12-bit conversion, this leads to exponentially increasing decision times per bit, potentially causing incorrect decisions or suspensions in the bit conversion process due to the critical dependency of the entire logic cycle on the comparator's output. The triple-tail comparator addresses this problem by greatly amplifying the input using the first two stages as a pre-amplifier and a half-latch stage.

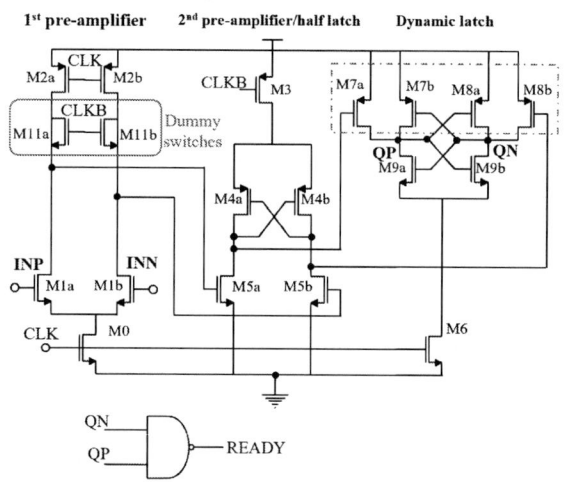

Fig. 7. Modified triple-tail comparator.

D. Bootstrap switch

To satisfy the requirements for half-Nyquist sampling of SAR ADC, the bootstrapped switch is employed

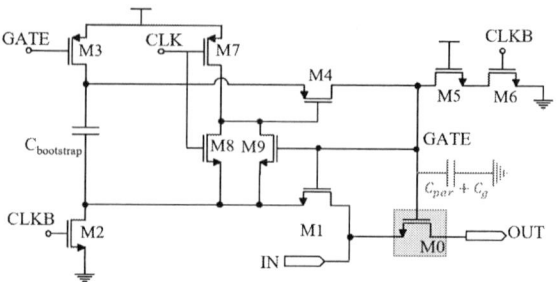

Fig. 8. Conventional bootstrap circuit.

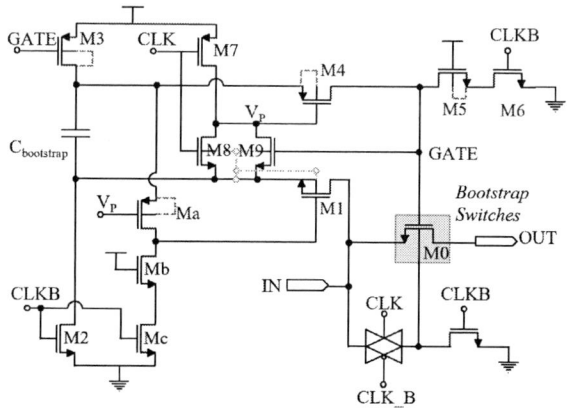

Fig. 9. Proposed bootstrap circuit.

due to its low on-resistance and minimal distortion. This configuration is achieved by driving the MOS switch gate to a voltage higher than its source using a constant voltage instead of a clock signal. However, the traditional bootstrap circuit, as illustrated in Fig. 8, switch M1, which connects the input with the bootstrap capacitor Cbootstrap, must be large for fast turn-on of the bootstrap switch, and the switch M0 must also be large to meet the sampling requirement. These two switches contribute a large parasitic capacitance. This results in a significant delay before VGATE settles to the desired level, extending the loop time and degrading the overall circuit performance. The high parasitic effects also reduce the bootstrap bandwidth and increase the resistance of M0, limiting the efficiency of the sampling operation at higher frequencies. Thus, in the proposed design shown in Fig. 9, this issue is alleviated by using a separate path for controlling the transistor M1, reducing the settling time and impedance impact of the track-and-hold circuit. This innovative design significantly improves sampling efficiency and mitigates performance losses at high operating frequencies.

979-8-3315-1550-8/25 $31.00 © 2025 IEEE

Additionally, a bulk cancellation technique is implemented by bootstrapping the MOS switch bulk to the input during sampling. This approach further mitigates distortion, enhancing circuit performance.

III. SIMULATION RESULT

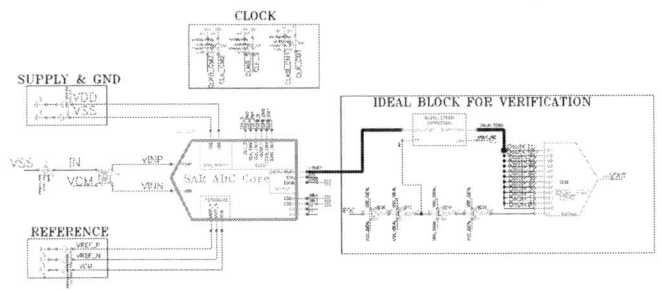

Fig. 10. SAR ADC dynamic test-bench setup.

The dynamic performance of the SAR ADC is evaluated by analyzing the spectrum of its output. The measurement setup, illustrated in Fig. 10, consists of two differential sine waves with a rail-to-rail amplitude and a frequency close to half the Nyquist frequency of the SAR ADC, which are applied to its input. The ADC output bits are then reconstructed into an analog signal using an ideal DAC. To receive a more practical result, transient noise is included in the analysis. The digital output is then converted back to an analog signal using an ideal DAC, which is subsequently used for spectral analysis. The SAR ADC waveform demonstrates the clock operation (upper of Fig. 11) and bit conversion sequence (lower of Fig. 11). The upper plot in Fig. 11 shows that the comparator clock (red line) intentionally avoids a full transition to 0 to reduce logic delay. The differential inputs, V+ and V-, represent the CDAC's top-plate voltage and also function as the comparator inputs. The SAR ADC operates correctly, as evidenced by the convergence of V+ and V- to the common-mode voltage level. The reconstructed output spectrum of the SAR ADC , obtained through a 4096-point FFT analysis, is illustrated in figure 12. The converter demonstrates a SNDR of 70.02 dB and a spurious-free dynamic range (SFDR) of 83.3 dB, yielding an ENOB of 11.33. During operation, the ADC average current consumptions is 2.85mA.

IV. CONCLUSIONS

This work presents a 12-bit, 100 MS/s asynchronous SAR ADC featuring an optimized delay path to enable

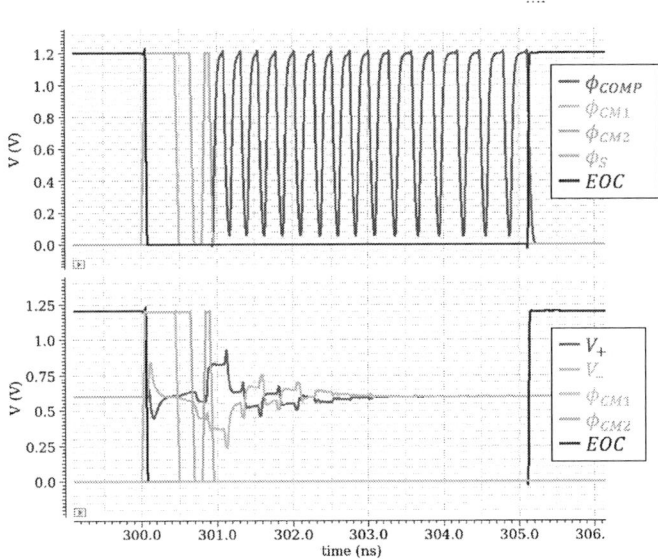

Fig. 11. Waveform of SAR ADC Clock and CDAC top plate

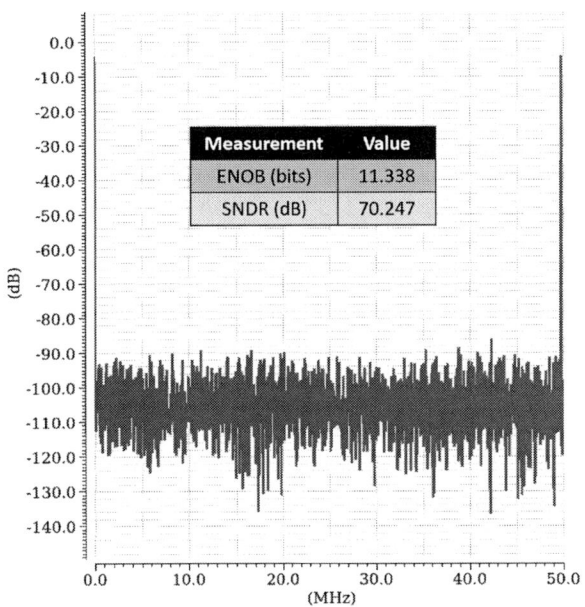

Measurement	Value
ENOB (bits)	11.338
SNDR (dB)	70.247

Fig. 12. Simulated 4096-point FFT output spectrum with 100MS/s sampling rate at 49.78MHz input frequency.

high-speed sampling. The proposed SAR logic architecture facilitates a high-speed clocking scheme for the comparator, necessitating the inclusion of redundancy bits to mitigate CDAC settling errors. To support higher bit-rate conversion, a triple-tail comparator is employed to accelerate comparison speed. A split-CDAC structure with dummy capacitors in the LSB branch minimizes overall capacitor count, reduces voltage swing

979-8-3315-1550-8/25 $31.00 © 2025 IEEE

in the LSB path, and enables the use of an integer-valued bridge capacitor. Simulation results demonstrate an ENOB of 11.3 bits at Nyquist rate using 4096-point FFT with a 49.78 MHz input signal. The ADC achieves a power efficiency of 11 fJ/conversion-step, consuming 2.85 mW from a 1.2 V supply. Post-layout simulations and silicon measurements are planned to further validate the design.

TABLE I
PERFORMANCE SUMMARY AND COMPARISON

	[7]	[10]	[11]	[12]	This Work*
Technology	CMOS 65nm	CMOS 65nm	CMOS 180nm	CMOS 40nm	CMOS 65nm
Supply (V)	1.2	1.2	1.8	0.9	1.2
Resolution (bits)	12	11	10	12	12
Sampling Rate (MS/s)	100	100	25	100	100
SNDR (dB) @ Nyquist	51.09	61.1	54	65	70.02
Power (mW)	0.8	1.6	3.83	2	2.85
FoM (fJ/conv.-step)	17.6	32	299.21	13.8	11
Architecture	SAR	SAR	SAR	SAR	SAR

*Schematic result only

V. ACKNOWLEDGMENT

This research is funded by Ho Chi Minh City University of Technology–VNU- HCM under grant number SVOISP-2024-ĐĐ-07. We acknowledge the support of time and facilities from Ho Chi Minh City University of Technology (HCMUT), VNU-HCM for this study.

REFERENCES

[1] T. Caldwell, D. Alldred and Z. Li, "A Reconfigurable $\Delta\Sigma$ ADC With Up to 100 MHz Bandwidth Using Flash Reference Shuffling," in IEEE Transactions on Circuits and Systems I: Regular Papers, vol. 61, no. 8, pp. 2263-2271.

[2] Texas Instruments, "ADS4125: 12-bit, 125 MSPS Low-Power Pipeline ADC," Datasheet, 2006.

[3] D. Xu et al., "A Noise Reduction 12-bit 125-MSPS SAR ADC with Modified Asynchronous Logic Regulation Technique," Journal of Circuits, Systems and Computers, vol. 30, no. 03, p. 2150040, Aug. 2020.

[4] A. T. Ramkaj, M. Strackx, M. S. J. Steyaert and F. Tavernier, "A 1.25-GS/s 7-b SAR ADC With 36.4-dB SNDR at 5 GHz Using Switch-Bootstrapping, USPC DAC and Triple-Tail Comparator in 28-nm CMOS," in IEEE Journal of Solid-State Circuits, vol. 53, no. 7, pp. 1889-1901, July 2018.

[5] J. Ma, J. Yu and Y. Hu, "An Analysis and Design Guideline of High-Speed and Low-Distortion Bootstrapped Switches," in IEEE Transactions on Circuits and Systems I: Regular Papers

[6] A. H. T. Chang, "Low-power high-performance SAR ADC with redundancy and digital background calibration," dspace.mit.edu, 2013. http://hdl.handle.net/1721.1/82177

[7] M. Li et al., "A 6.94-fJ/Conversion-Step 12-bit 100-MS/s Asynchronous SAR ADC Exploiting Split-CDAC in 65-nm CMOS," in IEEE Access, vol. 9, pp. 77545-77554, 2021, doi: 10.1109/ACCESS.2021.3079406.

[8] A. T. Ramkaj, J. C. Peña Ramos, M. J. M. Pelgrom, M. S. J. Steyaert, M. Verhelst and F. Tavernier, "A 5-GS/s 158.6-mW 9.4-ENOB Passive-Sampling Time-Interleaved Three-Stage Pipelined-SAR ADC With Analog–Digital Corrections in 28-nm CMOS," in IEEE Journal of Solid-State Circuits, vol. 55, no. 6, pp. 1553-1564, June 2020.

[9] T. Iizuka, H. Takenaka, H. Xu and A. A. Abidi, "Systematic Equation-Based Design of a 10-Bit, 500-MS/s Single-Channel SAR A/D Converter With 2-GHz Resolution Bandwidth," in IEEE Open Journal of the Solid-State Circuits Society, vol. 4, pp. 147-162, 2024.

[10] C. -H. Chan et al., "60-dB SNDR 100-MS/s SAR ADCs With Threshold Reconfigurable Reference Error Calibration," in IEEE Journal of Solid-State Circuits, vol. 52, no. 10, pp. 2576-2588, Oct. 2017, doi: 10.1109/JSSC.2017.2728784.

[11] H. N. Minh, D. N. Quoc and T. Hoang, "A design of 10-bit 25-MS/s SAR ADC using separated clock frequencies with high speed comparator in 180nm CMOS," 2015 International Conference on Advanced Technologies for Communications (ATC), Ho Chi Minh City, Vietnam, 2015, pp. 133-138, doi: 10.1109/ATC.2015.7388305.

[12] Xu Dai-guo, Pu-Jie, Xu Shi-liu, Zhang Zheng-ping, Zhang Jun-an, Wang Jian-an, A 12-bit 100-MS/s 83 dB SFDR SAR ADC with sampling switch linearity enhanced technique, IE-ICE Electronics Express, 2019, Volume 16, Issue 6, Pages 20190007, Released on J-STAGE March 25, 2019, Advance online publication March 08, 2019, Online ISSN 1349-2543, https://doi.org/10.1587/elex.16.20190007.

2025 10th IEEE International Conference on Integrated Circuits, Design, and Verification (ICDV)

QEA: An Accelerator for Quantum Circuit Simulation with Resources Efficiency and Flexibility

Van Duy Tran[1], Tuan Hai Vu[1], Vu Trung Duong Le[1], Hoai Luan Pham[1], and Yasuhiko Nakashima[1]

[1] Nara Institute of Science and Technology, 8916–5 Takayama-cho, Ikoma, Nara 630-0192, Japan.

Abstract—The area of quantum circuit simulation has attracted a lot of attention in recent years. However, due to the exponentially increasing computational costs, assessing and validating these models on large datasets poses significant obstacles. Despite plenty of research in quantum simulation, issues such as memory management, system adaptability, and execution efficiency remain unresolved. In this study, we introduce QEA, a state vector-based hardware accelerator that overcomes these difficulties with four key improvements: optimized memory allocation management, open PE, flexible ALU, and simplified CX swapper. To evaluate QEA's capabilities, we implemented and evaluated it on the AMD Alveo U280 board, which uses only 0.534 W of power. Experimental results show that QEA is extremely flexible, supporting a wide range of quantum circuits, has excellent fidelity, making it appropriate for standard quantum emulators, and outperforms powerful CPUs and related works up to 153.16× better in terms of normalized gate speed. This study has considerable potential as a useful approach for quantum emulators in future works.

Index Terms—quantum emulator, state vector, FPGA, SoC, and memory efficiency.

I. INTRODUCTION

Quantum algorithms have the potential to accelerate optimization, data processing, and feature extraction, which are essential in quantum computing [1]. However, development in this field is constrained by the restricted availability and capability of quantum hardware, requiring fast quantum simulation systems. As the number of qubits (#Qubits) increases, typical state-vector simulations such as naïve matrix multiplication (MM) become inefficient due to exponentially growing the #Qubits. Various approaches have been explored to address this challenge, including high-performance computing (HPC) systems that offer speed but demand power-intensive hardware such as DGX A100 and H100 [2], as well as software optimizations that minimize redundant operations but remain ineffective at large scale [3]. Alternative simulation techniques, such as tensor networks, the Heisenberg picture, and density matrices, as well as methods other than state-vector simulations, reduce the computational complexity to polynomial or linear time but are limited to specific circuit types. All the above works focus on general-purpose emulators or some fixed quantum algorithms, highlighting the need for more scalable and adaptable quantum simulation approaches.

Recently, substantial research has concentrated on designing quantum simulation hardware platforms (quantum emulators) employing Field-Programmable Gate Arrays (FPGAs), as a

potential solution [4]–[9]. This approach provides advantages in terms of programmability and practicality, as well as great performance and efficiency at lower costs. Existing studies, however, focus on specific applications with a restricted number of supported qubits, such as the Quantum Fourier Transform (QFT), Quantum Haar Transform (QHT), and Quantum Support Vector Machine (QSVM). As a result, their implementations lack the versatility required to enable a wider range of applications and more complicated quantum tasks. Furthermore, contemporary FPGA-based quantum emulators frequently encounter issues such as inefficient memory management, inefficient resource use, and slow execution speed, which are mainly due to non-optimized hardware designs.

To address these problems, we offer a hardware architecture named Quantum Emulation Accelerator (QEA) intended to support a wide range of applications with a high #Qubits while maintaining both flexibility and high performance at low hardware resources. Our approach combines hardware and software enhancements to efficiently simulate quantum circuits. On the software side, we still follow the state-vector simulator but enhance the performance of the overall process via gate-group-and-fusion and compact tensor product (TP) techniques. On the hardware side, we propose four key approaches, including:

- Optimized memory allocation management: A strategy for efficient memory utilization.
- Open processing element (PE): A processing element designed to allow data sharing among four processing elements, reducing execution time.
- Flexible arithmetic logic unit (ALU): A computation unit capable of switching between different operational modes to optimize data processing.
- Simplified CX swapper: A module designed for efficient computation of the CX gate.

To verify and evaluate QEA's performance, we implement it on the AMD Alveo U280 board and analyze its operational frequency, flexibility, and execution time.

II. BACKGROUND KNOWLEDGE

A. Quantum simulation algorithm

A quantum circuit can be simulated by a series of gates that act on a reference state $|\psi^{(0)}\rangle$ as $|\psi^{(m)}\rangle = \mathcal{U}|\psi^{(0)}\rangle$, where $|\psi^{(m)}\rangle$ and $\mathcal{U}(\boldsymbol{\theta})$ are target state and quantum operator,

979-8-3315-1550-8/25 $31.00 © 2025 IEEE

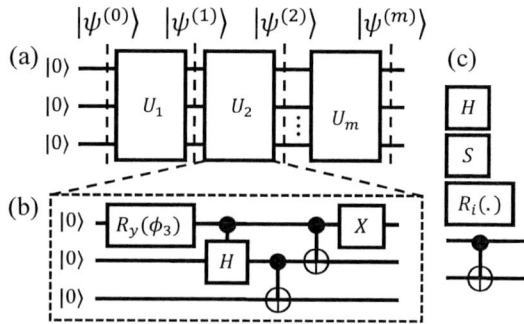

Figure 1: (a) Quantum circuit as unitary operator's chain (b) An example 3-qubit W state circuit [10] (c) Universal gate set $\{H, S, R_i(.), CX\}$ with $i \in \{x, y, z\}$ [11].

respectively. This quantum operator can be decomposed into MM between sub-operators but also into $\hat{m}$ gates (considering only single-qubit gates and two-qubit gates) application. These gates $\{g_j\}$ can be applied on $|\psi^{(0)}\rangle$ sequentially to achieve the same final state $|\psi^{(m)}\rangle$:

$$\mathcal{U} \equiv \prod_{j=1}^{m} U_j = (\prod_{j=1}^{\hat{m}} g_j), \ |\psi^{(j+1)}\rangle = U_j |\psi^{(j)}\rangle \quad (1)$$

with the state vector can be viewed as 2^n-dimensional complex array $\psi = [\alpha_0 \ \alpha_1 \ \dots \ \alpha_{2^N-1}]$ and $\{\alpha_j\}$ is called amplitudes of state vector.

B. Quantum gates

In (1), quantum gates are fundamental operations applied to qubits, represented as unitary matrices. In this work, we are considering the gate set Clifford + R_i with $i \in \{x, y, z\}$, which is known as a universal gate set followed by Gottesman-Knill theorem [11], any g_j will belong to the set $\{H, S, R_x(.), R_y(.), R_z(.), CX\}$ as Fig. 1 (c). This set is denoted as fixed gates $\{H, S, CX\}$, parameterized gates $\{R_x(.), R_y(.), R_z(.)\}$, sparse gates $\{S, R_z(.)\}$ and dense gates $\{H, CX, R_x(.), R_y(.)\}$ depend on its matrix representation. While H, S operator on one-qubit, CX applies a NOT operation to a target qubit conditioned on a control qubit.

By using Clifford + R_i, we can break down into any low-level or build up any high-level gates, such as control rotation $CR_x(.)$ can be decomposed into $\{S, CX, R_x(.), R_y(.), R_z(.)\}$. This process can be conducted by a transpiler [12], the transpiled state can recover the original state by $|\psi_{\text{tran}}\rangle = e^{i\phi}|\psi_{\text{origin}}\rangle$ where ϕ is the global phase born from the transpilation process.

C. Challenge

The challenge when emulating (1) for general purpose is the exponential scaling of both computation time and resources based on #Qubits. In detail, $\{U_j\}$ and ψ can be represented as $N \times N$ complex matrix ($N = 2^n$) and N-dimensional vector, respectively. Take example, storing such one matrix in case

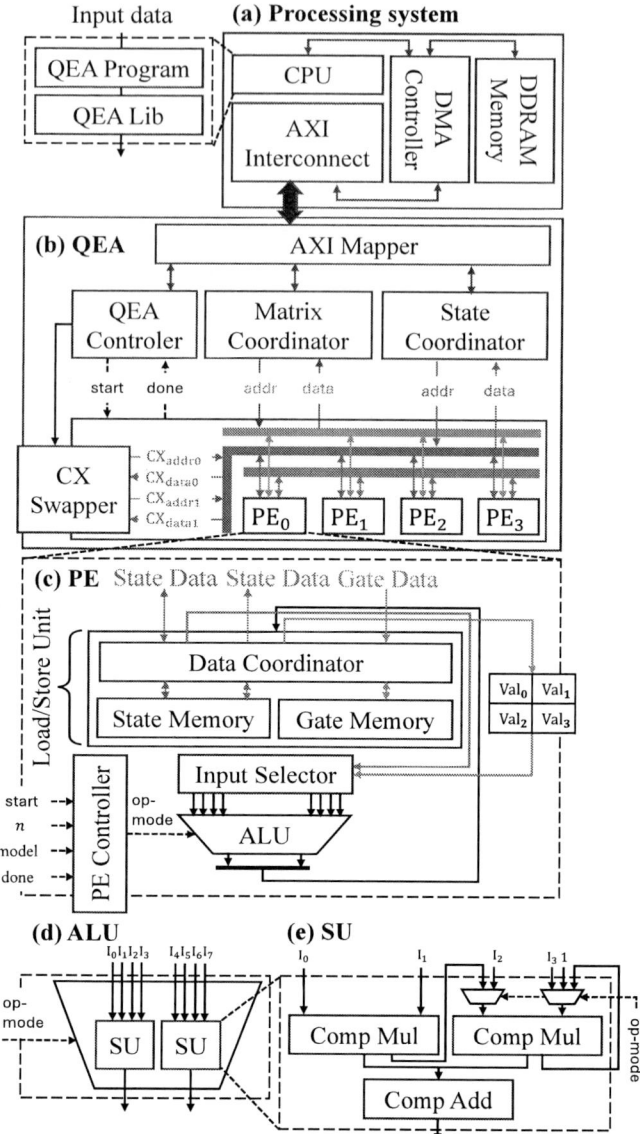

Figure 2: The detail of the QEA: (a) The processing system, (b) QEA core, (c) Processing Element (PE), (d) Arithmetic Logic Unit, (e) Special Unit (SU)

50 qubits require at least 8×10^{31} bit $\approx 10^{22}$ (GB) for single precision. 50 qubits state would take 7.2×10^{16} bit $\approx 9 \times 10^6$ (GB). Such huge required resources are impossible for more #Qubits even with the supercomputer. In almost all simulators, only ψ is being saved during the whole process. In QEA, $\{U_j\}$ can be applied directly to $|\psi\rangle$ without construction, the same properties as the wave-function approach [13].

III. PROPOSED HARDWARE ARCHITECTURE

A. Overview

Fig. 2 (a-c) shows an overview of QEA, which consists of two primary components: the processing system (PS) and

Algorithm 1 Matrix multiplication for a gate g at j qubit

Require: n, ψ, g, j, u (matrix representation of g), $N \leftarrow 2^n$
1: $\overline{g} \leftarrow 1 \ll (n - (j+1))$ ▷ Length of sub-generated group
2: $f \leftarrow 1$ ▷ Flag variable
3: **for** $i = 0$ to $N - 1$ **do**
4:    **if** $f = 1$ **then**
5:       **if** $type(g) == 0$ **then**
6:          $\psi[i] \leftarrow u[0][0] \times \psi[i]$ ▷ Sparse gate
7:       **else** $\psi[i] \leftarrow u[0][0] \times \psi[i] + u[0][1] \times \psi[i + \overline{g}]$
8:       **end if**
9:    **else**
10:       **if** $type(g) == 1$ **then**
11:          $\psi[i] \leftarrow u[1][1] \times \psi[i]$ ▷ Sparse gate
12:       **else** $\psi[i] \leftarrow u[1][0] \times \psi[i - \overline{g}] + u[1][1] \times \psi[i]$
13:       **end if**
14:    **end if**
15:    **if** $i \mod \overline{g} = \overline{g} - 1$ **then** $f \leftarrow \neg f$
16:    **end if**
17: **end for**
18: **return** ψ

Algorithm 2 CX gate application on state vector

Require: $n, \psi, j, k, N \leftarrow 2^n$
1: **for** $i = 0$ to $N - 1$ **do**
2:    $\overline{i}_2 \leftarrow bin(i, n)$ ▷ Binary representation of i in n bits
3:    **if** $\overline{i}_2[j] = 1$ **then**
4:       $j \leftarrow i \oplus (1 \ll k)$
5:       $swap(\psi[i], \psi[j])$
6:    **end if**
7: **end for**
8: **return** ψ

QEA. In the processing system, the QEA library allows users to create quantum programs and generate context data for execution on QEA. Direct Memory Access (DMA) is used to efficiently move large amounts of data between the user space and the PL, with a 256-bit AXI bus optimized for data communication. QEA, the main part of the architecture is composed of six important components:

- AXI Mapper: Manages the transfer of data between PS and QEA to ensure proper synchronization.
- QEA Controller: Controls execution management within the accelerator.
- Matrix Coordinator and State Coordinator: Assist the AXI Mapper by routing data to the state vector and gate context memories in each PE.
- The CX Swapper: Performs the computation of a CX gate and a state vector (described in Section III-E).
- The Processing Element Array (PEA): Consists of four PEs that perform computations to update the state vector when sparse and dense gates are applied.

B. Optimized Memory Allocation Management

Efficient memory allocation plays an essential role in quantum circuit simulation hardware design. As described in Section II-C, the exponential increase in #Qubits greatly raises memory demands. The accelerator in [8] uses numerous memory units for gate context, state vector, matrix data, and next state vector storage, resulting in significant memory overhead. Furthermore, the usage of 2×2 gate matrices to generate a $2^n \times 2^n$ matrix offers no computational speed-up, leading to inefficient memory utilization. To overcome these limitations, QEA uses efficient memory allocation management, as shown in Fig. 2 (b-d). QEA uses a single global

memory for gate context storage and two local memories for the state vector (State Memory) and gate data (Gate Memory) in each PE. The Matrix Coordinator delivers 2×2 gate matrix data directly to each Processing Element (PE) during quantum gate operations (except the CX gate), reducing the amount of gate data stored by PEs. Furthermore, the current state vector is distributed equally among all PEs, while the next state vector is directly updated to the current state vector, maximizing memory efficiency while maintaining computational speed.

Besides memory allocation, a fundamental problem is guaranteeing efficient data storage and computation when applying a 2×2 gate matrix to a 2^n-sized state vector. [8] suggests a Coordinate (COO) format for storing row/column indices and real and imaginary values, but it increases memory requirements. Because indices may be dynamically generated during execution (Algorithm 1), QEA uses a simpler format that stores only real and imaginary components, avoiding redundant data while retaining correctness. Furthermore, QEA employs a fixed-point 32-bit representation (2-bit integer, 30-bit fractional portion), resulting in faster computations than floating-point arithmetic while maintaining precision in quantum circuit simulations.

C. Open Processing Element

In a quantum circuit, data flow management is also essential in data computation and memory allocation, especially when multiple PEs are used to parallelize computing. As #Qubits grows, data access hazards occur, possibly affecting global execution time due to the distribution of state data among PEs. To reduce data hazards, each PE in the PEA allows for state data transfer across PEs. This approach enables state data to be shared between PEs, ensuring concurrent access for both the present PE and other PEs that require the same data. As a result, data access management is improved, and data access time decreases while overcoming the throughput constraints of State Memory within each PE. Fig. 2 (c) and (d) depict the mechanism in detail. Furthermore, because only one quantum gate is used in each phase, gate data can be previously loaded from Gate Memory, minimizing access latency and increasing overall quantum circuit simulation speed.

979-8-3315-1550-8/25 $31.00 © 2025 IEEE

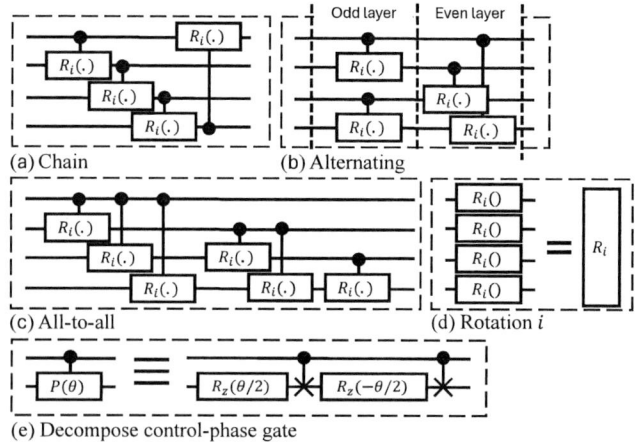

(a) Chain   (b) Alternating

(c) All-to-all   (d) Rotation i

(e) Decompose control-phase gate

Figure 3: Example of 4-qubit topologies (a-d) and (e) Decomposing of $CP(\theta)$ as $(I \otimes R_z(\theta/2))CX((I \otimes R_z(-\theta/2)))CX$.

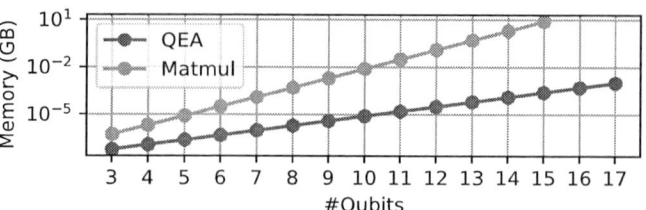

Figure 4: Comparison in memory usage (logarithm scale) between QEA and naive MM operation (Matmul).

D. Flexible Arithmetic Logic Unit

Besides the CX gate, there are two types of gates: sparse and dense gates. Sparse gates, which have fewer elements than dense gates, can be executed in around half the time, making them faster. Thus, creating an ALU that can switch between sparse and dense gate operations is critical for dramatically reducing execution time. To solve this, QEA integrates an ALU, as shown in Fig. 2 (e) and (f), which allows switching between sparse and dense gate computations. This ALU utilizes two complex multiplication operations and one complex addition operation within each Special Unit (SU). Furthermore, to optimize data access performance, the ALU has two SUs, ensuring efficient utilization of the State Memory.

E. Simplified CX Swapper

The two-qubit structure of the CX gate makes it a relatively difficult operation in a quantum circuit. When applying a CX gate to a state vector, the conventional method usually involves creating a full $n \times n$ CX matrix based on the state vector of n quantum bits and then performing matrix multiplication with the state vector. However, this approach is time-consuming because it requires generating the complete CX matrix and carrying out the multiplication. Fortunately, the main purpose of the CX gate is to switch values inside the state vector. Therefore, as shown in Algorithm. 2, a swap technique can handle the CX gate efficiently [14].

IV. VERIFICATION AND EVALUATION

A. Implementation and Verification

The Quantum Execution Architecture (QEA) was implemented and verified on the 16nm Alveo U280 FPGA board (Fig. 2) using the Vivado 2020.2 tool. The QEA requires 14,773 lookup tables (LUTs), 7,894 flip-flops (FFs), 578 Block RAMs (BRAMs), and 256 digital signal processors (DSPs)

while consuming only 0.543 W of power. With these resources, the QEA can best support quantum circuits of 3 to 17 (due to the limitation of BRAM on Alveo U280) and can be extended for higher #Qubits.

B. Benchmarking setting

To prove the generality of QEA, we evaluated 19 circuit templates [15], each labeled with a circuit ID from 1 to 19. The full circuit representation is presented in Fig. 3. All circuits are constructed from a combination of topologies, including chain, alternating, all-to-all [16], and rotation; as shown in Fig. 3. In circuit design, the chain topology arranges qubits in a linear sequence, where each qubit is coupled only to its immediate neighbors. The alternating topology extends this by introducing a bipartite structure, connecting qubits in two subsets with edges between them. In contrast, the all-to-all topology allows every qubit to interact directly with every other, forming a complete graph. The final, rotation i takes a role as parameter holder and provides no entanglement.

For comparison with other works, Quantum Fourier Transform (QFT) [17] is conducted. It is an important component in many quantum algorithms, such as phase estimation, order-finding, and factoring. QFT circuit is utilized from Hadamard (H), controlled phase ($CP(\theta)$) is decomposed as Fig. 3 (e), and SWAP gates equivalent to three sequential CX gates. These gates transform on an orthonormal basis $\{|0\rangle, \ldots, |N-1\rangle\}$ with the below action on the basis state $|j\rangle \rightarrow \frac{1}{\sqrt{N}} \sum_{k=0}^{N-1} e^{2\pi ijk/N}$. We apply the QFT on the zero state $|0\rangle$, which should return the exact equal superposition of all possible n-qubit states, means $\alpha_j = \alpha_k \ \forall j, k \in [0, N)$.

The evaluated metrics include memory usage, fidelity, mean-square error (MSE), and execution time defined clearly in [18]. The fidelity and MSE between two state vectors range from 0 to 1 with fidelity of 1 or MSE of 0 meaning two vectors completely overlap and vice versa. The execution time on the CPU (Qiskit) is measured from constructing the circuit to receiving the final state while the execution time of QEA is measured by running QEA with a maximum frequency of 250 MHz to calculate the final state from the initial state.

C. Comparison with powerful CPUs

In this section, the Intel(R) Core(TM) i9-10940X CPU at 3.30GHz will be used to simulate quantum circuits supported by Qiskit. Its results are compared with the QEA's results.

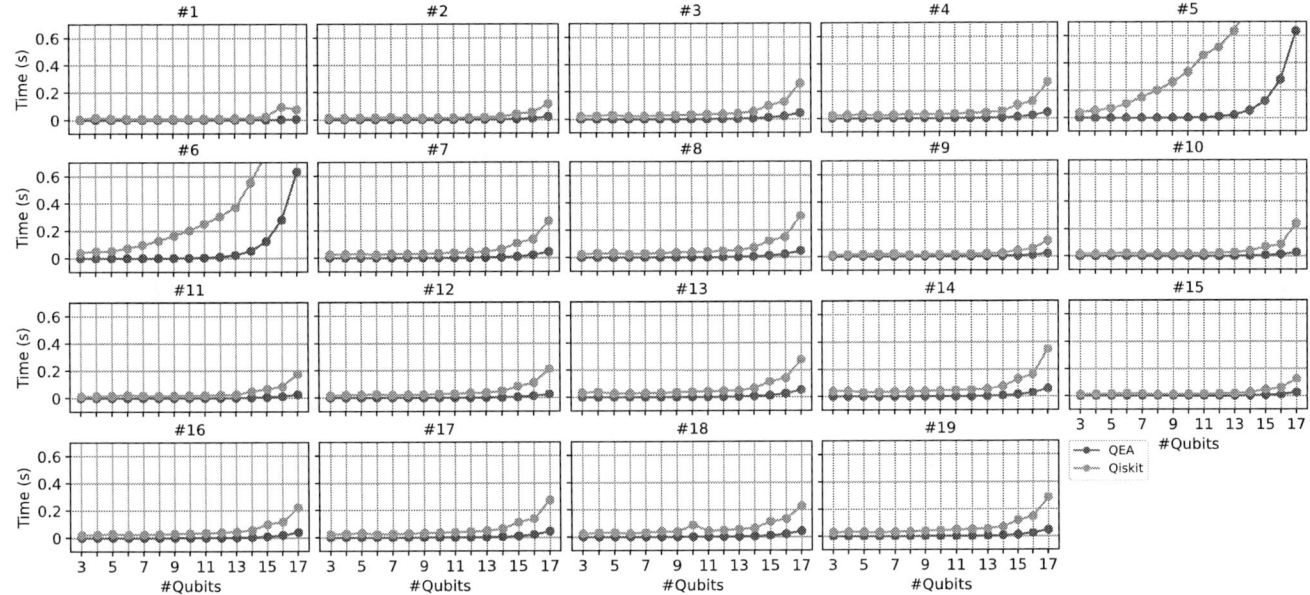

Figure 5: Comparison in execution time between QEA and a variety of random quantum circuits indexed from #1 to #19

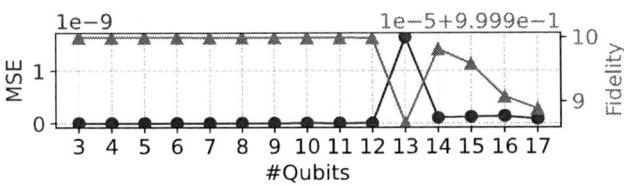

Figure 6: Average MSE and fidelity measurement between QEA and Qiskit for a set of parameterized quantum circuits indexed from 1 to 19.

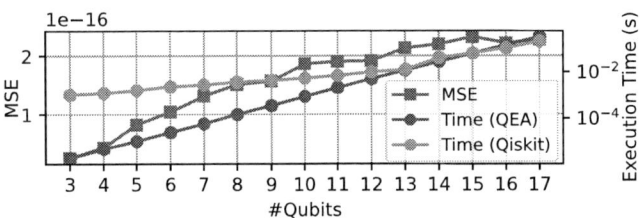

Figure 7: Comparison on QFT circuit. (a) Left y-axis: Mean square error between state vectors from Qiskit and QEA (b) Right y-axis: Execution time curve from Qiskit (Orange line) and QEA (Blue line).

To demonstrate the effectiveness and optimization in memory management and allocation, an evaluation was conducted between QEA and the naive matrix multiplication (Matmul), as shown in Fig. 4. The results show that as #Qubits rises, QEA improves the amount of memory used. At 7 and 13 qubits, QEA surpasses the naive Matmul operation by factors ranging from approximately $\approx 10^2$ to $\approx 10^4$, demonstrating the efficiency of proposed memory management optimization.

In addition to memory usage, the comparison in terms of execution time between QEA and Qiskit with a set of parameterized quantum circuits indexed from 1 to 19 is shown in Fig. 5. The QEA's results consistently achieve better execution times for circuits with a high #Qubits, while its performance is comparable to Qiskit's for circuits with fewer #Qubits. QEA also clearly shows its performance and effectiveness with applications using high #Qubits while Qiskit cannot. For QFT circuits comparison in Fig. 7, the execution time of Qiskit increases more slowly than that of QEA as #Qubits increases, this phenomenon is explained in [18]. Qiskit does not use a state vector-based simulator in the case of a simple structure as QFT, then it can achieve better performance than QEA for

high #Qubits. Note that this acceleration technique from Qiskit can not work on general circuits.

Fig. 6 and Fig. 7 (left y-axis) show a detailed comparison of QEA and Qiskit on the accuracy aspect. The comparison reveals that there are no notable differences between QEA and Qiskit. As a result, QEA meets the MSE and fidelity standards, demonstrating its capacity as a dependable quantum emulator.

D. Comparison with FPGA-based works

This section compares QEA to other related studies on FPGA to provide a more comprehensive evaluation of performance and efficiency. The QFT technique was utilized to compare with other studies since it is widely used in the creation of hardware designs. Table I compares QEA's hardware results to other FPGA-based works [4]–[7], [9] in terms of frequency, reconfigurable, #Qubits, and normalized gate speed due to differences in supported qubits, precision, hardware resources, and the number of gates in quantum circuits.

979-8-3315-1550-8/25 $31.00 © 2025 IEEE

Table I: Comparative analysis in post-implementation of QEA and existing FPGA-based emulators on QFT's performance.

Works	Device	Freq (MHz)	Reconfig[*]	Precision	#Qubits	Execution time (s)	#Gates [†]	NGS [††]
[4]	AMD Xilinx Zynq-7000	100	✓	32-bit FX	6	1.15×10^{-4}	10	1.8×10^{-7}
[5]	Arria 10AX115N4F45E3SG	233	✓	32-bit FP	16	1.84×10^{1}	528	5.33×10^{-7}
[6]	Xilinx XCKU115	160	✗	16-bit FX	16	2.70×10^{-1}	136	3.03×10^{-8}
[7]	2 × Intel Stratix 10 MX2100	299	✗	32-bit FP	30	4.47×10^{0}	465	8.95×10^{-12}
[9]	Xilinx XCVU9P	233	✗	18-bit FX	16	1.20×10^{-3}	-	-
This work	**AMD Alveo U280**	**250**	✓	**32-bit FX**	**17**	$\mathbf{3.29 \times 10^{-1}}$	**721**	$\mathbf{3.48 \times 10^{-9}}$

[*] The quantum emulator is fixed with a application and #Qubits.

[†] The number of gates (#Gate) in this work is higher than other work due to the no-use of control-rotation gates.

[††] The Normalized Gate Speed (NGS) (s / (gate × amplitude)) = Execution time / (#Gates × $2^{\#Qubits}$), smaller is better.

Table I shows that QEA surpasses other works in terms of performance and reconfigurable. QEA outperforms [4]–[6], [9] in terms of maximum frequency, with increases ranging from $1.07\times$ (250 vs. 233 MHz) to $2.5\times$ (250 vs. 100 MHz). QEA also supports a broader range of algorithms, which increases its versatility and allows it to handle a higher #Qubits. The NGS (Normalized Gate Speed) results for QEA additionally show a higher efficiency over previous works, with improvements ranging from $8.7\times$ (3.48×10^{-9} vs. 3.03×10^{-8} seconds) to $153.16\times$ (3.48×10^{-9} vs. 5.33×10^{-7} seconds). While QEA does not reach the best results in maximum frequency, supported qubits, and NGS when compared to [7], it excels in providing better adaptability over a larger spectrum of quantum circuits, rather than being limited to QFT only. In summary, the results demonstrate QEA's superior features when compared to alternative FPGA-based works.

V. CONCLUSION

In summary, this paper introduced QEA, a quantum emulator, to address current challenges in achieving high flexibility, memory efficiency, low hardware resources, and high performance. QEA was presented with four key ideas: optimized memory allocation management, open PE, flexible ALU, and simplified CX swapper. The complete verification and evaluation confirmed the exceptional properties of QEA in terms of low MSE, high fidelity, less memory usage, short execution time, and high flexibility compared to other comparable works in supporting a variety of applications with varying #Qubits from 3 to 17. In future work, QEA can be enhanced and integrated into a system using external data memories and multiple QEA cores to get higher performance with higher supported #Qubits.

ACKNOWLEDGMENT

This research is funded by the NAIST Scholar program. This work was supported by JST-ALCA-Next Program Grant Number JPMJAN23F5, Japan. The research has been partly executed in response to the support of JSPS, KAKENHI Grant No. 22H00515, Japan.

REFERENCES

[1] M. Schuld and N. Killoran, "Is quantum advantage the right goal for quantum machine learning?" *PRX Quantum*, vol. 3, p. 030101, Jul 2022.

[2] C.-C. W. et al, "Queen: A quick, scalable, and comprehensive quantum circuit simulation for supercomputing," 2024.

[3] V. Bergholm and et al, "PennyLane: Automatic differentiation of hybrid quantum-classical computations," 2022.

[4] A. Silva and O. G. Zabaleta, "Fpga quantum computing emulator using high level design tools," in *2017 Eight Argentine Symposium and Conference on Embedded Systems (CASE)*. IEEE, 2017, pp. 1–6.

[5] N. e. a. Mahmud, "Efficient computation techniques and hardware architectures for unitary transformations in support of quantum algorithm emulation," *Journal of Signal Processing Systems*, vol. 92, pp. 1017–1037, 2020.

[6] Y. Hong, S. Jeon, S. Park, and B.-S. Kim, "Quantum circuit simulator based on fpga," in *13th International Conference on Information and Communication Technology Convergence*. IEEE, 2022, pp. 1909–1911.

[7] W. et al, "A scalable emulator for quantum fourier transform using multiple-fpgas with high-bandwidth-memory," *IEEE Access*, vol. 10, pp. 65 103–65 117, 2022.

[8] T. X. H. e. a. Le, "Theoretical Analysis of the Efficient-Memory Matrix Storage Method for Quantum Emulation Accelerators with Gate Fusion on FPGAs," in *2024 IEEE 17th International Symposium on Embedded Multicore/Many-core Systems-on-Chip (MCSoC)*, 2024.

[9] S. Liang, Y. Lu, C. Guo, W. Luk, and P. H. Kelly, "Pcq: Parallel compact quantum circuit simulation," in *2024 IEEE 32nd Annual International Symposium on Field-Programmable Custom Computing Machines (FCCM)*. IEEE, 2024, pp. 24–31.

[10] D. Volya and P. Mishra, "State preparation on quantum computers via quantum steering," *IEEE Transactions on Quantum Engineering*, vol. 5, pp. 1–14, 2024.

[11] S. Aaronson and D. Gottesman, "Improved simulation of stabilizer circuits," *Phys. Rev. A*, vol. 70, p. 052328, Nov 2004.

[12] P. R. et al, "Highly optimized quantum circuits synthesized via data-flow engines," *Journal of Computational Physics*, vol. 500, p. 112756, 2024.

[13] Q. C. Nguyen, L. B. Ho, L. Nguyen Tran, and H. Q. Nguyen, "Qsun: an open-source platform towards practical quantum machine learning applications," *Machine Learning: Science and Technology*, vol. 3, no. 1, p. 015034, mar 2022.

[14] H. Hiroshi and D. Jun, "Optimization of quantum computing simulation with gate fusion," *QS*, vol. 2021, no. 23, pp. 1–7, 2021.

[15] S. e. a. Sim, "Expressibility and entangling capability of parameterized quantum circuits for hybrid quantum-classical algorithms," *Advanced Quantum Technologies*, vol. 2, no. 12, p. 1900070, 2019.

[16] T. e. a. Haug, "Capacity and quantum geometry of parametrized quantum circuits," *PRX Quantum*, vol. 2, p. 040309, Oct 2021.

[17] D. Coppersmith, "An approximate Fourier transform useful in quantum factoring," 2002.

[18] T. H. e. a. Vu, "FQsun: A Configurable Wave Function-Based Quantum Emulator for Power-Efficient Quantum Simulations," *arXiv preprint arXiv:2411.04471*, 2024.

979-8-3315-1550-8/25 $31.00 © 2025 IEEE

HW/SW Co-Design for a Variational AutoEncoder targeting Anomaly Detection on FPGA

Tuan-Phong Tran, Thien-Duy Ho, Tung-Bach Nguyen, Xuan-Tu Tran, Duy-Hieu Bui
VNU Information Technology Institute
144 Xuan Thuy Road, Cau Giay District, Hanoi, Vietnam
Corresponding author's email: hieubd@vnu.edu.vn

Abstract—**Variational AutoEncoder (VAE) is a generative model based on autoencoders, utilizing a latent space with a probabilistic distribution to generate new data. VAE is applied in various fields, such as image generation, data compression, and text generation. This paper proposes a VAE-based anomaly detection system implemented on an FPGA. The design applies co-design techniques, optimizing the model in software and accelerating the whole model in hardware using High-Level Synthesis. Our proposed model reduces the model size by a factor of 10, with a slight accuracy drop when compared to the reference model in [1]. The proposed model is then retrained with a 16-bit fixed-point representation instead of a 32-bit floating-point. The hardware implementation results on the Pynq-Z2 development board show that the proposed VAE accelerator runs 10 times faster than the software-only version running on a laptop computer while occupying a small area of 25,000 slices on the programmable logic.**

Keywords— Variational Autoencoder, High-Level Synthesis, Convolutional Neural Network, Artificial Intelligence, Edge AI

I. INTRODUCTION

Maintaining product quality is extremely important in the manufacturing industry. If anomalies are not detected in time, they can damage equipment and reduce operational efficiency, potentially negatively impacting employees' health and morale [2]. Automated anomaly detection systems help reduce costs by enabling early issue detection, which minimizes maintenance expenses and decreases production downtime caused by failures [3]. As a result, anomaly detection methods have become essential in various fields, including healthcare, manufacturing, and food quality monitoring.

Manual inspection methods rely on human observation and judgment, such as visual product inspection, manual sorting, or checking basic parameters using simple tools like scales or measuring devices. Although still widely used, these methods can be inaccurate and prone to human error, especially under high workloads or when high precision is required. Similarly, traditional machine learning approaches, such as classification models, require manual feature engineering, which is time-consuming and can lead to suboptimal results, particularly when dealing with highly dynamic and complex data.

Our project uses a variational autoencoder (VAE) as the core deep learning model to apply anomaly detection in hazelnuts. The goal is to detect defects or anomalies in hazelnuts that may indicate production issues, such as damage or irregularities in size, shape, or color. VAE models are well-suited for anomaly detection due to their ability to learn complex data distributions and generate new data points from these learned distributions. By training the VAE on a dataset of normal hazelnuts, the model learns to represent the typical characteristics of a hazelnut. It can identify deviations from this learned normal distribution when presented with new data, thereby detecting anomalies.

The software implementation of VAE presents several challenges. One of the main challenges is the high complexity of the original VAE model, which contains approximately three million parameters, making hardware deployment difficult. On the hardware side, optimization is challenging due to resource constraints, as VAEs typically have millions of parameters, making deployment on FPGAs or embedded systems difficult. Latency constraints are also critical in real-time applications, requiring an efficient inference process to ensure timely processing. Moreover, memory bandwidth poses another challenge, as managing large models in memory-limited environments requires careful optimization.

We applied a Hardware/Software co-design strategy, where the complete VAE model is implemented as a hardware IP core on the Pynq-Z2 FPGA development board. Only the generation of the epsilon coefficient in the latent space and the anomaly detection program, used to produce the final anomaly map, are executed in Python on the processing system. This co-design approach optimizes system performance and efficiency, enabling the deployment of large-scale VAE models on resource-constrained FPGA platforms while maintaining low latency and enhancing real-time processing capabilities. To address these challenges, we redesigned the reference VAE model in [1] to reduce the parameters from 3 million to approximately 300 thousand for hazelnut anomaly detection while maintaining the model accuracy. Our optimization facilitates easier hardware deployment without compromising performance. Combining model complexity reduction, synthetic anomaly data generation, and FPGA implementation, we successfully tackled key challenges in applying deep learning for defect detection in hazelnuts with low hardware resources.

The rest of this paper is organized as follows. Section II presents our proposed VAE model's architecture. The reference and proposed architecture will be evaluated regarding model size, performance, and accuracy. The proposed hardware architecture for the full model is depicted in Section III. After that, Section IV presents the hardware implementation and the system integration of the full model with the Zynq processing system on the Pynq-Z2 FPGA development board. Finally, there are some conclusions and perspectives in Section V.

II. PROPOSED VARIATIONAL AUTOENCODER ARCHITECTURE

We started with a reference VAE model on Kaggle [1], which uses the MVTec Anomaly Detection (MVTec AD) dataset. This model has over 3 million parameters with an

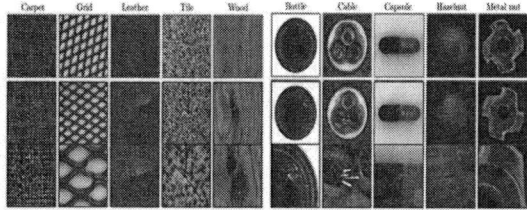

Fig. 1. MVTec Anomaly Detection dataset samples [5].

accuracy of approximately 90%. To optimize this model for hardware implementation, we used the HW/SW co-design approach to optimize the model in terms of the number of parameters and throughput while maintaining similar accuracy. Our proposed model is more suitable for hardware implementation on an FPGA. This section presents our proposed VAE model for hazelnut anomaly detection and the optimization techniques to implement the target model more efficiently on an FPGA.

A. Hazelnut anomaly detection dataset

The MVTec Anomaly Detection (MVTec AD) dataset [4] is a well-known anomaly detection benchmark designed to evaluate algorithms used in industrial applications, such as defect detection in manufacturing processes. This dataset, provided by MVTec Software GmbH, consists of 1,535 images categorized into object and texture types. These images are divided into training and test sets, with approximately 1,000 images used for training and 500 images designated for testing. The training set includes both normal (defect-free) and defective samples, while the test set contains images with previously unseen anomalies, allowing for a comprehensive evaluation of model performance. Some samples of the MVTec AD dataset are displayed in Fig. 1.

Regarding annotations, each image in the MVTec AD dataset is labeled with ground truth information about the location of defects. These annotations are provided as either pixel-wise segmentation masks for localizing defects or image-level labels indicating the presence of anomalies. This dual annotation format supports localization tasks (identifying the exact location of defects) and classification tasks (distinguishing between normal and defective objects).

B. The proposed Variational AutoEncoder architecture

1) Original architecture

Fig. 2 illustrates the architecture of the reference model in [1]. It contains three main parts: the encoder, the decoder, and latent space sampling. The entire VAE model consists of over 3.5 million parameters, with the encoder responsible for over 1.5 million parameters and the decoder responsible for 2.0 million. The latent space introduces a learnable sampling mechanism from the encoder's mean and variance. The model effectively compresses and reconstructs images while ensuring smooth transitions in the latent space through the reparameterization trick.

The encoder compresses input images in a 32-bit floating point (FP32) format into a latent representation through four convolutional layers. Each convolutional layer applies 3×3 filters to the input feature maps, followed by Batch Normalization to stabilize training and accelerate convergence. Finally, the Leaky ReLU activation function is used for non-linearity, enabling the network to learn complex patterns while mitigating the vanishing gradient problem. The

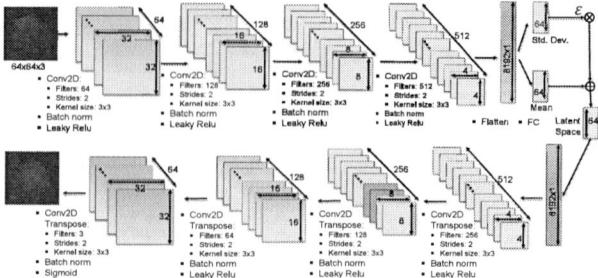

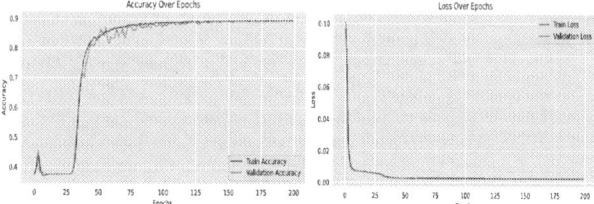

Fig. 2. The VAE reference architecture is proposed in [1].

Fig. 3. The acuracy and loss of the reference model after training and validation.

number of filters for the convolution layers is 64, 128, 256, and 512, respectively.

Instead of directly mapping to a latent representation, the encoder outputs two vectors, mean and log variance, defining a Gaussian distribution in the latent space. The sampling step ensures the model learns a continuous latent space, crucial for generating smooth interpolations.

The decoder reconstructs images from the sampled latent vector. It begins with a dense layer that expands the latent space into a shape matching the encoder's last feature map, followed by a reshaping operation. The decoder then employs transposed convolution layers to upsample the feature maps. The first transposed convolution layer restores spatial resolution to 8×8×256, followed by batch normalization and activation. The successive layers progressively increase the spatial dimensions to 16×16×128 and 32×32×64 while reducing the number of filters. The final layer generates an output of shape 64×64×3 using a sigmoid activation function, ensuring that pixel values remain in the valid range.

The model's accuracy on the training and test sets is shown in Fig. 3. Both training and validation results reach 90%. Meanwhile, the loss function for the training and test sets is approximately 0.001 and 0.002, respectively.

2) Proposed architecture

FPGA has limited logic elements, memory, and DSP blocks. Additionally, real-time applications require low latency, necessitating optimized inference processes for rapid decision-making. Excessive resource usage can also increase power consumption, impacting the performance of edge devices. Therefore, optimizing and fine-tuning the model architecture maximizes parallel processing capabilities, reduces memory bottlenecks, saves energy, and ensures hardware-friendly implementation on an FPGA.

The reference VAE model faces challenges related to resource constraints and suboptimal execution time when being implemented on an FPGA because it contains many parameters with floating-point operations. Additionally, in the anomaly detection phase, the output of the VAE model only

979-8-3315-1550-8/25 $31.00 © 2025 IEEE

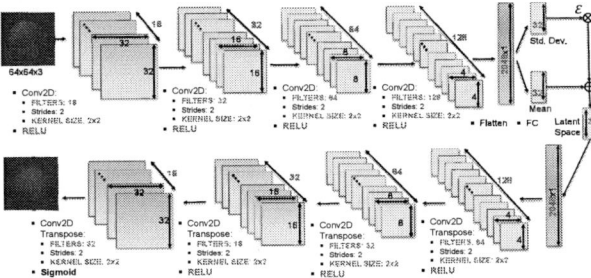

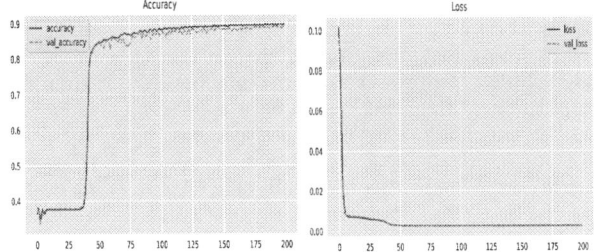

Fig. 4. The proposed VAE model's architecture.

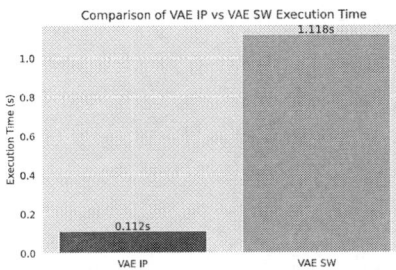

Fig. 6. Comparison of the parameter size between the reference model and the proposed model.

Since the reference model is implemented in software using the TensorFlow framework, processing floating-point data during training and performing calculations within the network is straightforward and efficient on computers. However, floating-point computation is costly in resource-constrained devices and hardware implementation on an FPGA. Therefore, we proposed to use quantization-aware training to convert all input images and model parameters, including weights and biases, to 16-bit fixed-point numbers [7] with 4 bits for the integer part and 12 bits for the fractional part. This reduces the proposed model's size and computation complexity. Fixed-point computation is equivalent to integer computation, which is less complex and more efficient than floating-point one when implemented on an FPGA. After converting to a fixed-point format, we updated the weights and biases accordingly. We tested the model again to compare its accuracy with floating and fixed-point representations. Fig. 6 shows the comparison of the reference model with our proposed one. The reference model is 30 times larger than the proposed model in terms of parameter size.

Fig. 5. The training and test accuracy of the proposed model on the hazelnut dataset.

provides a reconstructed image without highlighting the anomalous regions. The reference work in [1] relies on an existing library [6] that identifies anomalies from the reconstructed image to obtain an image with detected anomalies. We attempted to install this library on our PYNQ-Z2 board, which is used for this project, but found it infeasible due to its limited memory (only 512MB DRAM available). The library requires more than 1GB of DRAM to run. We replaced the anomaly detection method with a custom Python program to predict and detect anomalies. While this custom solution is less effective than the pre-existing library due to time constraints, it still ensures anomaly detection capability.

The proposed architecture of the model is illustrated in Fig. 4. To reduce the model size, we decreased the kernel size from 3×3 (in the original model) to 2×2 and reduced the number of filters in both the encoder and decoder blocks. A smaller kernel reduces the number of convolution operations, while fewer filters decrease the total number of parameters, optimizing bandwidth and processing speed. Batch normalization was also removed because it requires statistical computations (mean, variance) and complex arithmetic operations (division, square root), which are not suitable for hardware implementation on an FPGA. Additionally, Leaky ReLU was replaced with ReLU since FPGA implementations typically prioritize simple operations. ReLU only requires a comparison and assignment, whereas Leaky ReLU involves a multiplication with a small coefficient (αx) when $x \leq 0$, increasing latency and consuming more hardware on an FPGA.

After research and development, we successfully created a new VAE model with approximately 300 thousand parameters and nearly the same test accuracy of 89% as the original model. The training loss stays around 0.001. The proposed model maintains accuracy and improves effectiveness for hardware deployment. Fig. 5 shows the training and test accuracy of the proposed model with the hazelnut dataset.

III. HARDWARE AND SOFTWARE ARCHITECTURE

After optimizing the model to reduce its size and computational complexity using fixed-point representation, we applied a Hardware/Software co-design strategy to optimize the overall system. In the final implementation, the entire VAE model—including the encoder, sampling layer, and decoder—was implemented in hardware using High-Level Synthesis (HLS) on an FPGA. This section presents the hardware architecture of the complete VAE model and its integration with the Zynq processing system on the Pynq-Z2 development board.

A. Proposed hardware architecture

The proposed hardware architecture is presented in Fig. 7. The system is divided into three stages: the first stage is initialization, the second stage is data processing, and the final stage is storage and result retrieval. In the first stage, the input image data is loaded into DDR3 memory along with the weights and biases of the VAE model. The epsilon coefficients, generated by a Python program, are also loaded into DDR3 memory for the sampling process. Next, in the second stage, the Cortex-A9 transfers the IP configuration, the address of the image data, weights, and biases in DDR3 memory to the VAE IP and activates the hardware accelerator. The VAE IP uses the AXI4 master interface to load data into the Input Mem and Weights/Bias Mem in the programmable logic via the AXI4 bus. The MAC block then performs the convolution operation, which is the main operation of all layers in the model. A Finite State Machine (FSM) controls the entire processing flow. In the final stage, the processed data is stored in the Output Mem and then transferred to

979-8-3315-1550-8/25 $31.00 © 2025 IEEE

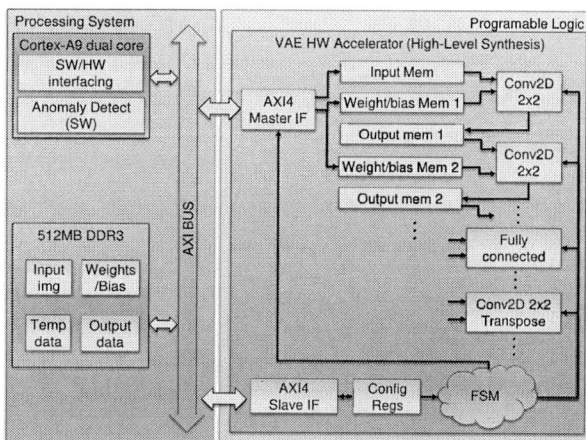

Fig. 7. The proposed hardware architecture for VAE IP.

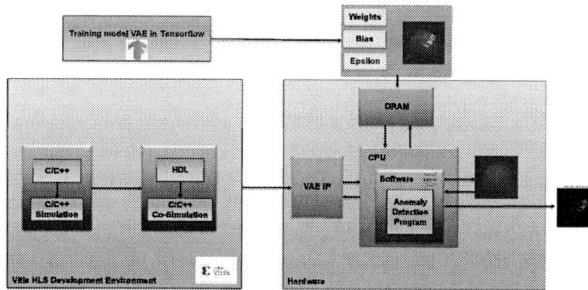

Fig. 8. Our design flow to implement the VAE model on Pynq-Z2

DRAM in the processing system through the AXI4 Master Interface. The processing system continues to run the anomaly detection program in Python. Once the output is transferred to the processing system, a Python program running on Jupyter Notebook calculates the pixel-wise Mean Squared Error (MSE) between the original and reconstructed images using the following formula:

$$\text{Error Map} = \frac{1}{C} \sum_{C=1}^{C} \left(I_{original,C} - I_{reconstructed,C} \right)^2$$

Where C is the number of color channels; $I_{original,C}$ is the pixel value of the original image at channel C, and $I_{reconstructed,C}$ is the corresponding pixel value in the reconstructed image. The resulting error map indicates pixel-wise differences and effectively visualizes anomalies. Values that exceed a predefined threshold are marked as abnormal regions. This process generates an anomaly map, which is visualized as a heatmap where brighter areas correspond to higher reconstruction errors and potential defects in the input image.

The proposed hardware architecture has been implemented in C++ with optimization for High-Level Synthesis with Vitis HLS 2021.2. The AXI4 master and slave interface is automatically inferred through the HLS pragmas. The proposed hardware architecture has been tested in C++, co-simulated with the generated hardware module from HLS, and generated into a Vivado IP that can be integrated into a system-on-chip.

B. Our design flow

Fig. 8 presents our design flow to implement the proposed VAE model on the Pynq-Z2 development board. First, the

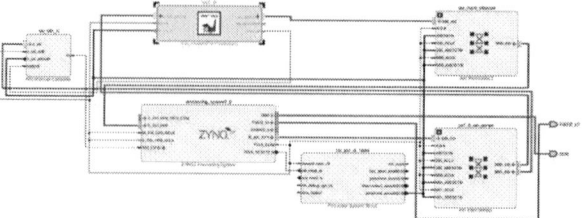

Fig. 9. Integration of the full VAE module with the Zynq processing system on the Pynq-Z2 development board.

input image dataset is quantized into a fixed-point format with 16 bits, including 4 bits for the integer and 12 bits for the fractional part. After training, the weights and biases in the fixed-point format are extracted into text files to be used with the HLS design flow. The model is then translated from the deep learning framework into C++ source code, enabling smooth integration into embedded systems. Next, the Vitis HLS tool converts the C++ code into Verilog or VHDL and generates the VAE IP core. Finally, the IP core is integrated into the Zynq-7000 SoC on the Pynq-Z2 board. The image data, already converted to the fixed-point format, is loaded into DRAM along with the kernel weights and biases via the Zynq processor. The extracted data from DRAM is then reconstructed into an image using Python running on the Pynq-Z2 development board.

C. High-Level Synthesis and Pynq-Z2 development board

High-Level Synthesis (HLS) is a technology that converts high-level C/C++ code into hardware description languages such as VHDL or Verilog for execution on an FPGA. One of its main advantages is the significant reduction in development time, as writing C/C++ code is much faster than developing RTL code in Verilog or VHDL. Additionally, HLS allows for easier optimization, enabling quick modifications to improve algorithm performance. It also integrates seamlessly with tools like Xilinx's Vivado, which supports exporting RTL code for FPGA implementation. However, HLS also has some drawbacks, such as potentially suboptimal hardware compared to hand-written RTL designs, leading to resource utilization and performance inefficiencies. Moreover, the generated RTL code may not be fully optimized or easily controllable, making fine-tuned hardware design more challenging.

For our project, we chose HLS due to the time limitation in implementing the proposed model using HDL. HLS provides a powerful approach to implementing complex algorithms and computations on hardware using high-level languages like C++, making it a suitable choice for accelerating development while maintaining efficiency. To deploy our design, we use the Pynq-Z2 FPGA development board, which is based on the Xilinx Zynq-7000 dual-core processor and is designed to run PYNQ – a framework that simplifies FPGA programming with Python. The Pynq-Z2 features a Zynq XC7Z020-1CLG400C FPGA, which integrates an ARM Cortex-A9 processor with FPGA fabric, along with 512MB DDR3 RAM, microSD storage, and multiple interfaces such as HDMI, USB, Ethernet, PMOD, and Arduino headers. It is a versatile platform for various applications, including real-time image and video processing, machine learning, IoT, and embedded systems. Additionally, the ease of programming with Python makes it an ideal choice for education and research.

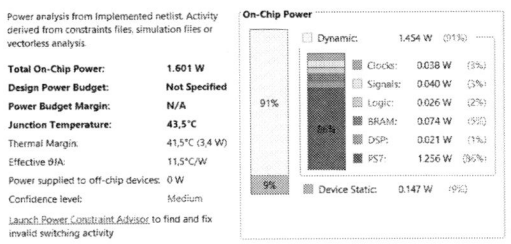

Resource	Utilization	Available	Utilization %
LUT	9694	53200	18.22
LUTRAM	476	17400	2.74
FF	11748	106400	11.04
BRAM	81.50	140	58.21
DSP	22	220	10.00

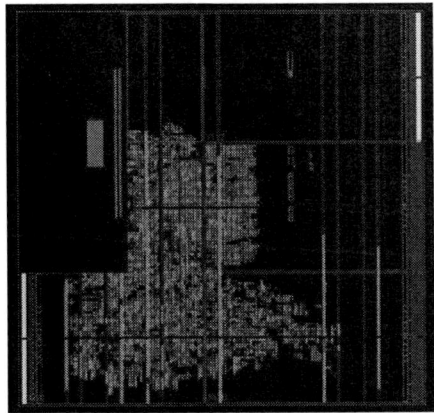

Fig. 10. Resources utilization and power analysis from implemented netlist.

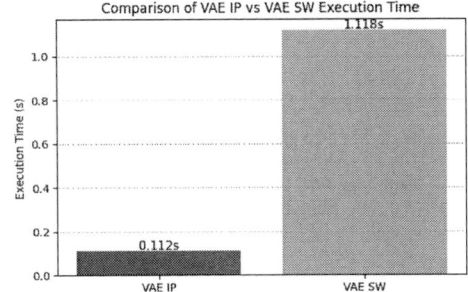

Fig. 12. Comparison of the execution time between hardware VAE model and the software VAE model on the Pynq-Z2 development board.

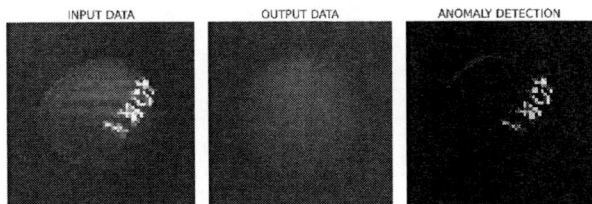

Fig. 13. Reconstructed images from VAE IP and anomaly detection.

complex loops, and optimize memory usage in C++ source code for HLS. The design is then synthesized into RTL (Verilog/VHDL) for deployment on FPGA using Vitis HLS. Finally, after obtaining the RTL code, the design is integrated in Vivado to generate the bitstream to test on the FPGA development board for evaluation.

The proposed VAE architecture for anomaly detection has been successfully implemented using Vitis HLS 2021.2 and integrated into Zynq-7000 SoC using Vivado 2021.2. It has been successfully synthesized, implemented, and validated on the Pynq-Z2 development board. The software is implemented in Python using the Pynq platform.

As shown in Fig. 10, the design uses 11,748 flip-flops (11%), 9,694 LUTs (18.22%), and 22 DSP slices (10%) of the FPGA resources for convolutional computations. It also consumes 81.5 BRAM tiles (58%) and 476 LUTRAMs (2.7%) of these resources on the FPGA kit. These results demonstrate that the VAE hardware implementation achieves moderate resource usage overall, with the most significant consumption being BRAM, which could be further optimized in future versions. Fig. 11 shows the FPGA resource allocation after placement and routing.

B. Runtime between VAE IP hardware and software

After implementing the VAE IP into an FPGA, we evaluate the system performance by running the controlling software and the implemented hardware on the Pynq platform on the Pynq-Z2 development board. The Pynq platform allows programming the programmable logic and control of the IPs through its Python interface. This will enable us to use Python to build the software driver for the VAE IP and perform the anomaly detection for the hazelnut dataset.

Fig. 12 illustrates the execution time comparison between the software-only VAE model running on a laptop with an AMD Ryzen 7 5800H processor and our proposed VAE accelerator running on the Pynq-Z2 development board. The results show a significant improvement in execution speed with hardware acceleration. Specifically, the software-only VAE model takes 1.118 seconds to process one input image,

Fig. 11. The proposed VAE IP after placement and routing on FPGA.

D. Integration of the VAE IP into Zynq-7000 SoC in Vivado

Fig. 9 shows the system architecture of the VAE IP with the Zynq-7000 via the AXI4 interface. The proposed VAE IP has two interfaces: one master interface for direct memory access and one slave interface for configuration. The two interfaces are connected to the AXI4 bus. The Zynq processor controls the VAE IP and can directly access the DRAM memory for the input data, weights, and bias. This arrangement reduces the workload on the processor during the VAE IP execution.

Additionally, the system's interrupt mechanism in the connection between the Zynq7000 SoC and the IP improves communication efficiency between the IP core and the Zynq processor (Cortex-A9). Instead of the processor continuously polling the IP's status, the IP proactively sends an interrupt to the processor when the VAE computations are complete. This reduces the processor workload and enhances system performance by allowing the processor to respond only when an interrupt is triggered.

IV. HARDWARE IMPLEMENTATION RESULTS AND EVALUATION

A. Synthesis results

During the hardware design process using HLS, we first develop a software model to simulate and verify the algorithm. Once the algorithm is confirmed to work correctly, the next step is to convert this model into C/C++ code suitable for synthesis in the HLS tool. We use integer data types, minimize while our proposed VAE accelerator requires 0.112 seconds. This results in a speedup of approximately 10 times compared to the software running on a laptop.

Fig. 13 shows the final results of the model using the output from the hardware decoder, completing the end-to-end inference fully in hardware. The generated images are consistent with those produced by the software-only implementation, confirming that the hardware encoder and decoder preserve essential latent representations and reconstruct high-quality outputs. This demonstrates the reliability and correctness of our proposed VAE accelerator.

C. Comparison with related works

TABLE I. COMPARISON WITH RELATED WORKS

	This work	[9]	[10]
Platform	PYNQ-Z2	U250	U250
Model	VAE	AE	VAE
Frequency (Mhz)	100	300	200
Data format	16-bit fixed-point	16-bit fixed-point	16-bit fixed-point
# DSPs	**22**	2,221	113
# LUTs	**9,694**	516,000	15,570
# Flipflops	11748	-	-
# BRAMs	81.5	-	12
Latency	112.4ms	343ns	5.6ns
Throughput	8.9 img/s	2.9M img/s	178.6M img/s

The comparison of our proposed design with the related works in [9] and [10] is shown in Table I. It summarizes the key parameters, including the operating frequency, the data format, hardware resource usage, and execution latency. While all three implementations aim to accelerate deep learning inference on FPGAs, they differ in design objectives and deployment contexts. The designs in [9] and [10] are optimized for high-performance inference on data center-grade platforms such as the Xilinx Alveo U250. In contrast, our work targets a low-cost and resource-constrained FPGA, the PYNQ-Z2, making it more suitable for lightweight applications at the edge.

Despite operating at a lower frequency and with more limited hardware capabilities, our design meets real-time inference requirements for anomaly detection while consuming significantly fewer resources. With only 22 DSPs and less than 10,000 LUTs, it presents a compact and efficient solution, emphasizing low area cost and power consumption. Moreover, using fixed-point arithmetic and an HLS-based design approach facilitates integration with embedded systems, enabling fast development and prototyping. Although the latency is higher than the other works, it remains acceptable for the intended application scenario, emphasizing cost-effectiveness and deployability for edge-AI applications.

V. CONCLUSIONS AND FUTURE WORKS

Variational AutoEncoder (VAE) models present several challenges when deployed on FPGAs. The encoder and decoder rely on Conv2D and Conv2DTranspose operations,

while latent space sampling introduces additional complexity due to the need for stochastic operations. This work proposes a hardware/software co-design approach for implementing a VAE for hazelnut anomaly detection. Compared to a reference model with 3.5 million parameters, our optimized version uses only around 300 thousand parameters—about 10 times fewer. We also convert the model to fixed-point arithmetic to accelerate hardware execution and improve throughput. The encoder and decoder are fully deployed on an FPGA using High-Level Synthesis (HLS) in the current implementation. At the same time, only the epsilon-scaling step in the latent sampling process is performed in software to reduce hardware complexity. The proposed VAE accelerator has been successfully integrated into the Zynq-7000 SoC with low hardware resource usage. Testing shows that the hardware-accelerated VAE achieves 10× speedup over its software counterpart while preserving anomaly detection accuracy. The system is deployed on the PYNQ-Z2 development board using the PYNQ framework. Future work includes further optimizing BRAM usage and potentially implementing the remaining sampling operations in hardware for a fully standalone FPGA design.

ACKNOWLEDGMENT

This work has been supported by Vietnam National University, Hanoi (VNU) under Project No. QG.24.57 (QG-HW4AI).

REFERENCES

[1] Jiří Raška, "MVTect - HazelNut - Variational AutoEncode II." Retrieved from https://www.kaggle.com/code/jraska1/mvtect-hazelnut-variational-autoencode-ii

[2] P. Bergmann, K. Batzner, M. Fauser, D. Sattlegger, and C. Steger, "Regularization-based continual learning for anomaly detection in discrete manufacturing," *arXiv preprint*, arXiv:2101.00509, 2021.

[3] M. Stephen and K. Sheriffdeen, "AI-Enabled Anomaly Detection in Industrial Systems: A New Era in Predictive Maintenance," *ResearchGate*, 2024.

[4] Paul Bergmann, Kilian Batzner, Michael Fauser, David Sattlegger, Carsten Steger: The MVTec Anomaly Detection Dataset: A Comprehensive Real-World Dataset for Unsupervised Anomaly Detection; in: International Journal of Computer Vision 129(4):1038-1059, 2021, DOI: 10.1007/s11263-020-01400-4.

[5] Yajie Cui, Zhaoxiang Liu, Shiguo Lian, "A Survey on Unsupervised Anomaly Detection Algorithms for Industrial Images." Arxiv 2022. URL: https://arxiv.org/abs/2204.11161

[6] Van Looveren et al. Alibi Detect: Algorithms for outlier, adversarial, and drift detection. URL: https://github.com/SeldonIO/alibi-detect

[7] Francof2a, "Fixed Point Precision Neural Network for MNIST dataset," GitHub, [Online]. URL: https://github.com/francof2a/fxpmath/blob/master/examples/Fixed_Point_Precision_Neural_Network_for_MNIST_dataset.ipynb [Accessed: Mar. 2, 2025]

[8] Fang, J., Shafiee, A., Abdel-Aziz, H., Thorsley, D., Georgiadis, G., & Hassoun, J. (2020). Post-Training Piecewise Linear Quantization for Deep Neural Networks. *arXiv preprint arXiv:2002.00104*.

[9] Z. Que, E. Wang, U. Marikar, E. Moreno, J. Ngadiuba, H. Javed, B. Borzyszkowski, T. Aarrestad, V. Loncar, S. Summers, M. Pierini, P. Y. Cheung, and W. Luk, "Accelerating Recurrent Neural Networks for Gravitational Wave Experiments," *in Proc. ACM/SIGDA Int. Symp. on Field-Programmable Gate Arrays (FPGA)*, 2021, pp. 245–255.

[10] Z. Que, M. Zhang, H. Fan, H. Li, and W. Luk, "Low Latency Variational Autoencoder on FPGAs," *in Proc. Int. Conf. Field-Programmable Technology (FPT)*, 2021, pp. 1–8.

979-8-3315-1550-8/25 $31.00 © 2025 IEEE

Efficient ECG Beat Classification Using Inception Network on Software and FPGA Platforms

Nam Le Nguyen Nhat and Thi Diem Tran

Department of Computer Engineering, University of Information Technology—VNUHCM
Email: diemtt@uit.edu.vn

Abstract—Convolutional Neural Networks (CNNs) have become a dominant solution for Electrocardiogram (ECG) beat classification, owing to their superior feature extraction capabilities. However, traditional CNN-based accelerators often encounter limitations such as large parameter counts, limited configurability, and inefficient hardware utilization. In this work, we propose a lightweight one-dimensional CNN (1D-CNN) inspired by the InceptionNet architecture, specifically designed for efficient and accurate ECG beat classification. The model integrates customized Inception blocks with parallel convolutional layers of varying kernel sizes to capture multi-scale temporal features. To reduce model complexity and enhance efficiency, 1×1 convolutions are used for dimensionality reduction, resulting in a compact network with only 7,740 parameters. Additionally, we present a hardware accelerator that exploits spatial data reuse across multiple processing units and employs dual-port RAM for concurrent data access, improving throughput and memory efficiency. When implemented on hardware with parallel and pipelined execution at 100 MHz, the system achieves a classification accuracy of 99.32%, sensitivity of 98.50%, specificity of 99.58%, and a positive predictive value (PPV) of 98.30% across five ECG classes with the MIT-BIH database, demonstrating its suitability for real-time deployment in resource-constrained environments.

Index Terms—ECG, beat classification, Inception Network, FPGA, SoPC, latency

I. INTRODUCTION

Cardiovascular disease (CVD) remains one of the leading causes of mortality worldwide, posing a significant threat to global public health. Early detection and continuous monitoring are critical for reducing its fatal consequences [1]. Electrocardiogram (ECG) signals play a vital role in the diagnosis and assessment of various heart conditions by capturing the electrical activity of the heart over time. These signals provide valuable insights into the cardiac cycle and help identify abnormalities such as arrhythmias, myocardial infarctions, and ischemia. Key features of ECG signals include the P wave, QRS complex, and T wave, each representing different phases of the heart's electrical conduction system. Analyzing the duration, amplitude, and morphology of these components allows clinicians and automated systems to detect irregular heart rhythms and other pathological patterns with high accuracy.

ECG signal classification is a fundamental task in automated cardiac diagnosis, typically categorized into beat classification and rhythm classification. Beat classification focuses on identifying individual heartbeats, such as normal beats, premature ventricular contractions (PVCs), or atrial premature beats, while rhythm classification involves analyzing longer sequences of heart activity to detect arrhythmias like atrial fibrillation or ventricular tachycardia. Traditional machine learning algorithms, such as support vector machines (SVM), k-nearest neighbors (KNN), and decision trees, have been widely applied to these tasks using handcrafted features extracted from the ECG signal [2], [3]. More recently, deep neural networks (DNNs), including convolutional neural networks (CNNs) and recurrent neural networks (RNNs), have demonstrated superior performance by automatically learning hierarchical representations from raw or minimally preprocessed ECG data. These deep learning models can capture both temporal dependencies and morphological variations in the signals, making them well-suited for robust and accurate classification in real-world healthcare applications.

Recent advancements in electrocardiogram (ECG) classification have utilized various neural network architectures and machine learning techniques to enhance arrhythmia detection. Yin [4] introduced a spiking neural network (SNN) for ECG diagnosis, demonstrating promising energy efficiency and achieving a classification accuracy of 97.86%, making it suitable for neuromorphic hardware implementations. Zhang et al. [5] proposed a low-power application-specific integrated circuit (ASIC) based on an artificial neural network (ANN), which attained 98.68% accuracy while consuming less than 0.5 mW of power—highlighting its potential for wearable devices. Similarly, the study in [6] developed an ultra-lightweight binary neural network (BNN) that achieved 97.5% accuracy while significantly reducing model size, making it well-suited for edge deployment. Other research [7], [8] applied a feature-induced long short-term memory (LSTM) network, reaching the highest reported accuracy of 98.96% and exhibiting robustness to inter-patient variability, albeit at the cost of increased computational complexity. Additionally, the work in [9] employed morphological feature extraction and two-lead ECG signals in combination with classical machine learning classifiers, yielding an accuracy of 94.26%, particularly effective in inter-patient testing scenarios. In contrast, Biradar and Thippeswamy [10] presented a simple ANN model in 2020, achieving 98.35% accuracy, serving as a baseline in terms of architectural simplicity. While deep learning models such as LSTM and feature-induced networks offer high accuracy and generalization, their practical deployment is often hindered by power and memory constraints. Conversely, binary and spiking models provide superior energy efficiency but may compro-

979-8-3315-1550-8/25 $31.00 © 2025 IEEE

mise on accuracy. Overall, these studies reflect a growing trend toward lightweight, real-time, and energy-efficient ECG arrhythmia classifiers. However, a major challenge remains in balancing classification accuracy, model interpretability, and deployment feasibility for resource-constrained environments.

Convolutional Neural Networks (CNNs) have demonstrated significant potential in arrhythmia classification tasks due to their ability to automatically extract spatial features from ECG signals. Odugoudar and Walia [11] proposed a convolutional neural network (CNN)-based system that achieved 97.8% classification accuracy using a 1D representation of ECG signals, highlighting the effectiveness of convolutional feature extraction in time-domain data. Ullah et al. [12] transformed ECG signals into 2D spectral images and applied deep CNNs, reaching an accuracy of 99.11%, and recall of 98.55%. This approach benefits from frequency-domain representations, enhancing pattern recognition capabilities for complex arrhythmias. The work [13] introduced a hybrid lightweight architecture combining 1D CNN and LSTM networks, achieving 98.22% accuracy with reduced computational cost. Their model capitalizes on CNN's feature extraction and LSTM's temporal learning for improved beat-wise classification, making it suitable for edge computing scenarios. Meanwhile, Tippannavar et al. [14] utilized a 2D CNN for heart disease classification from ECG data, reporting an overall accuracy of 99.47%. Their method leverages spatial representations but requires higher memory due to the 2D transformation of input signals. Despite their strong performance, these methods face several limitations. CNN-based models operating on 2D inputs often involve significant preprocessing and higher memory consumption, which may hinder deployment on resource-limited devices. Hybrid CNN-LSTM architectures, though computationally efficient, may still struggle with generalization across patients if not adequately trained on diverse datasets.

In this work, we propose a one-dimensional Convolutional Neural Network (1D-CNN) inspired by the InceptionNet architecture for efficient ECG beat classification. The model incorporates customized Inception blocks that utilize parallel convolutional layers with varying kernel sizes to extract features across multiple temporal resolutions. To improve computational efficiency, 1×1 convolutions are employed for dimensionality reduction, which significantly reduces the number of parameters and helps mitigate overfitting. The resulting model achieves high classification performance while maintaining a lightweight structure with only 7,740 parameters, making it well-suited for real-time applications and deployment in resource-constrained environments. When implemented on hardware using parallel and pipelined execution strategies and operating at 100 MHz, the system achieves an accuracy of 99.32%, sensitivity of 98.50%, specificity of 99.58%, and a positive predictive value (PPV) of 98.30% across five ECG classes. Section II details the proposed software and hardware architectures, while Section III presents the model evaluation, performance results, and hardware resource utilization.

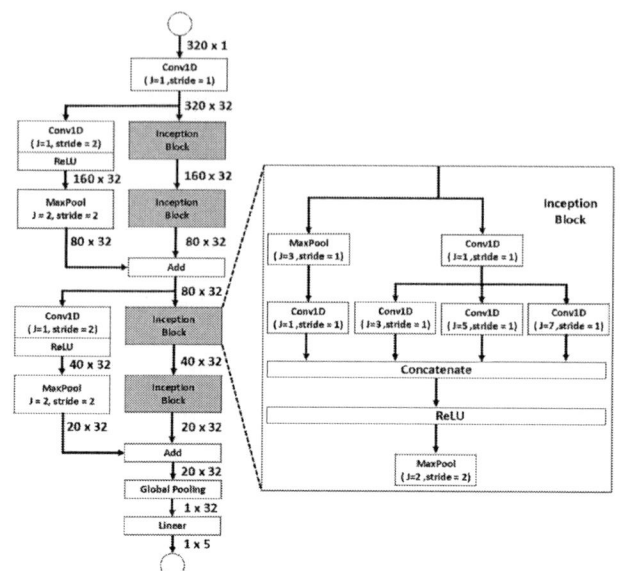

Fig. 1. Proposed Network Architecture for ECG Beat Classification

II. METHODOLOGY

A. Proposal Structure: Inception Network-Based Software System

The proposed Mini-InceptionNet is a lightweight one-dimensional convolutional neural network (1D-CNN) designed for ECG beat classification. As shown in Fig. 1, the input signal with a shape of 320 × 1 is first passed through a 1D convolutional layer with kernel size 1 and stride 1 to preserve temporal resolution, then flows through several convolutional layers and four customized Inception blocks, which are interleaved with ReLU activations, residual connections, and downsampling operations using either strided convolutions or max pooling. Each Inception block consists of five parallel branches: one branch performs max pooling, while the remaining four apply 1D convolutions with kernel sizes of 1, 3, 5, and 7, respectively; all branches use stride 1 and are concatenated along the feature dimension, followed by a ReLU activation and optional downsampling via max pooling with a stride of 2. This multi-branch design enables the network to efficiently capture temporal patterns at multiple resolutions. After the final Inception block, global max pooling is used to reduce the temporal dimension, and a fully connected layer maps the result to an output of shape 1 × 5, corresponding to the five target ECG beat classes. Despite its hierarchical and parallel structure, the entire model contains only 7,740 trainable parameters, making it highly suitable for real-time and resource-constrained deployment scenarios.

B. Proposal Structure: Inception Network-Based Hardware System

Convolutional layers serve as the fundamental components of a one-dimensional Convolutional Neural Network (1D CNN). These layers apply a set of learnable filters (also

979-8-3315-1550-8/25 $31.00 © 2025 IEEE

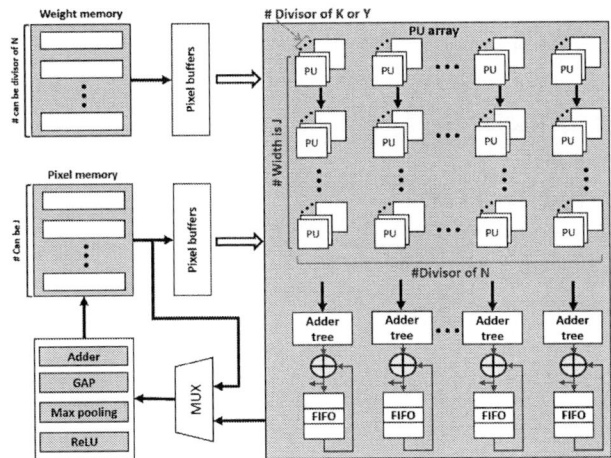

Fig. 2. Proposed Hardware Architecture for ECG Beat Classification

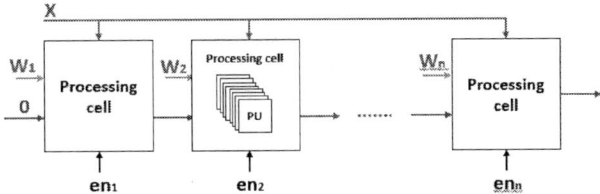

Fig. 4. Multi-Stage Processing Unit Design

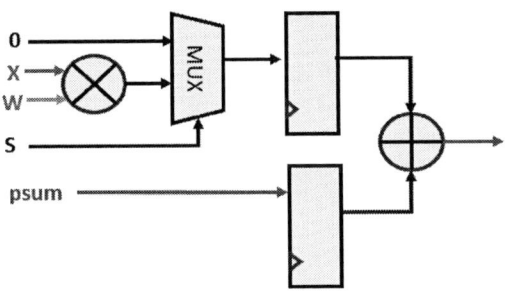

Fig. 3. Processing Unit Design

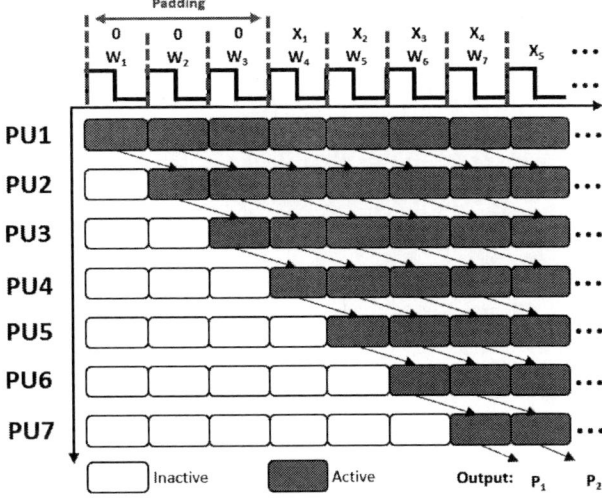

Fig. 5. Computing Pipeline of PU

referred to as kernels) to the input data, generating output feature maps by computing weighted sums over local receptive fields. Mathematically, the convolution operation is defined as:

$$Z[n, y] = \sum_{k=0}^{K-1} \sum_{j=0}^{J-1} W[n, k, j] \cdot X[k, y \times s + j] + b[n], \quad (1)$$

where $Z[n, y]$ denotes the output feature map. Here, $n \in [0, N)$ indexes the output channels, and $y \in [0, Y)$ indexes the spatial positions in the output. The variable $k \in [0, K)$ corresponds to the input channels, while $j \in [0, J)$ indexes the kernel positions. The parameter s represents the stride of the convolution, and $b[n]$ denotes the bias term associated with each output channel.

Fig. 2 illustrates the proposed hardware architecture for embedded ECG classification, optimized for both resource efficiency and high inference performance. To exploit data-level parallelism and minimize memory access latency, the system adopts a spatial reuse strategy, in which data fetched from memory is shared across multiple processing units (PUs) and processed in parallel. The PU array is structured such that its width corresponds to the kernel size J, and the total number of PUs is selected as a divisor of either the input channels K or the output positions Y. This configuration ensures that the hardware can efficiently parallelize computations across vari-

ous CNN layers. To maximize memory bandwidth utilization, the design employs dual-port RAM, allowing simultaneous access to weight and pixel data without contention. In addition, ping-pong buffers are integrated between the memory and the PU array for both weights and inputs, ensuring uninterrupted data streaming and reducing idle cycles. Each column of the PU array is followed by an adder tree, which aggregates partial sums produced by the PUs. These intermediate results are temporarily stored in FIFO buffers, which allow for synchronization and flexible interfacing with the next pipeline stage. A multiplexer (MUX) then routes the data to the appropriate post-processing modules, which may include ReLU activation, max pooling, global average pooling (GAP), or element-wise addition. Fig. 3 presents the architecture of a processing unit designed to perform multiplication between two input signals: X, representing a pixel value, and W, representing a weight. The operation is governed by a control signal S. When $S = 1$, the unit multiplies X and W, and the result is accumulated with the partial sum (psum) received from the previous processing unit. If $S = 0$, the unit outputs a value of zero. Moreover, the architecture supports output generation at every clock cycle, providing high throughput and suitability for real-time processing applications. Fig. 4 illustrates the multi-stage processing unit designed for the concurrent computation of filters up to size 7×1 with multiple inputs processed simultaneously. In this architecture, each processing unit (PU) computes a partial result, which is then forwarded as input to

TABLE I
SOFTWARE-BASED COMPARISON OF DIFFERENT METHODS APPLIED TO THE MIT-BIH DATABASE

Content	[4]	[5]	[6]	[7]	[9]	[10]	[12]	[13]	[14]	[11]	This work
Database	MIT-BIH	MIT-BIH	MIT-BIH	MIT-BIH	MIT-BIH	MIT-BIH	MIT-BIH	MIT-BIH	MIT-BIH	MIT-BIH	MIT-BIH
Method	SNN	ANN	BNN	Bi-LSTM	KNN	ANN	2D-CNN	CNN+LSTM	2D-CNN	1D-CNN	CNN
Parameter	-	-	29202	-	-	-	3678792	64034	$\sim 132 \times 10^6$	-	7740
Preprocess	WT	Filter	No Preprocessing	WT	Filter	Filter	2D Image	Filter	No Preprocessing	WT	Filter
ACC	*97.86%	*98.68%	97.5%	*98.96%	*94.26%	98.35%	99.11%	98.22%	*99.47%	97.8%	**99.32%**
SEN	*94.65%	96.66%	97.9%	*97.04%	*85.65%	95.06%	97.91%	98.23%	*98.14%	97%	**98.50%**
SPEC	*98.66%	*99.18%	99.8%	*99.35%	*96.41%	99.01%	99.61%	99.64%	*99.69%	97.32%	**99.58%**
PPV	*94.65%	96.67%	96.4%	*97.40%	*85.65%	95.06%	98.58%	98.26%	*98.14%		**98.30%**

* Calculated from the confusion matrix

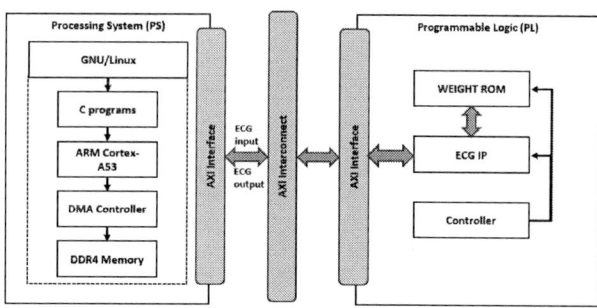

Fig. 6. System-on-Chip Architecture

the next PU. Each processing cell in the architecture consists of eight processing units arranged across seven distinct stages, thereby optimizing computational throughput and resource utilization. With this design, it is capable of computing filters of sizes 1×1, 3×1, 5×1, and 7×1. At the final stage of the multi-stage processing unit, an adder tree is used to sum all the partial results. The architecture is designed with an adder tree that has $N = 8$ inputs.

When executing, the convolution is pipelined as shown in Fig. 5. For a 7×1 filter, during the first 7 clock cycles, the weights for each Processing Unit (PU) are loaded serially and remain unchanged during the subsequent computations. Once all weights are loaded, the corresponding PU becomes active and starts generating outputs from the 8th clock cycle. For other filter configurations, the signal S can be selected to perform computations, with the corresponding filter.

III. SETUP EXPERIMENTS AND RESULTS

A. Software and hardware Implementation

The Mini-InceptionNet model was trained and evaluated using the MIT-BIH Arrhythmia Database [20], which comprises 46 annotated ECG recordings classified into 19 beat types by Physionet. This study focuses on five clinically relevant beat classes: normal (NOR), left bundle branch block (LBBB), right bundle branch block (RBBB), premature ventricular contraction (PVC), and atrial premature beat (APB). Each heartbeat was segmented by extracting 159 samples before and 160 samples after the R-peak, ensuring the full inclusion of the P wave, QRS complex, and T wave. Over 100,000 beat segments were collected and partitioned into training (70%), validation (15%), and testing (15%) sets. The model

was implemented using Google TensorFlow and trained on an NVIDIA GeForce RTX 3090 GPU for 100 epochs using the Adam optimizer, with a batch size of 20. An initial learning rate of 0.0001 was used for the first 40 epochs and reduced to 0.00001 thereafter to stabilize convergence.

For hardware deployment, the system was implemented on the Xilinx Kria KV260 Vision AI Starter Kit, as shown in Fig. 6. The System-on-Chip (SoC) architecture was designed using Vivado 2021.2 and consists of two main components: the Processing System (PS) and the Programmable Logic (PL). The PS includes an ARM Cortex-A53 CPU running a GNU/Linux OS, which executes C programs and communicates with the PL via AXI interfaces. The PL integrates custom ECG processing logic, implemented in Verilog HDL, including modules such as the ECG IP core, a weight ROM for storing model parameters, and a controller for managing operations. Data transfer between DDR4 memory and the accelerator is handled efficiently using a DMA controller.

The performance of the ECG beat classification system is assessed using four standard evaluation metrics: accuracy (ACC), sensitivity (SEN), specificity (SPEC), and positive predictive value (PPV). These metrics are defined as follows:

$$ACC = \frac{TP + TN}{TP + TN + FP + FN} \qquad (2)$$

$$SEN = \frac{TP}{TP + FN} \qquad (3)$$

$$SPEC = \frac{TN}{TN + FP} \qquad (4)$$

$$PPV = \frac{TP}{TP + FP} \qquad (5)$$

where a true positive (TP) is if there is a beat detected in ± 75ms window from beat annotation, and the other is considered a false positive (FP), no detection around beat annotation around that range, it is regarded as a false negative (FN).

B. Results

Table I provides a comprehensive summary of software-based ECG classification methods applied to the MIT-BIH database. The performance metrics—accuracy (ACC), sensitivity (SEN), specificity (SPEC), and positive predictive

TABLE II
HARDWARE-BASED COMPARISON OF DIFFERENT METHODS APPLIED TO THE MIT-BIH DATABASE

Content	[15]	[16]	[17]	[18]	[19]	This work
CNN Model / Model	VGG-16	VGG-16	VGG-16	ANN	CNN	Inception / CNN
CNN Type	2-D CNN	2-D CNN	2-D CNN	-	1-D CNN	1-D CNN
Purpose	ECG, *other	ECG, *other	ECG, *other	ECG	ECG	ECG
FPGA	Virtex7 VX690T	Zynq XC7Z045	Zynq XC7Z045	Artix7	Artix-7	KV260
Clock (MHz) / Frequency	150	100	150	98.209	25.5	100
CNN Size (GOP)	30.76	30.76	30.76	-	-	1.975×10^{-3}
Parameters	50.15M	50.15M	50.15M	-	9385	7740
Precision	16-bit Fixed	16-bit Fixed	16-bit Fixed	24-bit Fixed	22-bit Fixed	18-bit Fixed
DSP Utilization	2833	784	780	214	121	448
LUT / Resource Utilization (kLUT)	561.427k	155.886k	183k	7598	128.960k	26.987k
BRAM Utilization	1248	909**	486	-	16.5	26
Throughput (GOP/s)	354	229.55	137	-	-	38.12
Hardware Resource Efficiency (GOP/s/kLUT)	0.631	1.473	0.749	-	-	1.4125

* The application includes other tasks beyond ECG

** BRAM: The labeled BRAM type is 18K, others' is BRAM 36K

Fig. 7. Throughput and latency comparison among FPGA, CPU, and GPU platforms.

value (PPV)—are evaluated across various approaches. The proposed CNN model in this work outperforms prior methods in nearly all aspects, achieving the highest overall accuracy of 99.32%, with 98.50% sensitivity, 99.58% specificity, and 98.30% PPV. In earlier work, the SNN-based method [4] achieved an accuracy of 97.86%, while ANN [5] improved it slightly to 98.68%. Both exhibited moderate sensitivity and specificity, ranging from 94.65%–98.66%. BNN [6], though parameter-efficient with 29,202 parameters, reached 97.5% accuracy and 97.9% sensitivity. The Bi-LSTM method [7] achieved a slightly higher accuracy of 98.96%, with sensitivity of 97.04% and specificity of 99.35%. The KNN method [9] lagged behind with an accuracy of only 94.26%, sensitivity of 85.65%, and specificity of 96.41%, indicating limitations in handling ECG classification. A conventional ANN approach [10] showed improved results (98.35% ACC), but still trailed deep learning-based techniques. The 2D-CNN method in [12] performed exceptionally well, with 99.11% accuracy, 97.91% sensitivity, and 99.01% specificity, but required 3.6 million parameters. The CNN+LSTM model [13] achieved 98.22% accuracy and 98.23% sensitivity, representing a good trade-off between depth and temporal modeling. In [14], another 2D-CNN model reported a high accuracy of 99.47% and

exceptional specificity (99.69%), though the parameter count (132 million) is significantly higher than others, making it impractical for embedded or real-time applications. The 1D-CNN model in [11] reached 97.8% accuracy and 97% sensitivity, lower than recent hybrid models. In short, the proposed CNN model not only surpasses these methods in classification performance but also maintains a lightweight design with only 7,740 parameters, making it highly suitable for real-time and resource-constrained environments such as edge devices or wearable health monitors.

Table II presents a comparative analysis of hardware-based implementations of various CNN models for ECG classification using the MIT-BIH database. The proposed design, based on a lightweight 1-D CNN with an Inception structure, demonstrates superior hardware efficiency and minimal resource utilization, while still achieving competitive throughput. Specifically, the proposed implementation achieves a throughput of 38.12 GOP/s using only 26.987k LUTs, resulting in a hardware resource efficiency of 1.4125 GOPskLUT. In comparison, prior methods based on the VGG-16 architecture, such as [15], [16], and [17], exhibit significantly higher resource consumption. For example, [15] reports a throughput of 354 GOPs, but requires 561.427k LUTs, 2833 DSPs, and 1248 BRAMs, leading to a lower efficiency of 0.631 GOPskLUT. While [16] improves efficiency to 1.473 GOP/s/kLUT, it still demands 784 DSPs and 909 BRAMs, which are considerably higher than the 448 DSPs and 26 BRAMs used in our design. Other lightweight models, such as those in [18] and [19], employ 1-D CNNs with fewer parameters (e.g., 9,385 in [19]), but either suffer from low throughput or consume a large number of LUTs (e.g., 128.96k in [19]), making them less suitable for resource-constrained platforms. In contrast, the proposed model uses only 7,740 parameters, operates at 18-

bit fixed-point precision, and has an extremely compact CNN size of 1.975×10^{-3} GOP. These characteristics highlight its suitability for real-time ECG processing on embedded systems, particularly on low-power platforms like the Xilinx Kria KV260. Overall, the design offers an excellent trade-off between performance, efficiency, and hardware practicality.

Fig. 7 illustrates the throughput and latency performance of the ECG inference system implemented on three platforms: FPGA (KV260), CPU (Intel i7-9750H), and GPU (NVIDIA GTX 1650). The results clearly highlight the superiority of the FPGA implementation. In terms of throughput, the FPGA achieves 38.12 GOP/s, significantly outperforming the CPU and GPU, which reach only 1.0189 and 1.1669 GOP/s, respectively. This corresponds to a speedup of approximately 37.4× over the CPU and 32.7× over the GPU, emphasizing the effectiveness of the FPGA's parallel and pipelined architecture for lightweight CNN workloads. In terms of latency, the FPGA again demonstrates the best performance with an inference delay of just 0.231 ms. In contrast, the CPU and GPU exhibit latencies of 6.315 ms and 6.012 ms, respectively. This translates to a latency reduction of about 96.3% compared to the CPU and 96.2% compared to the GPU. Such ultra-low latency makes the FPGA platform highly suitable for real-time ECG signal processing in embedded and wearable healthcare applications.

IV. CONCLUSION

This work presents a lightweight 1D-CNN model inspired by the InceptionNet architecture for efficient ECG beat classification. By integrating parallel convolutional layers with varying kernel sizes and employing 1×1 convolutions for dimensionality reduction, the model effectively captures multi-scale temporal features while maintaining low complexity. With only 7,740 parameters, the model achieves high accuracy (99.32%) and excellent sensitivity (98.50%), specificity (99.58%), and PPV (98.30%), making it highly suitable for real-time and resource-constrained applications. The hardware implementation on an FPGA at 100 MHz demonstrates low latency and efficient resource utilization through pipelined and parallel execution strategies. While the model shows strong performance on the MIT-BIH database, its generalizability to other ECG datasets and noisy real-world signals remains to be further validated. Future work will focus on enhancing robustness against signal artifacts, integrating adaptive filtering techniques, and extending the system for multi-lead ECG analysis and arrhythmia detection in ambulatory settings.

ACKNOWLEDGMENTS

This research was supported by The VNUHCM-University of Information Technology's Scientific Research Support Fund.

REFERENCES

[1] M. B. Yilmaz and H. Gunes, "The ever-growing burden of cardiovascular disease," in *Epigenetics in Cardiovascular Disease*, pp. 3–17, Elsevier, 2021.

[2] G. Goovaerts, S. Padhy, B. Vandenberk, C. Varon, R. Willems, and S. Van Huffel, "A machine-learning approach for detection and quantification of qrs fragmentation," *IEEE journal of biomedical and health informatics*, vol. 23, no. 5, pp. 1980–1989, 2018.

[3] R. J. Martis, U. R. Acharya, and L. C. Min, "Ecg beat classification using pca, lda, ica and discrete wavelet transform," *Biomedical Signal Processing and Control*, vol. 8, no. 5, pp. 437–448, 2013.

[4] Z. Yin, "Electrocardiogram diagnosis based on spiking neural networks," in *Applied and Computational Engineering*, vol. 96, pp. 7–14, 2024.

[5] C. Zhang, J. Chang, Y. Guan, Q. Li, X. Wang, and X. Zhang, "A low-power ecg processor asic based on an artificial neural network for arrhythmia detection," in *Applied Sciences*, vol. 13, p. 9591, MDPI, 2023.

[6] N. Pu, Z. Wu, A. Wang, H. Sun, Z. Liu, and H. Liu, "Arrhythmia classifier based on ultra-lightweight binary neural network," in *arXiv*, 2023.

[7] B. Ganguly, A. Ghosal, A. Das, D. Das, D. Chatterjee, and D. Rakshit, "Automated detection and classification of arrhythmia from ecg signals using feature-induced long short-term memory network," in *IEEE Sensors Letters*, vol. 4, p. 6001604, IEEE, 2020.

[8] A. Sampath and R. Sumithira, "Sparse based recurrent neural network long short term memory (rnn-lstm) model for the classification of ecg signals," *Applied Artificial Intelligence*, vol. 36, pp. 1–28, 2022.

[9] H. Zakaria, E. Nurdiniyah, A. Kurniawati, D. Naufal, and N. Sutisna, "Morphological arrhythmia classification based on inter-patient and two leads ecg using machine learning," in *IEEE Access*, vol. PP, pp. 1–1, IEEE, 2024.

[10] S. Srivastava, H. Bhardwaj, A. Dixit, and N. K. Shinde, "Ecg pattern analysis using artificial neural network," *SSRG International Journal of Electronics and Communication Engineering*, vol. 7, no. 5, pp. 1–4, 2020.

[11] A. Odugoudar and J. Walia, "Ecg classification system for arrhythmia detection using convolutional neural networks," in *arXiv*, 2023.

[12] A. Ullah, S. M. Anwar, M. Bilal, and R. M. Mehmood, "Classification of arrhythmia by using deep learning with 2-d ecg spectral image representation," in *Remote Sensing*, vol. 12, p. 1685, MDPI, 2020.

[13] Y. Obeidat and A. Alqudah, "A hybrid lightweight 1d cnn-lstm architecture for automated ecg beat-wise classification," in *Traitement du Signal*, vol. 38, pp. 1281–1291, 2021.

[14] S. Tippannavar, R. Harshith, S. Jain, and R. Shashidhar, "Ecg based heart disease classification and validation using 2d cnn," in *Proceedings of the International Conference on Contemporary Computing and Informatics (IC3I)*, 2022.

[15] C. Zhang, Z. Fang, P. Zhou, P. Pan, and J. Cong, "Caffeine: Towards uniformed representation and acceleration for deep convolutional neural networks," in *Proceedings of the 2016 ACM/IEEE 43rd Annual International Symposium on Computer Architecture (ISCA)*, 2016.

[16] Q. Xiao, Y. Liang, L. Lu, S. Yan, and Y.-W. Tai, "Exploring heterogeneous algorithms for accelerating deep convolutional neural networks on fpgas," in *2017 54th ACM/EDAC/IEEE Design Automation Conference (DAC)*, (Austin, TX, USA), pp. 1–6, 2017.

[17] K. Guo, L. Zeng, J. Yu, Y. Wang, H. Yang, L. Shi, S. Yin, and G. Yuan, "Angel-eye: A complete design flow for mapping cnn onto embedded fpga," *IEEE Transactions on Computer-Aided Design of Integrated Circuits and Systems*, vol. 37, pp. 35–47, Jan. 2018.

[18] Z. Hadjer, M. Kedir-Talha, K. Meddah, and S. Slimane, "Fpga-based system for artificial neural network arrhythmia classification," *Neural Computing and Applications*, 2020.

[19] A. Jaramillo-Rueda, L. Vargas-Pacheco, and C. Fajardo, "A computational architecture for inference of a quantized-cnn for detecting atrial fibrillation," *Ingeniería y Ciencia*, vol. 16, no. 32, pp. 135–149, 2020.

[20] G. B. Moody and R. G. Mark, "The impact of the mit-bih arrhythmia database," *IEEE engineering in medicine and biology magazine*, vol. 20, no. 3, pp. 45–50, 2001.

Analysis of Plant Electrical Signals on an IoT Platform

Xuan-Bach Nguyen-Duy[1,2]
1. University of Information Technology
Ho Chi Minh City, Vietnam
bachndx@uit.edu.vn
2. Vietnam National University
Ho Chi Minh City, Vietnam

Bao-Chau Pham-Ngoc[1,2]
1. University of Information Technology
Ho Chi Minh City, Vietnam
chaupnb@uit.edu.vn
2. Vietnam National University
Ho Chi Minh City, Vietnam

Anh-Vu Dinh-Duc[1,2]
1. International University
Ho Chi Minh City, Vietnam
ddav@hcmiu.edu.vn
2. Vietnam National University
Ho Chi Minh City, Vietnam

Abstract—The waves emitted by plants, also known as plant bioelectrical signals, are associated with electrophysiological phenomena within the plant. These signals can reflect the plant's state, health condition, and responses to environmental factors. This study integrates traditional sensors—such as temperature, humidity, light, sound, and imaging—with Galvanic Skin Response (GSR) to decode the significance of the plant's emitted frequencies. The research leverages an IoT-based data collection platform, analyzing the data to classify and evaluate these signals, thereby providing insights into the correlation between plant-emitted frequencies and plant conditions. Future phases of this research will include the application of supervised ML algorithms, such as Random Forest or SVM, for automated classification of plant conditions based on GSR and environmental signals.

Index Terms—IoT, Bio-Signal, GSR, Electrical Signal, Wireless
.

I. INTRODUCTION

To monitor plant growth, current methods rely on multiple tools and sensors, such as temperature, humidity, and light sensors. If a single sensor could replace all of these, it would simplify installation, streamline data collection, and significantly reduce investment costs.

Plants absorb water for growth, development, and survival. As water moves throughout the plant's body, ion collisions occur within its internal structure [1]. This study utilizes Galvanic Skin Response (GSR) to measure these wave signals. The objective is to place plants in various conditions and scenarios to observe changes in their emitted frequencies and derive meaningful evaluations.

Several experimental studies have suggested that plants exhibit emotions. For instance, they experience stress in noisy environments but appear more comfortable when exposed to music. Research has shown that sound vibrations can influence the rearrangement of microfibers, increase soluble sugar levels, and regulate the transcription of specific genes, thereby promoting plant growth [4]. Additionally, some duckweed species exposed to soft music exhibited protein content approximately 1.6 times higher after seven days compared to untreated plants [2].

There are also hypotheses suggesting that plants can produce sounds to communicate with one another [3]. Plants can generate acoustic signals due to the phenomenon of cavitation, particularly under drought conditions. When tension is created within the xylem sap due to transpiration, air is drawn into conduits through pit membranes, forming gas bubbles. The expansion of these bubbles disrupts water transport, causing vibrations in the vascular tissue. This results in the emission of broadband acoustic signals, primarily in high-frequency ranges, with variations across different plant species.

Moreover, as external environmental factors change, the bioelectrical signals emitted by plants also fluctuate accordingly [2][6]. Electrical signals in plants arise due to various external stimuli and play a crucial role in their adaptation to changing environmental conditions [1]. This forms the basis for collecting and classifying plant electrical signals in this study.

This paper aims to investigate and evaluate plant frequency responses under similar conditions.

II. METHODOLOGY

A. Hardware Device

The IoT system is designed with integrated sensor hardware to collect data and transmit it to the cloud for storage and analysis [5][7][8] .The proposed system consists of:

- ESP32 serves as the central System-on-Chip (SoC) that integrates Wi-Fi, Bluetooth, and a dual-core microcontroller, enabling sensor interfacing, data processing, and wireless transmission.
- DHT22 sensor for temperature and humidity monitoring.
- CO_2 sensor for air quality measurement.
- Light sensor to track illumination levels.
- Soil moisture sensor for monitoring water content.
- GSR sensor for bio-signal detection.
- Microphone for bioacoustic analysis.
- Camera for visual monitoring.

Figure 1 illustrates the schematic diagram of the circuit. The ESP32 serves as the central MCU, interfacing with various sensors and other components such as the power supply, LCD display, and push buttons. Figure 2 shows the finalized circuit after assembly. The fully enclosed product is shown in Figure 3, with the antenna and soil moisture sensor connected.

979-8-3315-1550-8/25 $31.00 © 2025 IEEE

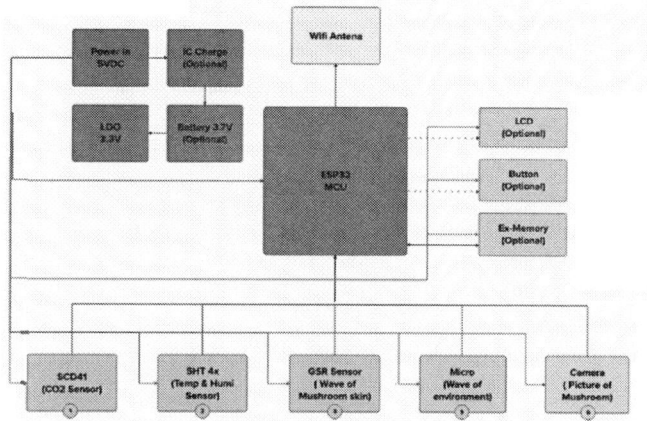

Fig. 1: Circuit Block Diagram.

Fig. 4: Box with Mushroom.

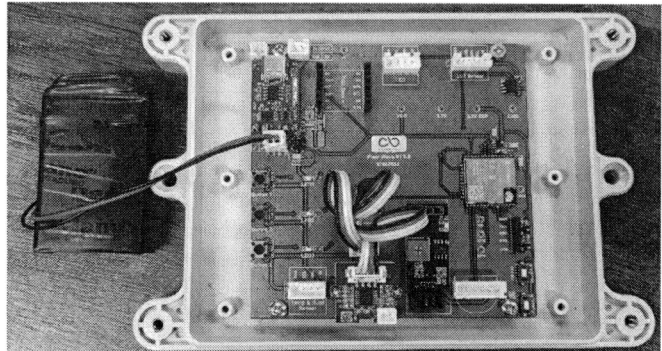

Fig. 2: Circuit Board.

Fig. 3: Complete Equipment.

B. Hardware Setup

The study was tested with Mushrooms grown in boxes that support water spraying to create humidity as Figure 4. Circuit devices and sensors were also mounted in the box to collect the living environment of the Mushrooms. In particular, the GSR sensor was attached to the surface of the Mushrooms to collect the electrical signals emitted from the Mushrooms as Figure 5 and Figure 6.

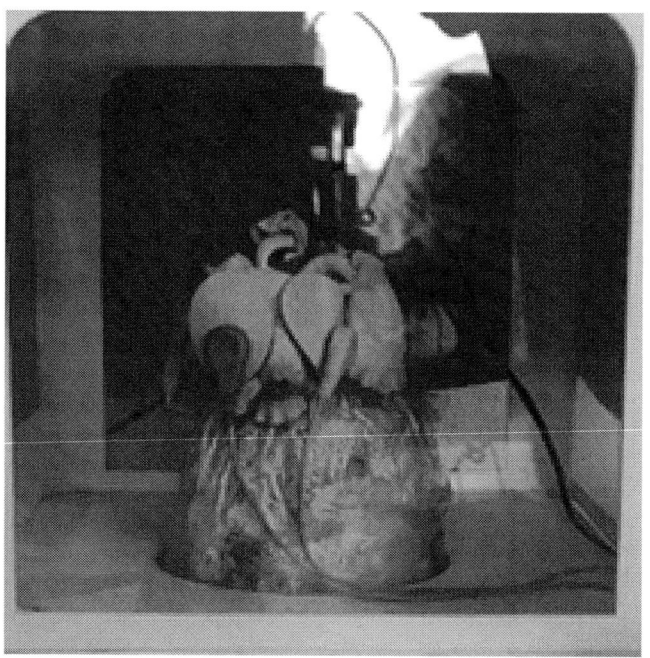

Fig. 5: GSR Sensor stick on Mushroom.

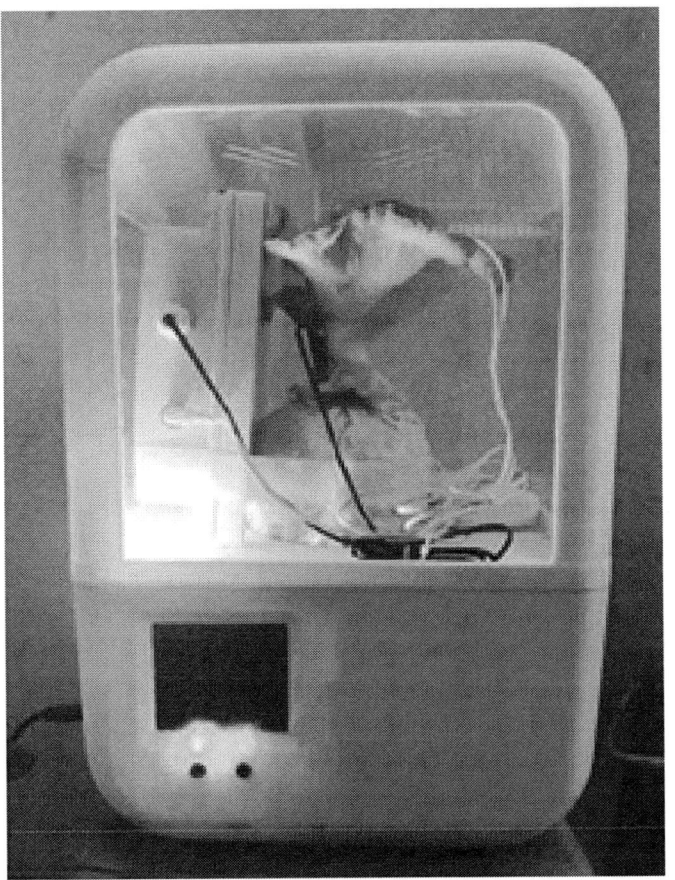

Fig. 6: Full Hardware Devices.

C. Data Collection and Transmission

ESP32 connects WiFi router and collects data sensors every 10 minutes and transmit them via MQTT to a cloud-based database. The system uses a mobile app for real-time monitoring and alerts. The bio-signal data, including GSR readings and sound intensity variations, is recorded and analyzed for correlations with environmental changes.

This study evaluates the electrical signals and emitted frequencies of mushrooms under four different conditions:

Case 1: Standard conditions: temperature 25–32°C, humidity 85–90%, light intensity 100–300 lux, CO_2 less than 1000 ppm, and adequate water supply.

Case 2: Same as Case 1 but without watering.

Case 3: Same as Case 1 but placed in darkness.

Case 4: Same as Case 1 but placed in darkness and without watering.

Data collection will be conducted continuously for 4–10 days for each case.

All collected data will be aggregated, classified, and compared to determine how different experimental conditions correspond to the electrical signals and frequency variations of mushrooms.

III. RESULTS AND DISCUSSION

Device sends data to cloud with raw data as Figure 7.

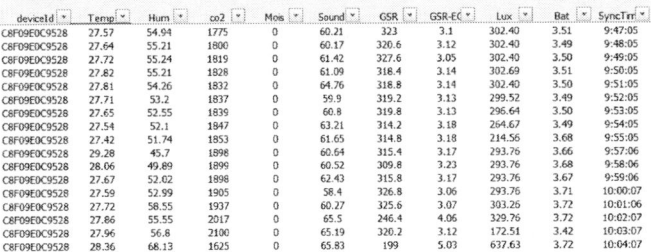

Fig. 7: Raw data.

Case 1: Sensors Data Chart shows from Figure 8 to Figure 12

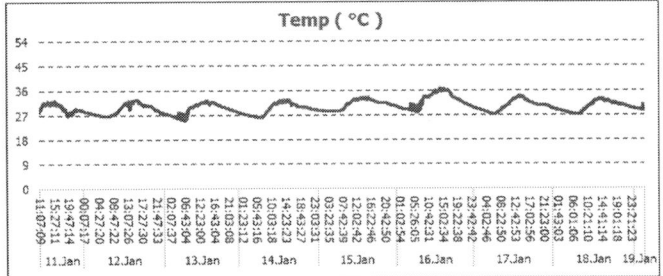

Fig. 8: Temperature Data Chart.

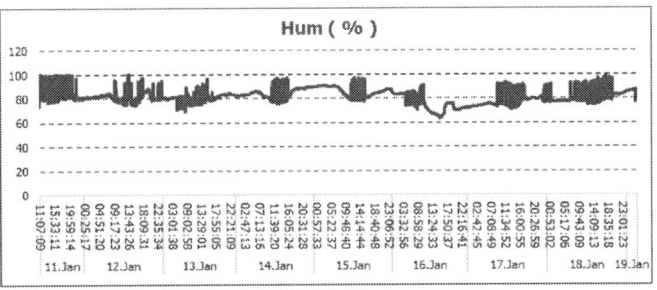

Fig. 9: Humidity Data Chart.

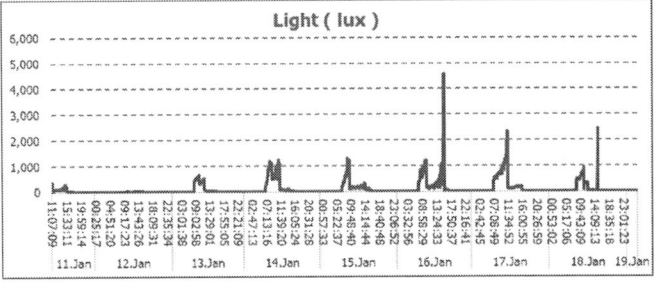

Fig. 10: Light Data Chart.

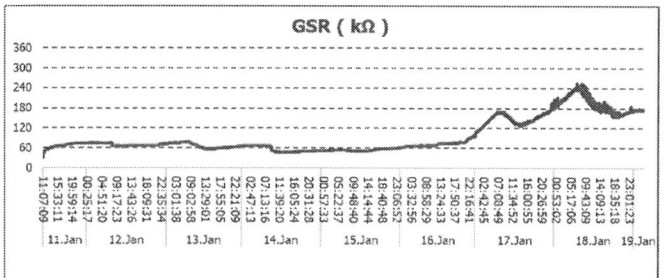

Fig. 11: GSR Data Chart.

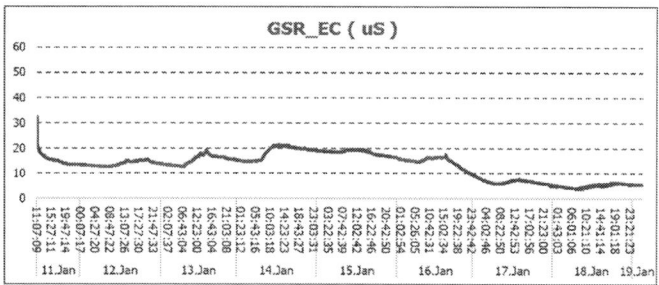

Fig. 12: GSR-EC Data Chart.

Case 2: When not watered, differences compared to Case 1 can be observed in certain measured data, as presented from Figure 13 to Figure 15.

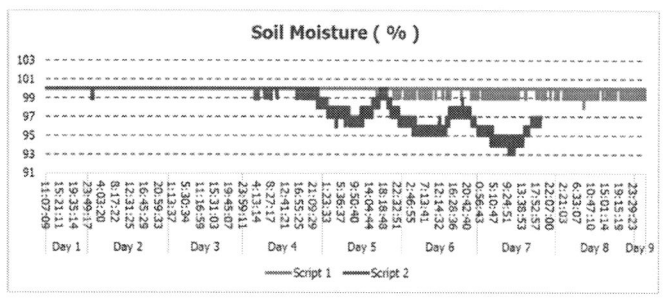

Fig. 13: Moisture of Case 1 and Case 2.

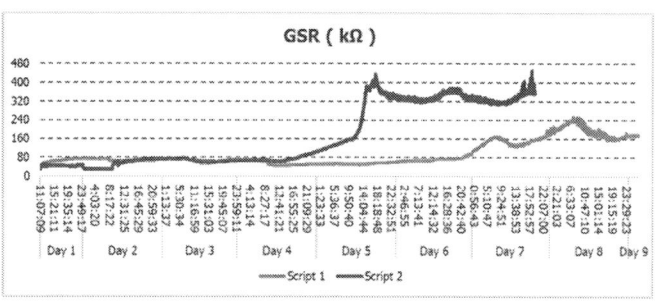

Fig. 14: GSR of Case 1 and Case 2.

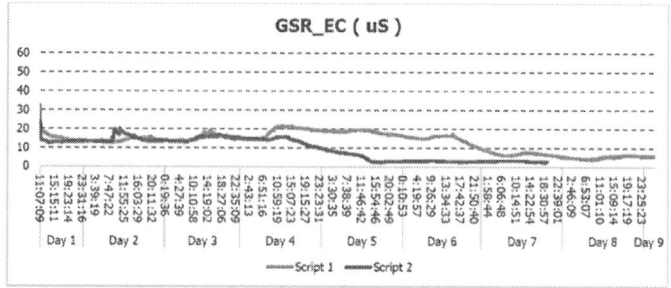

Fig. 15: GSR-EC of Case 1 and Case 2.

Case 3: When placed in darkness, differences compared to Case 1 can be observed in certain measured data, as presented from Figure 16 and Figure 17.

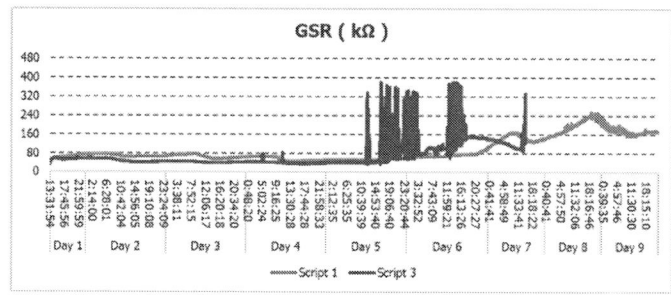

Fig. 16: GSR of Case 1 and Case 3.

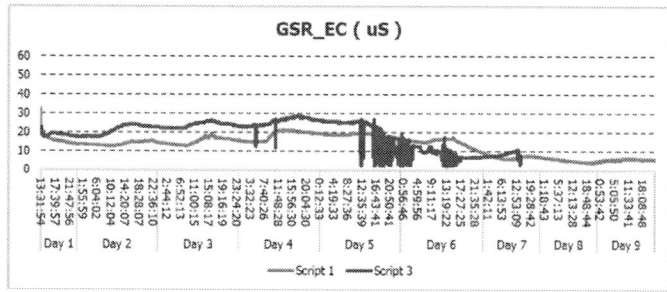

Fig. 17: GSR-EC of Case 1 and Case 3.

Case 4: Combined with placing in the shade and not watering, differences compared to Case 1 can be observed in certain measured data, as presented from Figure 18 to Figure 20.

979-8-3315-1550-8/25 $31.00 © 2025 IEEE

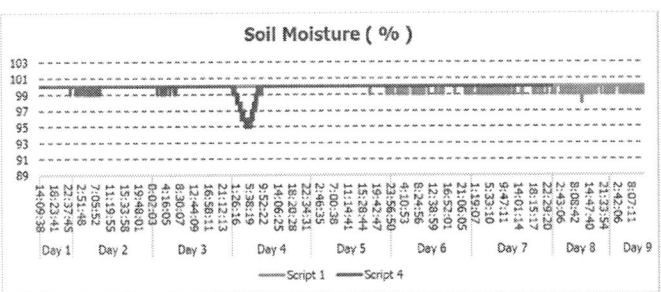

Fig. 18: Moisture of Case 1 and Case 4.

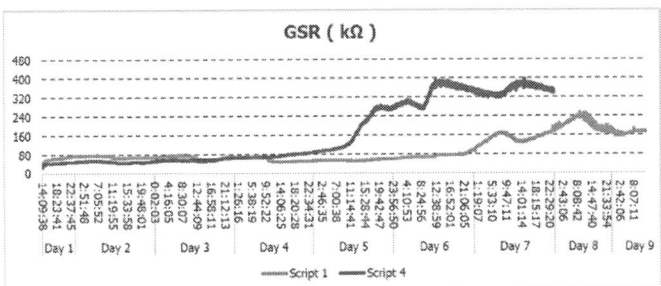

Fig. 19: GSR of Case 1 and Case 4.

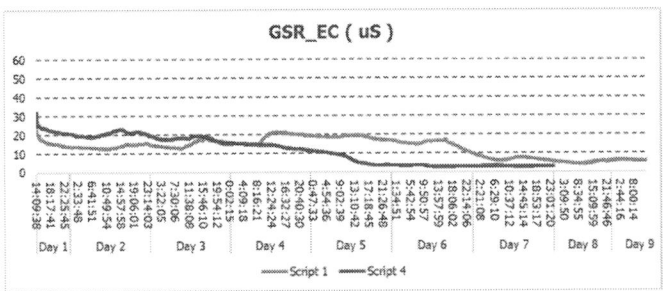

Fig. 20: GSR-EC of Case 1 and Case 4.

Due to the use of low-cost sensors, signal stability and sensor drift were initially evaluated. Preliminary observations showed stable readings for temperature and humidity within ±0.5°C and ±2% RH respectively. However, GSR values fluctuated based on contact quality and environmental noise. Future work will include calibration against reference-grade instruments and sensor redundancy to validate signal reliability.

Although direct benchmarks are limited due to the novelty of GSR-based plant sensing, prior studies in [2][6] showed similar signal volatility under stress stimuli. Compared to controlled experiments using specialized electrophysiology systems, our setup yielded consistent patterns despite hardware limitations, supporting feasibility of low-cost platforms for field applications.

IV. CONCLUSION AND FUTURE WORK

In case no water, the nutrient/moisture level in the substrate of the mushrooms significantly decreased from the fourth day onward.

At this stage, the GSR value showed an increase of 2–5 times compared to normal conditions. Meanwhile, GSR-EC exhibited a decreasing trend toward zero.

In case darkness, the light sensor consistently returned a value of zero throughout the experiment. The substrate moisture showed a significant declining trend over time. The GSR values began to increase and fluctuate significantly from the fifth day, coinciding with a prolonged decrease in substrate moisture. GSR-EC also fluctuated, but with a lower amplitude compared to GSR.

In case no water and darkness, despite the lack of water, water loss was lower in darkness compared to the no-water case under normal lighting.

As a result, the primary impact came from light deprivation, leading to findings that were similar to the no-water case.

These initial observations indicate that GSR and GSR-EC signals can serve as potential indicators of water and light stress in plants. The clear pattern differences observed across experimental cases form a foundation for subsequent phases of research. In the next stage, the collected multi-sensor data—particularly time-series variations of GSR, GSR-EC, and environmental parameters—will be utilized to train supervised machine learning models. These models aim to automatically classify plant stress conditions and predict deviations from normal growth environments. This direction is expected to enhance the decision-making capabilities of the system and support intelligent early warning mechanisms in future IoT-based plant monitoring applications.

A. Preliminary Conclusions

Strong fluctuations in GSR indicate significant light deficiency. A steady and substantial increase in GSR suggests severe water deficiency. The research team is continuing data collection and expanding the study to include the effects of machine noise, music exposure, and multi-plant interactions to assess potential communication between plants..

ACKNOWLEDGMENT

This research is funded University of Information Technology -Vietnam National University Ho Chi Minh City

REFERENCES

[1] Maxim A.Mudrilov, Maria M.Ladeynova, Darya V.Kuznetsova, Vladimir A.Vodennev, "Ion Channels in Electrical Signaling in Higher Plants", vol. 88, pp. 1467-1487, 2023.

[2] Jessica K.S Pachu, Francynes C.O.Macedo, Jose B Malaquias, Francisco S.Ramalho, Ricardo F.Oliveria, Wesley A.C Godoy, Angelica S.Salustino, "Electrical signalling and plant response to herbivory: A short review" Plant Signaling and Behavior Journal, vol. 18, 2023.

[3] Jin-Soo Son, Seonghan Jang, Nicolas Mathevon, Choong-Min Ryu "Is plant acoustic communication fact or fiction?"New Phytologist Journal, 2024.

[4] Anindita Roy Chowdhury, Anshu Gupta, "Effect of Music on Plants - An Overview" International Journal of Integrative sciences, Innovation and Technology, vol. 4, 2015.

[5] J. Smith, A. Khan, "IoT in Precision Agriculture: A Survey," IEEE Internet of Things Journal, vol. 7, no. 4, pp. 3124-3138, 2023.

[6] P. Garcia, M. Lee, "Plant Bio-Signals and Their Application in Smart Farming," Journal of Agricultural Technology, vol. 5, no. 2, pp. 145-160, 2022.

[7] R. White, L. Brown, "Sensor-Based Smart Agriculture," IEEE Sensors Journal, vol. 18, no. 8, pp. 3456-3469, 2022.

[8] H. Lee, S. Kim, "Advances in IoT-Based Smart Farming," IEEE IoT Systems, vol. 9, no. 5, pp. 100-120, 2023.

2025 10th IEEE International Conference on Integrated Circuits, Design, and Verification (ICDV)

A Transformer Feedback Oscillator

Weiwen Lin[1,2], Zhiqun Li[1,2,*], Zhennan Li[1,2], Yan Yao[3], Bofan Chen[1,2], Muhammad Hashim[1,2], Yassin Abdullah[1,2]

1 State Key Laboratory of Millimeter Waves, Southeast University, China;
2 Institute of RF- & OE-ICs, Engineering Research Center of RF- & OE-ICs, Southeast University, Nanjing 210096, Jiangsu, China;
3 Nanjing Electronic Equipment Institute, China;
* Correspondence: zhiqunli@seu.edu.cn

Abstract—This paper presents an LC voltage-controlled oscillator for millimeter-wave communication. The resonator of the oscillator is optimized in the design, and the gate voltage amplitude of the negative resistance transistor is slightly higher than the drain with the help of a transformer, which makes the transistor to turn on and turn off more quickly results the improve of the phase noise. The tuning range is expanded with switched capacitors. The varactor bias structure makes the capacitance change range being set appropriately within the Vtune adjustment range. The design is based on 40nm CMOS process. A compromise is get between power consumption and performance. The output frequency range is about 8.22-9.949 GHz, and the FTR is 19%. When the output frequency is 9.949 GHz, the phase noise at 1MHz offset is -115.68 dBc/Hz, the FoM value is -181.31 dBc/Hz, the FoM_T is -186.88 dBc/Hz, and the power consumption is 27.072mW.

Keywords—Oscillator, Phase noise, Resonator.

I. INTRODUCTION

Oscillator is an important part of wireless communication system, which provides important local oscillator signals for transmission and reception. However, with the increasing demand of communication and the shortage of channel resources, a good quality of local oscillator signals for wireless communication systems has become a hot topic in the field of radio frequency research. In order to satisfy the demands of high-speed data transmission and complex modulation technology, various kind of oscillator appeared. These works aim to meet the application requirements of low phase noise, low power consumption and small area[1–3].

A noise circulating oscillator demonstrates better phase noise, due to the stacking of nmos and pmos, the supply voltage is relatively high, which is not suitable for low-voltage application [4]. Although a class F oscillator based on 0.12 μm SiGe BiCMOS process shows good phase noise, the power consumption is as high as 75 mW, and is work at a high supply voltage of 2.5V [5]. Multi-core multi-mode oscillator shows wide band frequency, due to many cores, the chip area is larger, and compared with the single-core, the mode

switching control circuit needs to be designed carefully, which increases the complexity of the system [6].

This article proposed an oscillator with a transformer feedback structure based on 40nm CMOS process. The second part of the article describes the structure of the oscillator and analyzes it. The third part gives the simulation results. And it also gives the performance comparison of the oscillator with similar works. The fourth part draws a conclusion.

II. OSCILLATOR ANALYSIS AND DESIGNING

The varactor changes the capacitance by changing the two terminal voltage results the frequency change of oscillator. The C-V characteristic curve shows that the varactor is not linear. When the capacitance is changed by voltage, there are regions with different slopes. If the varactor can be adjusted to the region where the slope of the C-V characteristic changes acutely, one tuning curve can overlap a wider frequency range. Therefore, in this design, the varactor is connected in series with fixed capacitors, and the varactor is biased to work in the region with the sharply capacitance change range to improve the utilization efficiency. The connection is shown in Figure 1.

The varactor uses a resistor bias to make it work in the suitable range. The simulation result in Figure 2 shows that the value of Vtune is changed from 0 to 0.9 volts for different Vbias. Vbias is set to 0.2 V, which can better ajust varactor for a wider range of changing with Vtune.

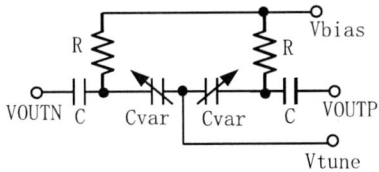

Fig. 1. Structure of biased varactor.

979-8-3315-1550-8/25 $31.00 © 2025 IEEE

79

For LC oscillators, the frequency range of one tuning curve is usually small. If switch capacitor array can be used, the tuning range will be widened. This design uses a 6-bit switch capacitor array as shown in Figure 3. The switch that controls the connection and disconnection of the capacitor is nmos transistor. The gate of nmos is controlled by the control signal, the source and drain voltages are controlled by the resistor at the output of the inverter. Since the gate and source of the MOS transistor are opposite logic levels when the control signal is 0 or 1, the transistor can be fully turned on or off. Taking the capacitor controlled by SW0 in Figure 3 as an example, when SW0 is high, the output of the inverter is low, and the connection node between resistor R and capacitor C0 also becomes low, which can fully turn on MN0. Conversely, when SW0 is low, MN0 can be fully turned off. By adjusting the value of the switched capacitor per bit, the tuning curves overlap each other, and the frequency range of oscillation can be fully covered.

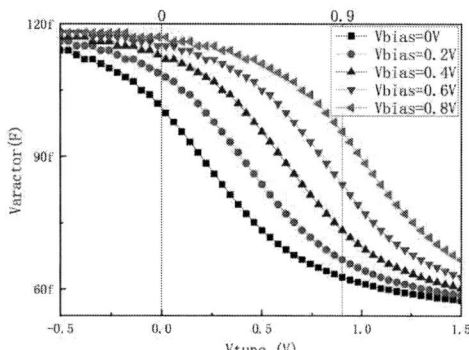

Fig. 2. C-V characteristic of varactor with different bias voltages.

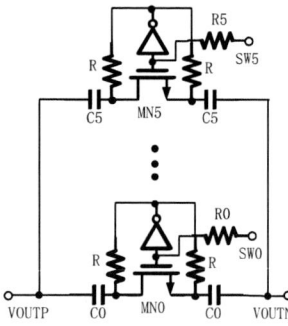

Fig. 3. Structure of 6-bit switch capacitor array.

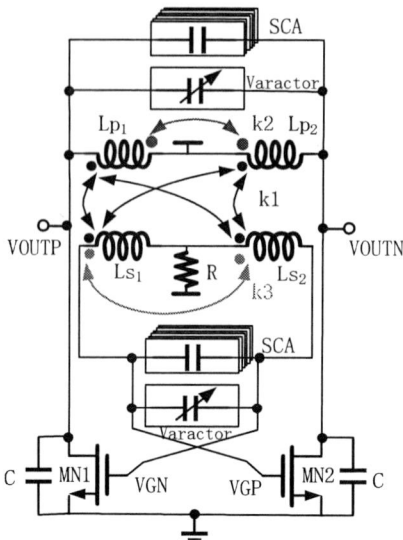

Fig. 4. Structure of oscillator core.

The schematic of the proposed VCO is shown in Figure 4. The primary coil which composed of Lp1 and Lp2 is connected to the drain of the two negative resistance transistors. The secondary coil which composed of Ls1 and Ls2 is connected to the gate. The size of the primary and the secondary coil is appropriately adjusted so that the voltage amplitude of the gate is slightly higher than the drain. The center tap of the primary coil is directly connected to the power supply. The varactor and switch capacitor array in the box in Figure 4 are simplified presents, for the detail of which are shown in Figures 1 and 3, respectively. When designing the layout, the primary and the secondary coil are put closed to each other for magnetic coupled.

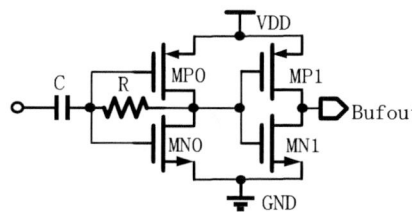

Fig. 5. Structure of buffer.

The output of the VCO core is connected to the buffer. The design of the buffer is shown in the Figure 5. The DC signal of the oscillator is filtered out by a capacitor, and then go through a buffer. The buffer isolates the load from the core circuit, which can prevent frequency changing.

III. SIMULATION RESULT

The simulation is being done to predict and approximately estimate the performance of the design.

The impedance characteristics of resonator is simulated by set the switch capacitances at lowest and highest frequency tuning curves, respectively. The simulation results are shown in the Figure 6, where the negative resistance shows the ability of oscillating and working frequency.

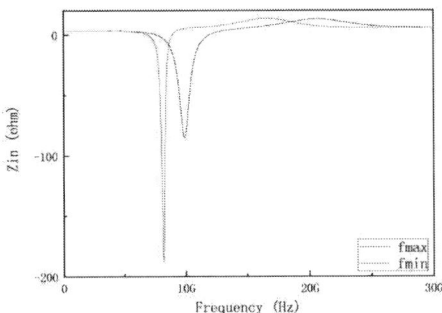

Fig. 6. Real part of resonator impedance.

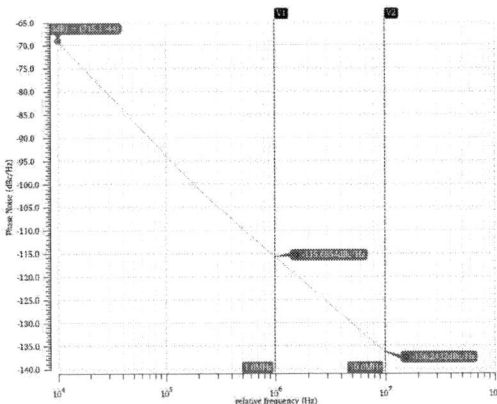

Fig. 7. Simulated phase noise at carrier frequency 9.949 GHz.

Figure 7 shows the simulated phase noise of the oscillator at 9.949 GHz carrier. The simulation result shows that the phase noise at 1MHz offset is about -115.68 dBc/Hz.

The oscillating transient waveform is shown in the Figure 8. The amplitude of the gate transient voltage of the negative resistance transistor is slightly larger than the drain, so that the time of turning on and turning off of the transistor can be relatively shorter. Shortening the transient time of turning on and turning off can

reduce noise injection and improve phase noise.

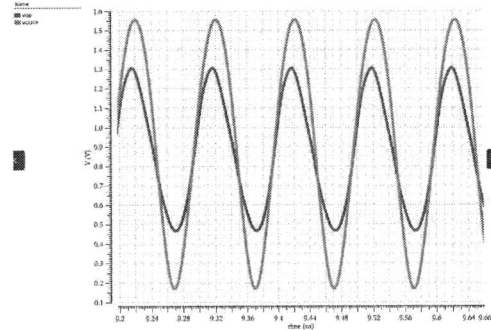

Fig. 8. Transient voltage waveform of the drain and gate of the negative resistance transistors at carrier frequency 9.949 GHz.

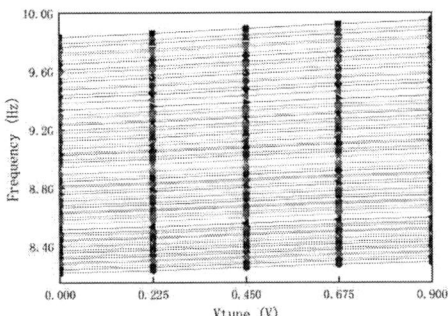

Fig. 9. Simulated tuning range of the proposed oscillator.

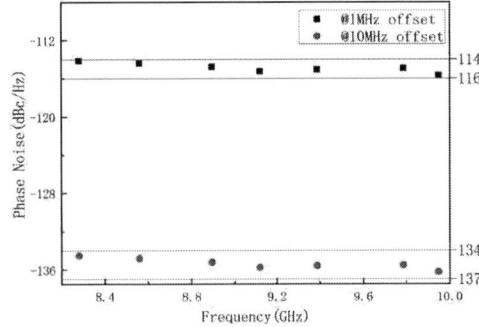

Fig. 10. Simulated phase noise at 1MHz and 10MHz frequency offset over the oscillating range.

The tuning curve estimated by pss simulation is shown in Figure 9. It can be seen that the oscillation frequency range is about 8.22-9.949 GHz. The FTR is

19%.

Figure 10 shows the phase noise at specific frequency after running pss and pnoise simulation.

The calculation formula of FoM is denoted as Equation (1). FoM_T takes into account the frequency range of the oscillator and can better compare the performance comprehensively. The calculation formula is denoted as Equation (2).

$$FoM = PN - 20\log(\frac{f_0}{\Delta f}) + 10\log(P_{DC}/1mW) \quad (1).$$

$$FoM_T = PN - 20\log(\frac{f_0}{\Delta f}FTR/10) + 10\log(P_{DC}/1mW) \quad (2).$$

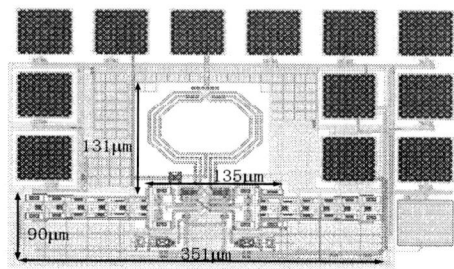

Fig. 11. Layout designing of the proposed oscillator.

TABLE I
PERFORMANCE SUMMARY AND COMPARISON WITH OTHER VCOS

Parameter	[7]	[8]	[9]	This work
Tech. (nm)	65	65	65	40
Freq. (GHz)	19	10.6	11.2	9.949
Power (mW)	46	330	2.2	27.072
PN@1MHz (dBc/Hz)	-111	-132	-107.73	-115.68
FTR (%)	10	7.5	9.6	19
FoM (dBc/Hz)	-181	-187	-185.3	-181.31
FoM_T (dBc/Hz)	-186.1	-184.5	-184.9	-186.88
Core Area (mm²)	0.09	0.45	0.07	0.05

The layout design of the VCO is shown in Figure 11. For differential signals, a symmetrical design solution is used to reduce mismatch.

Table I summarizes the simulation results of the proposed VCO and compares it with other works.

IV. CONCLUSION

A 8.22-9.949 GHz frequency range oscillator is designed in this paper. Fixed capacitors are in series with the properly biased varactor to improve the utilization efficiency of varactor. The transformer structure makes the oscillator gate voltage higher than drain results better phase noise. It is suitable for wide frequency application as well as requiring better phase noise local oscillation signal.

ACKNOWLEDGMENT

This work was supported by the Jiangsu provincial Key R&D Program under Grant BE2022052-2. (Corresponding author: Zhiqun Li).

REFERENCES

[1] Shu, Y., Qian, H. J., & Luo, X. (2020). 17.4 A 18.6-to-40.1GHz 201.7dBc/Hz FoM_T Multi-Core Oscillator Using E-M Mixed-Coupling Resonance Boosting. 2020 IEEE International Solid- State Circuits Conference - (ISSCC), 272–274.
[2] Li, Y., Wu, X., Gu, J., Gu, Q. J., Xu, Z., & Yu, X. (2022). A K-Band CMOS Standing Wave Oscillator Using Digital-Controlled Artificial Dielectric Differential Transmission Lines. IEEE Microwave and Wireless Components Letters, 32(10), 1195–1198.
[3] Franceschin, A., Riccardi, D., & Mazzanti, A. (2022). Ultra-Low Phase Noise X-Band BiCMOS VCOs Leveraging the Series Resonance. IEEE Journal of Solid-State Circuits, 57(12), 3514–3526.
[4] Ji, X., Wang, Y., Xia, X., & Guo, Y. (2021). A Capacitively Coupled Noise Circulating VCO. IEEE Microwave and Wireless Components Letters, 31(10), 1127–1129.
[5] Wagner, E. C., & Rebeiz, G. M. (2018). A 9.4–11.7 GHz VCO in 0.12 μm SiGe BiCMOS with −123 dBc/Hz Phase Noise at 1 MHz Offset for 5G Systems. 2018 IEEE Radio Frequency Integrated Circuits Symposium (RFIC), 16–19.
[6] Jain, S., Jang, S.-L., & Tchamov, N. T. (2016). Oscillation Mode Swapping Dual-Band VCO. IEEE Microwave and Wireless Components Letters, 26(3), 210–212.
[7] Sun, Y., Deng, W., Jia, H., He, Y., Wang, Z., & Chi, B. (2023). A Compact and Low Phase Noise Square-Geometry Quad-Core Class-F VCO Using Parallel Inductor-Sharing Technique. IEEE Journal of Solid-State Circuits, 58(10), 2861–2873.
[8] Zhang, S., Deng, W., Jia, H., Liu, H., Sun, S., Guan, P., & Chi, B. (2023). A 100 MHz-Reference, 10.3-to-11.1 GHz Quadrature PLL with 33.7-fs_rms Jitter and -83.9 dBc Reference Spur Level using a -130.8 dBc/Hz Phase Noise at 1MHz offset Folded Series-Resonance VCO in 65nm CMOS. 2023 IEEE Custom Integrated Circuits Conference (CICC), 1–2.
[9] Amin, Md. T., Yin, J., Mak, P.-I., & Martins, R. P. (2015). A 0.07 mm² 2.2 mW 10 GHz Current-Reuse Class-B/C Hybrid VCO Achieving 196-dBc/Hz FoM_A. IEEE Microwave and Wireless Components Letters, 25(7), 457–459.

2025 10th IEEE International Conference on Integrated Circuits, Design, and Verification (ICDV)

Synthesis of cosecant squared pattern antenna arrays using the methods of stacked beams

Le Nhu Thai
Viettel Aerospace Institute
Hanoi, Vietnam
thailn2@viettel.com.vn

Vu Thanh Cong
Viettel Aerospace Institute
Hanoi, Vietnam
congvt6@viettel.com.vn

La Tuan Anh
Viettel Aerospace Institute
Hanoi, Vietnam
anhlt131@viettel.com.vn

Nguyen Hoai Son
Viettel Aerospace Institute
Hanoi, Vietnam
sonnh13@viettel.com.vn

Le Thi Hang
Viettel Aerospace Institute
Hanoi, Vietnam
hanglt@viettel.com.vn

Abstract— **This article presents two methods for creating a Cosecant squared pattern (CSP). Regarding the first method, the method of synthesizing radiation patterns according to amplitude and phase distribution is presented. For the second method, use the phase beam scanning method in array antenna theory. The article also presents the results of an array of research and simulations of 16 elements Patch antenna array to demonstrate theoretical research results.**

Keywords— *cosecant squared pattern, array antenna, microstrip patch antenna*

INTRODUCTION

Antennas with a constant height pattern or CSP are specially designed for air-surveillance radars. These permit an adapted distribution of the radiation in the beam and cause a more ideal space scanning. This antenna pattern can get the required elevation coverage where the received power is independent of the radar range for a constant height target. It is a means of achieving a more uniform signal strength at the input of the receiver as a target moves with a constant height within the beam [1].

Moreover, the CSP radar exhibits exceptional resilience to certain forms of interference and clutter, enhancing its efficacy in environments where mitigating such factors is critical for accurate target detection. This resilience is attributed to the unique characteristics of the antenna pattern, which helps to suppress unwanted signals and focus on relevant targets within the surveillance area. Because of these advantages, CSP radar is widely used in surveillance radars and target detection, especially low-flying targets, for example: The Coast Watcher 100 system (France), YLC-4 system (Russia).

Research article synthesized CSP for array antennas with stacked beam method (beams of the form (sin(x))/x) [2] and used beam scanning principle in phased array antenna theory [3]. According to [4], CSP is obtained using a beam forming network based on a hybrid architecture, in which the main corporate feed linefeeds identical 11-element traveling wave slotted arrays. In addition, CSP synthesized by changing the shape of the reflector is widely used and is considered a popular and simple method [5,6].

I. SYNTHESIZE RADIATION PATTERN ACCORDING TO AMPLITUDE AND PHASE DISTRIBUTION

According to the theory of phased array antennas, we know the essence of the principle of creating a CSP or any shape pattern is just the sum of the beams together according to amplitude and phase (stacked beam). The description of M beams decomposition is shown in Fig.1.

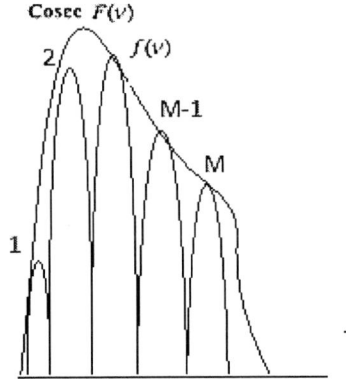

Fig. 1. Describe the method of synthesizing cosecant squared pattern

$$F(\theta) = z_0 \sum_{m=1}^{M} \frac{p^m}{m} \frac{\sin[\pi(z-m)]}{\pi(z-m)} + A_0 \frac{\sin(\pi z)}{\pi z} \qquad (1)$$

According to the stacked beam principle explained in Fig.1 and using the sub-beam synthesis equation (1), we obtain a CSP by synthesizing from M beams of the form $\frac{\sin(x)}{x}$ shown in Fig.2.

Fig. 2. The CSP by synthesizing from beams of the form $\frac{\sin(x)}{x}$.

The amplitude distribution function of the phased array antenna is determined as follows:

$$A(z) = \sum_{m=1}^{M} a_m \exp\left(\frac{-j2\pi mz}{L}\right) \qquad (2)$$

where:

979-8-3315-1550-8/25 $31.00 © 2025 IEEE

$$a_m = \frac{1}{M} \int_{-L/2}^{L/2} F(\theta) \overset{*}{f(\theta)} \quad (3)$$

From 2 and 3, the amplitude and phase distribution function for CSP is determined as follows:

$$A(n) = \sqrt{\left[A0 - \frac{1}{2}z_0 \ln(1 - 2p\cos(y) + p^2)\right]^2 + \left(z_0 \arctan \frac{p\sin(y)}{1-p\cos(y)}\right)^2} \quad (4)$$

$$\psi(n) = -\arctan \frac{z_0 \arctan\left(\frac{p\sin(y)}{1-p\cos(y)}\right)}{A_0 - 0.5z_0 \ln(1-2p\cos(y)+p^2)} \quad (5)$$

where: $Y(n) = \frac{2\pi}{D1}(n-1)$

with D1- the distance between the two elements.

Using (4) and (5), we obtain the amplitude and phase results of M=16 elements, shown in Fig. 3.

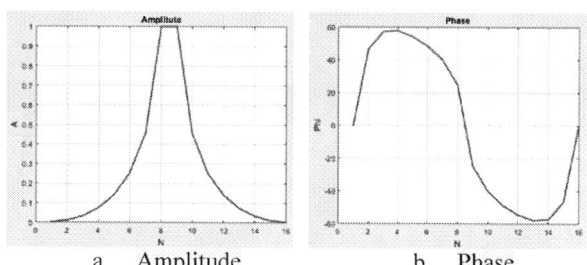

a. Amplitude   b. Phase

Fig. 3. The amplitude and phase distribution for 16 elements.

Corresponding to the values of amplitude and phase of the 16-element array shown in Fig.3, and using (6) we obtain a CSP in Fig.4.

$$RP(\theta) = \sum_{i=1}^{M} A(n).\exp^{i\psi(n)} \quad (6)$$

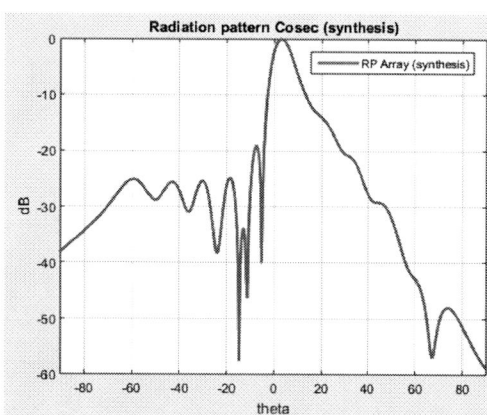

Fig. 4. CSP synthesis results of the first method.

II. SYNTHESIS OF CSP USING MULTI-BEAM STACKING METHOD IN BEAM SCANNING OF PHASED ARRAY ANTENNA THEORY

The essence of this method is to use the beam scanning principle in phased array antenna theory. Specifically, by changing the phase difference between elements of the antenna array, it will scan a certain angle, and then the phase difference will be determined by the following (7):

$$\Delta\psi = kd\sin(\theta) \quad (7)$$

where:

$$k = \frac{2\pi}{\lambda}$$

with d- distance between elements

Fig. 5 shows the method of obtaining the expected beam based on the phase scanning method. Realize that to obtain the expected CSP, we need a main beam offset at *theta_min* angle and a few adjacent side beams with smaller amplitudes.

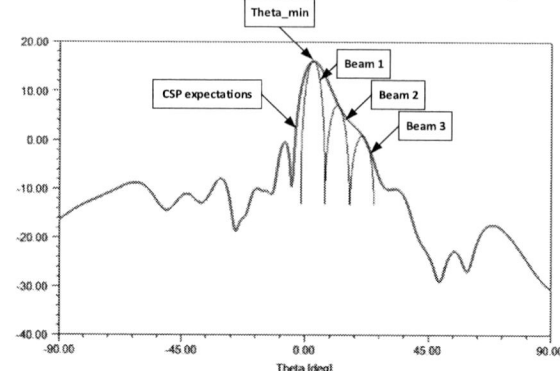

Fig.5. Describe the method of synthesizing CSP in phased array antenna theory

Fig. 6 shows the CSP simulation results with *theta_min*=2.5 degrees and the amplitude and phase parameters are shown in Table 1 using the sub-beam synthesis equation and phase optimization.

TABLE I. AMPLITUDE AND PHASE DISTRIBUTION OF CSP WITH THETA_MIN=2.5 DEGREES.

M	Amplitude	Phase
1	1	0
2	1	-11
3	1	-22
4	1	-33
5	1	-44
6	1	-55
7	1	-66
8	1	-77
9	1	-88
10	1	-100
11	1	-120
12	1	-140
13	1	-180
14	1	-200
15	1	-300
16	1	-400

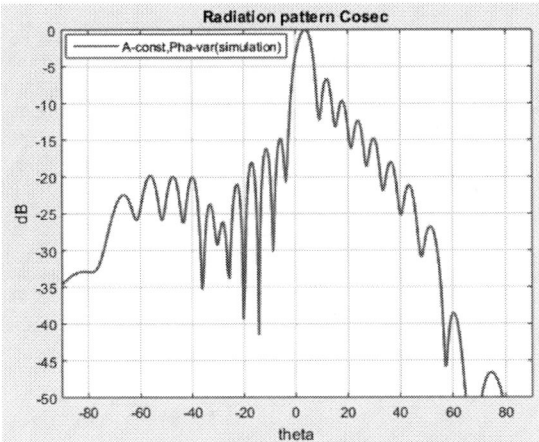

Fig. 6. Fig. 4. CSP synthesis results of the second method.

III. SIMULATION RESULTS

A. Antenna elements

To evaluate the CSP synthesis results of the above two methods, the evaluation on an antenna array consisting of 16 patch elements is used. The structure of an element is shown in Fig. 7.

Fig. 7. Structure of the element antenna

The results of studying the reflection coefficient and radiation pattern of an antenna element are shown in Fig.8.

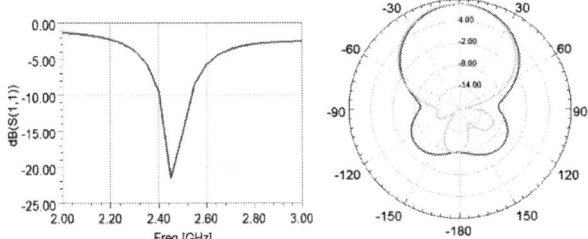

Fig. 8. The reflection coefficient and radiation pattern of an antenna element.

B. Antenna array with CSP

For the method of synthesizing CSP based on amplitude and phase distribution, we have simulation results of a CSP array of 16 elements in Fig. 9, corresponding to the amplitude and phase distribution presented in Table 2.

TABLE II. AMPLITUDE AND PHASE DISTRIBUTION OF FIRST METHOD

M	Amplitude	Phase
1	0.0041	0
2	0.0113	46.7928
3	0.0337	57.3411
4	0.0745	57.7168
5	0.1412	54.3814
6	0.2509	48.6931
7	0.4543	40.2752
8	1	24.8407
9	1	-24.8407
10	0.4543	-40.2752
11	0.2509	-48.6931
12	0.1412	-54.3814
13	0.0745	-57.7168
14	0.0337	-57.3411
15	0.0113	-46.7928
16	0.0041	0

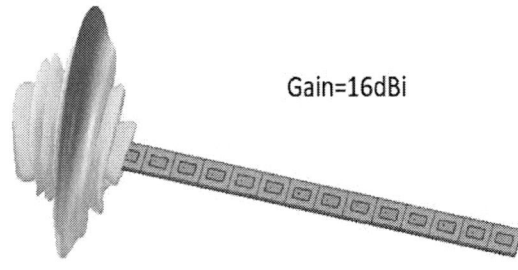

Fig. 9. The simulation results of the 3D radiation pattern of the 16 elements array for first method.

To synthesis of CSP using multi-beam stacking method in beam scanning of phased array antenna theory, we obtained the CSP simulation results of an array of 16 patch elements (Fig. 10) with phase and amplitude of shown in Table 1.

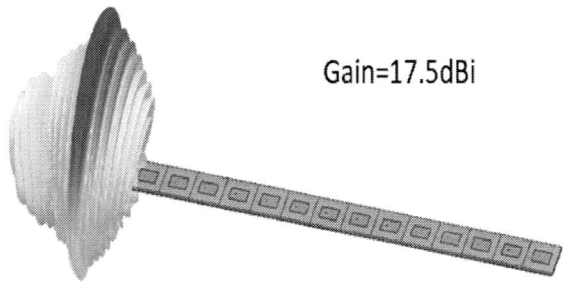

Fig. 10. The simulation results of the 3D radiation pattern of the 16 elements array for second method.

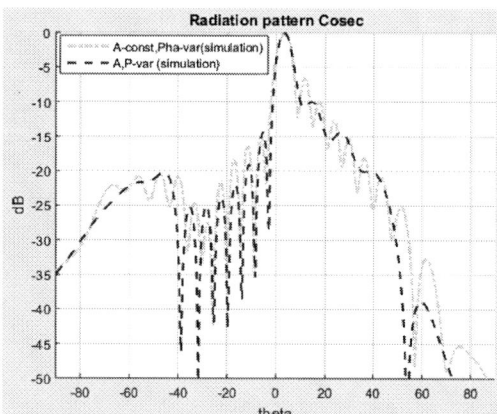

Fig. 11. Comparison of the simulation results of the radiation pattern of two methods.

The comparison results between the two methods are shown in Fig. 11. According to Fig. 9,10 and 11, it can be seen that the simulation results using the two methods mentioned above are quite similar to each other. Notice that for the results in Fig. 10, we have more gain than the results of the method shown in Fig. 9, but the smoothness of the beam is a bit worse.

IV. CONCLUSION

In this paper, we propose two methods that can synthesize CSP beams using amplitude and phase distributions and uniquely optimize phase distribution according to the beam scanning principle in phased array antenna theory. The research results are proven by simulation results showing the feasibility of the two presented methods.

REFERENCES

[1] https://www.radartutorial.eu/06.antennas/Cosecant%20Squared%20Pattern.en.html.

[2] И.А. Кирпичева, А.В. Останков, А.И. Рябчунов. Оптимизация шаблона для повышения эффективности синтеза антенной решетки с косекансной диаграммой направленности. Радиотехника и Связь. Т. 16. № 2. 2020

[3] Noach Amitay, Victor Galindo. Theory and analys of phased array antenna. New york: Wiley, 1972. Ch.1

[4] Antonio Morini, Davide Mencarelli, Marco Farina, Luca Pierantoni, Vincenzo Malaspina. Cosec2 hybrid travelling/resonant antenna formaritime surveillance applications. IET Microw. Antennas Propag., 2020, Vol. 14 Iss. 4, pp. 223-232

[5] Okan Yurduseven, Okan Mert Yucedag, Ahmet Serdar Turk. Cosecant-squared parabolic reflector antenna design for air and coastal surveillance radars. Engineering, Physics. 2010.

[6] Ali Akbar Dastranj, etc. Cosecant-squared pattern synthesis method forbroadband-shaped reflector antennas. ET Microw. Antennas Propag., 2014, Vol. 8, Iss. 5, pp. 328–336

2025 10th IEEE International Conference on Integrated Circuits, Design, and Verification (ICDV)

Nonlinear Capacitance Compensation Low Noise Amplifier and Mixer with UWB Anchor Antenna

Wen-Cheng Lai*

Dept. of Electrical Eng., Ming Chi Univ. of Technology, New Taipei City, Taiwan, R.O.C.
*Email: wenlai@mail.mcut.edu.tw; wenlai@mail.ntust.edu.tw

Abstract—A low-noise amplifier of nonlinear capacitance compensation using InGaP / GaAs HBT technology is demonstrated for ultrawide-band applications. An advanced methodology is proposed to achieve ultrawide-band gain and input impedance matching through extremely wide-band frequencies simultaneously. This article proposes low power low noise amplifier and charge-injection mixer with circular antenna that complies with UWB technical specifications, and presents simulated raw data of different models and measured results of a circular antenna with a diameter of 14mm.

Keywords—low-noise amplifier, nonlinear capacitance compensation, charge-injection mixer, circular anchor antenna

I. INTRODUCTION

According to the Federal Communications Commission (FCC) definition of ultra-wideband (UWB), bandwidth need between 3.1 ~ 10.6 GHz. With the continuous development and progress of radio frequency receiver frontend and antenna, the transmission rate of the UWB communication is more than 100Mbps, which is faster than IEEE 802.11 standard. Since the bandwidth of RF front-end circuit is ultra-wide, designing InGaP/GaAs HBT low-noise amplifier (LNA) of nonlinear capacitance compensation becomes a high challenge to meet the characteristics of high current gain and cut-off frequency. Furthermore, signals that meet the requirements of 10dB bandwidth need to better than 500MHz (1) or fractional bandwidth (FBW) need greater than 0.2 (2). This article provides frontend diagram of UWB RF as following in Fig. 1.

$$(f_H - f_L) > 500MHz \qquad (1)$$

$$(f_H - f_L) > 500MHz \qquad (2)$$

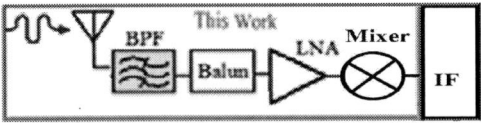

Fig. 1. The proposed frontend diagram of UWB RF.

II. CIRCUIT DESIGN IN DETAIL

The low noise amplifier of nonlinear capacitance compensation with passive balun consists of an input matching stage, a main amplification stage, and a buffer stage, respectively as shown in Fig. 2.

A. Input Matching Stage

The input impedance of the ultrawide-band low noise amplifier is because the preamplifier is passive balun from bandpass filter usually needs to be matched to 50 ohms. This architecture is very difficult input matching that use inductive feedback or use resistance feedback method of noise generated by itself. This proposal achieves a capacitive feedback input impedance matching [1]-[6] as including

inductor, capacitor, common emitter amplifier for composing 1st stage of partial gain amplifier. Impedance matching circuit of passive balun evaluated S parameters and layout micrograph placement as displayed in Fig. 3. The Fig. 4 introduces circuit of bandpass filter as below [7]-[10].

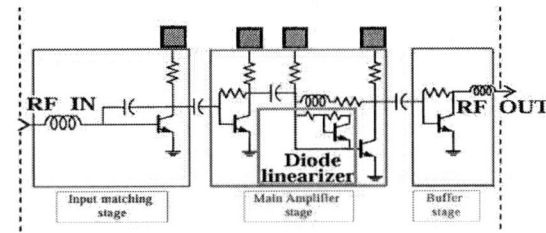

Fig. 2. The proposed LNA of nonlinear capacitance compensation architecture.

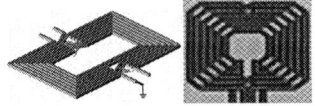

Fig. 3. The proposed circuit of passive balun and microphoto.:

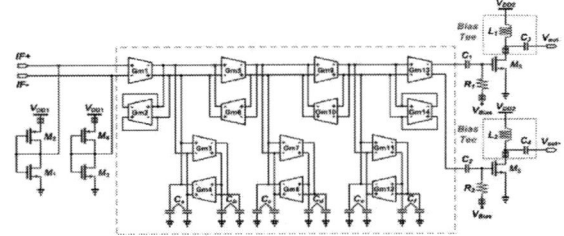

Fig. 4. Circuit of bandpass filter.

B. Main Amplifier Stage

The provided main amplifier stage has power gain enhancement and flat strategy for UWB frequency range covering. Therefore, this design proposes the resistor shunt-shunt feedback and adds amplification with inductor feedback that make the gain flatter and increases the bandwidth by 3dB [11]-[13]. The Fig. 5 illustrates nonlinear capacitance compensation idea comes from traditional diode linearizer (HBT process) that provides gain expansion and negative phase deviation characteristic. The advantage of this solution lies in its area-efficient, low power depletion, moderate linearity improvement and compensate the non-linearity due to the dramatically change of gate capacitance.

C. Buffer Stage

The placement of buffer stage located next stage as main amplifier to use shunt-shunt feedback with a wider frequency response. Because the frequency response of emitter follower is weakness at high frequencies.

979-8-3315-1550-8/25 $31.00 © 2025 IEEE

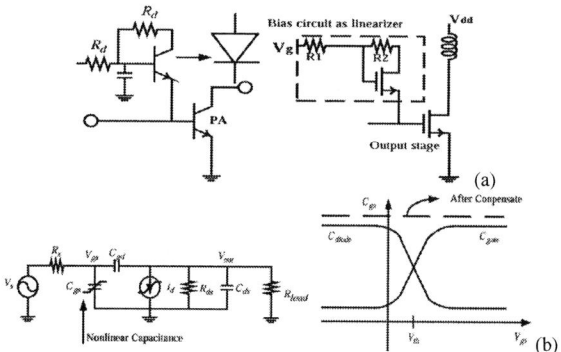

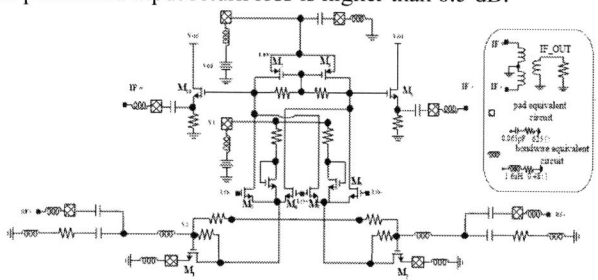

Fig. 5. The proposed nonlinear capacitance compensation (a) circuit (b) small signal analysis.

D. Charge-Injection Mixer

The proposed circuit of broadband CMOS double-balance mixer for UWB receiver is presented in Fig. 6. The broadband mixer is fabricated with the 0.18 μm 1P6M standard CMOS process. Measurement of the CMOS mixer is performed input return loss is higher than 8.5 dB.

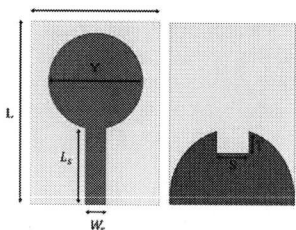

Fig. 6. Circuit of double-balance mixer for UWB receiver.

III. UWB ANCHOR ANTENNA

In [14] provides a circular shape antenna design. printed on a dielectric substrate and fed by a 50 microstrip line. present V shape antenna with the half circle with rectangle slot back ground. In Fig. 7 depicts an antenna that combines the two antenna shapes. S_{11}< -10dB bandwidth in the range of 4.06 to 10.97 GHz. At 4.94 GHz, the return loss reached 61.34 dB. This corresponds to band group #1, band #3 within the UWB spectrum allocation. This letter also used a chamber to measure the radiation pattern, obtaining both 3D and 2D experimental results [15]-[22].

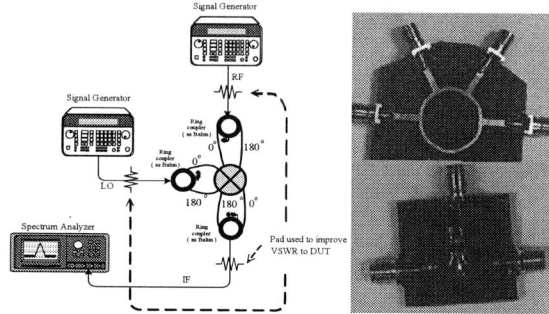

Fig. 7. The proposed Circular monopole of UWB anchor antenna with patch size.

The proposed circular monopole UWB patch antenna with dimensions of 28 x 23 x 1.6 mm³. The antenna radiates through the circular patch on the top and achieves a more compact design through a rectangular slot in the ground plane. The feedline has an impedance of 50 ohms and dimensions of Ls x Ws. These work features alter the current

distribution and affect the inductance and capacitance of the antenna, resulting in wide bandwidth characteristics. The advantage of this design lies in its simpler and more planar structure. The final antenna art was simulated using HFSS, and the detailed dimensions are listed in Table I.

TABLE I.   DIMENSIONS OF UWB PATCH ANTENNA.

Parameter	Dimensions (mm)
Substrate height, h	1.6
Substrate width, W	23
Substrate length, L	28
Y	14
Ls	11.64
Ws	3
T	3.6
S	4.2

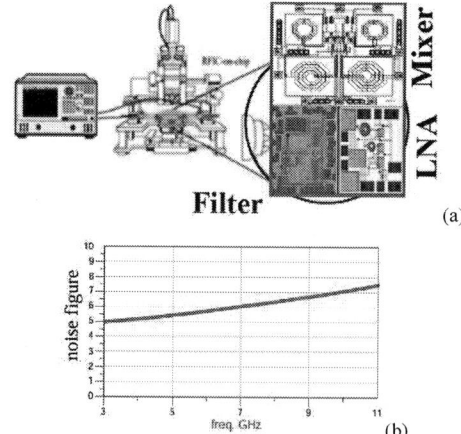

Fig. 8. Measured charge-injection mixer.

IV. MEASURED RESULTS

The measured double-balance mixer with microstrip ring coupler and transformer as using mini-circuit TC1-1T as displayed in Fig. 8. The Fig. 9 (a) and (b) shows noise figure results at insertion loss < -10dB at probing conductive measurement. It can be seen from the simulation and experimental measurement results that the S parameters are not very accurate. The reason is that the inaccurate transistor model leads to inaccurate S parameters, especially the power gain (S_{21}), which is quite different. In terms of input loss, simulation and measurement are very margin. The experimental results report that the input loss is less than -10dB from 2GHz to 15GHz, the 3dB bandwidth of 8GHz, and power gain of 17dB from 2GHz to 15GHz, 9.7GHz. The result is a very flat and high-gain ultra-wideband amplifier.

Fig. 9. The proposed noise figure results of LNA on wafer station.

This article simulated the S_{11} reflection loss using HFSS as shown in Fig. 10. The reflection loss is below -10 dB from 3.575 GHz to 13.349 GHz, with the lowest reflection loss of -24.66 dB occurring at 4.25 GHz. This paper found that the measured reflection loss is below -10 dB from 4.07 GHz to 10.817 GHz, with the lowest point of -61.299 dB occurring at 4.948 GHz. There is a discrepancy between the simulated and measured S_{11}, which is due to imperfect impedance matching, the omission of the SMA connector during simulation, and differences in the measured condition.

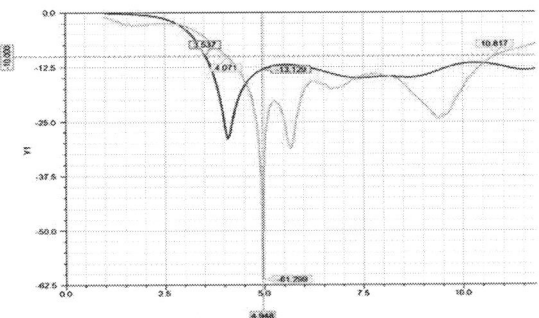

Fig. 10. The S_{11} reflection loss simulation.

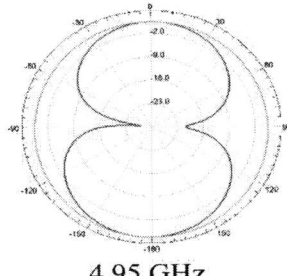

4.95 GHz

Fig. 11. Received simulation of radiation patterns for 4.95GHz.

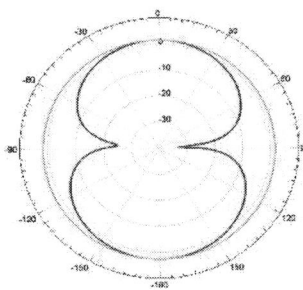

4.23 GHz

Fig. 12. Received simulation of radiation patterns for 4.23GHz

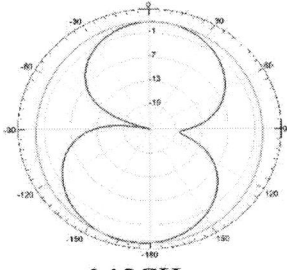

6.12GHz

Fig. 13. Received simulation of radiation patterns for 6.12GHz.

The radiation patterns observed at operating frequencies of 4.2 GHz, 4.96 GHz, and 6.125 GHz were analyzed as shown in Fig. 11, Fig. 12 and Fig. 13, respectively. The simulation reported that the S_{11} reflection coefficient was optimal at 4.21 GHz, and the 2D radiation pattern during the measurements as shown in Fig. 14, Fig. 15 and Fig. 16, respectively.

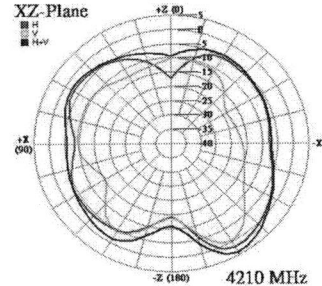

Fig. 14. Anechoic measured 4.21GHz results at shielding room.

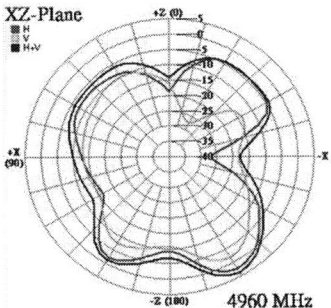

Fig. 15. Anechoic measured 4.96GHz results at shielding room..

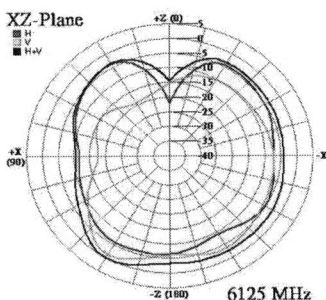

Fig. 16. Anechoic measured 6.12GHz results at shielding room..

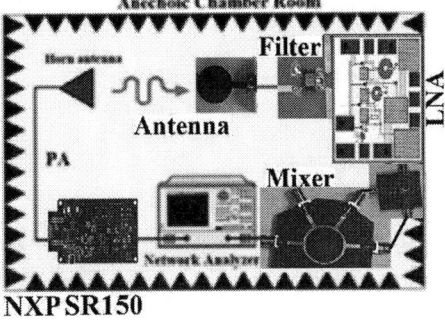

Fig. 17. Received 3D measurement at anechoic chamber.

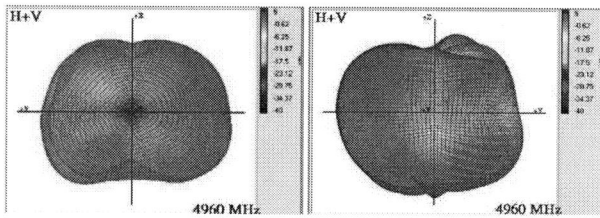

Fig. 18. The 3D patterns of received radiation at 4.2 GHz.

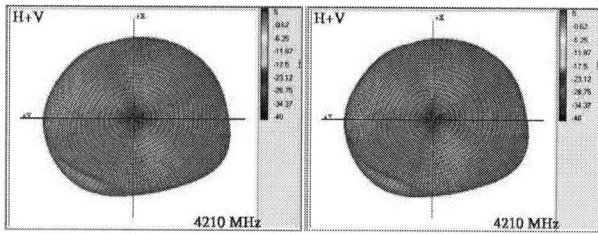

Fig. 19. The 3D patterns of received radiation at 4.96 GHz.

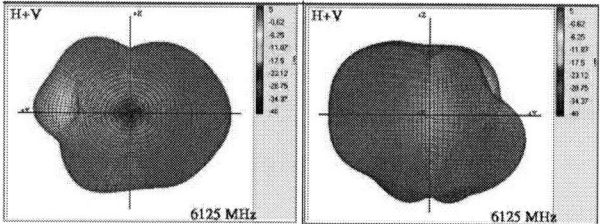

Fig. 20. The 3D patterns of received radiation at 6.125 GHz.

V. CONCLUSIONS

For the low noise amplifier, three stages were used obtaining a good input/output matching. The design and measurement of the 14mm circular UWB antenna demonstrate its compliance with UWB low noise amplifier of nonlinear capacitance compensation. The antenna achieved a bandwidth of 4.07 GHz to 10.817 GHz, with the lowest return loss of -61.299 dB at 4.948 GHz, which aligns with standard band group #1, band #3 in the UWB spectrum. Simulations and measurements revealed some discrepancies, likely due to conditional parameters and impedance mismatches. Radiation patterns were analyzed at frequencies of 4.21 GHz, 4.96 GHz, and 6.12 GHz, with the 3D measurements obtained using an anechoic chamber as shown in Fig. 17, Fig. 18, Fig. 19 and Fig. 20, respectively. This design setup power ampilfer of UWB evaluation module (NXP Trimension® Ultra-Wideband Development Kits: SR150-Based) for transmitter condition at anechoic chamber. The overall performance of the antenna performs it is suitable for UWB applications, though further refinement could enhance its consistency between simulation and physical measurement.

REFERENCES

[1] Hosaein Hashemi, *et al.*, "con-current multiband Low-noise Amplifiers-theory,Design,and Apllications," *IEEE Trans. on Microwave Theory and Techniques*, vol. 50, no.1, Jan. 2002.

[2] W.-C. Lai," Integrated Circuits of RF Front-End with Antenna for Millimeter-Wave Receiver," *15th Global Symposium on Millimeter-Waves & Terahertz*, 2024.

[3] S. Wu and B. Razavi, "A 900-MHz/1.8-GHz CMOS receiver for dual bandapplications," *IEEE JSSC*, vol.33, no.12, pp.2178-2185, Dec. 1998.

[4] W.-C. Lai, "Mixing Wake-up Receiver of Low Noise Amplifier with Back-gated QVCO," *Int'l Conf. on Electrical, Computer and Energy Techn.*, 2021.

[5] J.-F. Huang, *et al.*, "1.5V 6–10 GHz Broadband CMOS LNA and Transmitting Amplifier for DS - UWB Radio Received," *IEICE Trans. on Electron*, vol. E94-C no.11 pp.1807-1810, Nov. 2011.

[6] W.-C. Lai, "RF Receiver Front-End Chip Design with Antenna for Ultra-wideband Applications," *IEEE 21st Int'l Conf. on Communication Techn.*, 2021.

[7] T. Islam, *et al.*, "A 3.5 GHZ Low-Noise Amplifier with TR Switch Package," *18th Int'l Microsystems, Packaging, Assembly and Circ. Techn. Conf.*, 2023.

[8] C.-H. Hsu, *et al.*, "A 18-24GHz Low Noise Amplifier with Filter," *IET Int'l Conf. on Eng. Tech. and App.*, 2022.

[9] W.- C. Lai, "Chip Design of an All- Digital Frequency Synthesizer with Reference Spur Reduction Technique for Radar Sensing," *Sensors*, 22 (7), 2570, Mar. 2022.

[10] T. Islam, *et al.*, "Ku-band low noise amplifier with TR switch," *IET Int'l Conf. on Eng. Tech. and App.*, 2023.

[11] W.- C. Lai, *et al.*, "A Triple- Band Voltage- Controlled Oscillator Using Two Shunt Right- Handed 4th- Order Resonators," *Journal of Semiconductor Technology and Science*, vol. 16, no. 4, pp. 506-510, Aug. 2016.

[12] J. -C. Wang, *et al.*, "Editorial for the Special Issue on Recent Advances in Microwave Components and Devices," *Micromachines*, 15(2), 180 , Jan. 2024.

[13] W.- C. Lai, *et al.*, "An X-Band GaN HEMT Oscillator with Four-Path Inductors," *Applied Computational Electromagnetics Society Journal*, vol. 35, no. 9, Sept., 2020.

[14] J. Liang, *et al.*, "Study of a Printed Circular Disc Monopole Antenna for UWB Systems," *IEEE Tran. on Antennas and Propagation*, vol. 53, no. 11, Nov. 2005.

[15] W.- C. Lai, *et al.*, "Integrated RFIC On-Chip and GPS Antenna with Human Body for Wrist and Wearable Communication Applications," *Applied Computational Electromagnetics Society Journal*, vol. 31, no. 9, pp. 1084-1091, Sept. 2016.

[16] J. -F. Huang , *et al.*, "Design of A Compact Printed Double-Sided Dual-Band Dipole Antenna by Fdtd for Wifi Application," *Microwave and Optical Technology Letters*, vol.55, no. 8, pp. 1845–1851, Aug. 2013.

[17] W.- C. Lai," RF Front-End CMOS Receiver with Antenna for Millimeter-Wave Applications," *6th Int'l Conf. on Integrated Circ. and Microsystems*, 2021.

[18] O. P. Kumar, *et al.*, "A Compact Dual-Band Notched UWB Antenna for Wireless Applications," *Micromachines*, 13(1), 12, 2022.

[19] W.- C. Lai, "Integrated Homodyne Receiver Chip Design with Dual-Band Antenna," *IEEE 14th Int'l Conf. on ASIC*, 2021.

[20] Yanfei Mao, *et al.*, "A 24 GHz End-Fire Rod Antenna Based on a Substrate Integrated Waveguide," *Sensors*, 25(5), 1636, Mar. 2025.

[21] J. -F. Huang , *et al.*, "" Design of a Printed Dipole Array Antenna with Wideband Power Divider and RF Switches," *Microwave and Optical Technology Letters*, vol.55, no. 10, pp. 2410–2413, Oct. 2013.

[22] W.- C. Lai," RF Receiver Dedicated Short Range Communication Chip Design with Antenna, *SBMO/IEEE MTT-S Int'l Microwave and Optoelectronics Conf.*, 2021.

Dynamic Queue Management and Packet Loss Mitigation in P4-Enabled Data Planes

Bui Ngoc Thanh Binh, Tran Nguyen Tuan Kiet, and Nguyen Viet Ha

University of Science, Ho Chi Minh City, Vietnam
Vietnam National University, Ho Chi Minh City, Vietnam
Email: 19207049@student.hcmus.edu.vn, 19207084@student.hcmus.edu.vn, nvha@hcmus.edu.vn

Abstract—**Modern network infrastructures, including data centers, 5G edge nodes, and large-scale IoT networks, frequently handle dynamic and sporadic data traffic. Meanwhile, the traditional static-threshold-based congestion control often leads to excessive packet loss and inefficient resource allocation. This paper introduces a P4-based adaptive queue management approach that dynamically adjusts congestion thresholds, leverages QoS-driven recirculation, and enables rapid fault recovery. The proposed method adapts congestion thresholds based on queue depth, selectively reprocesses high-priority packets, and generates real-time summaries when packet losses exceed critical levels. Experimental results on the BMv2 software switch, across traffic loads from 50 Mbps to 300 Mbps, demonstrate that the proposed approach reduces packet loss by 59% compared to traditional static-threshold congestion control and 44.6% relative to RED, while achieving a recovery rate of 67% faster and a throughput 27% higher than the traditional method. Although the approach incurs an additional latency of 1.8 ms over the traditional method and 1.9 ms over RED due to recirculation overhead, it significantly improves packet retention and congestion management, effectively mitigating packet loss.**

Index Terms—**P4, SDN, queue management, adaptive threshold, network optimization.**

I. INTRODUCTION

The rise of cloud computing, artificial intelligence, and the Internet of Things (IoT) has placed unprecedented demands on network infrastructure, requiring high throughput, low latency, and stringent Quality of Service (QoS). Traditional network architectures, which rely on static threshold-based loss detection, rigid routing, and reactive congestion control, struggle to adapt to fluctuating traffic. This leads to prolonged queuing delays and critical packet loss in applications such as real-time streaming and edge computing.

To address these challenges, programmable data planes—particularly those enabled by the P4 language—offer dynamic, real-time packet processing for traffic control. Unlike fixed-function ASIC-based switches, P4-enabled switches provide software-defined logic for congestion handling and adaptive queue management. Studies have demonstrated that P4-based networking significantly mitigates congestion and accelerates failure recovery, highlighting the flexibility of software-based packet pipelines over static hardware flows.

This research is funded by Vietnam National University, Ho Chi Minh City (VNU-HCM) under grant number B2023-18-01.

Our approach integrates three congestion-mitigation techniques in P4. First, adaptive threshold adjustment dynamically monitors queue depth, lowering the loss-detection threshold to reduce false positives under low traffic and raising it to trigger early congestion alarms under heavy load. Second, selective packet recirculation reroutes high-priority packets via a dedicated RECIRC_PORT when congestion thresholds are exceeded, preventing unnecessary drops and enabling network recovery from transient bottlenecks. Finally, digest-triggered recovery tracks transmitted packets against an expected count, generating a digest to assist the controller in taking prompt remedial actions upon detecting significant loss.

We implement and evaluate our approach on the BMv2 software switch, which is widely used to rapidly prototype P4-based data plane innovations. While BMv2 enables flexible experimentation, real-world deployment would leverage hardware acceleration to minimize latency. Our technique remains compatible with such platforms, making it suitable for future FPGA-based implementations where predictable performance and energy efficiency are essential.

Extensive BMv2 tests with traffic loads of 50M, 100M, and 300M demonstrate that our adaptive threshold mechanism improves queue utilization, lowers queuing latency, and increases overall throughput, outperforming traditional static congestion control. This adaptable queue management system offers a scalable, energy-efficient solution for modern infrastructures, including autonomous vehicles, live streaming, and IoT networks, while laying the foundation for advanced FPGA-based deployments

II. METHODOLOGY

Programmable data planes have emerged as a key enabler for high-performance and adaptive networking. Unlike traditional network devices that rely on fixed-function ASICs, programmable switches allow fine-grained control over packet processing logic, enabling dynamic adjustments to congestion control, queue management, and forwarding policies. The P4 language has been widely adopted for designing such systems, offering a high-level abstraction to define packet processing rules directly at the switch level. In this work, we leverage P4-enabled data planes to implement an adaptive queue management mechanism that dynamically responds to congestion events. Our approach integrates three key methods: *adaptive threshold adjustment*, *selective packet recirculation*, and *digest-triggered recovery*. These methods collectively reduce packet loss, optimize queue utilization, and enhance overall network performance. Although AQM techniques have been widely studied for cloud data centers, many traditional approaches lack adaptability for P4-programmable networks.

A. Adaptive Threshold Adjustment

Traditional congestion detection mechanisms rely on static thresholds, which can be ineffective under fluctuating network traffic [3]. To address this limitation, we implement an adaptive threshold adjustment scheme that dynamically modifies the packet loss detection threshold T_n based on the current queue depth Q. In our P4-based design, the queue depth Q represents the number of packets waiting in the switch buffer and is stored in meta.queue_depth. On Tofino switches or Xilinx FPGA, Q is retrieved from telemetry data, whereas on BMv2 switches it is estimated using packet arrival/departure counters and timestamps. By monitoring Q, the system dynamically adjusts the loss detection threshold, thereby improving congestion control in real-time. The threshold is updated iteratively as:

$$T_n = T_{n-1} + \Delta T, \tag{1}$$

where the adjustment factor ΔT is defined as:

$$\Delta T = \begin{cases} -(Q - Q_{high}), & \text{if } Q > Q_{high}, \\ +(Q_{low} - Q), & \text{if } Q < Q_{low}, \\ 0, & \text{otherwise.} \end{cases} \tag{2}$$

The upper and lower queue depth thresholds are set as $Q_{high} = 100$ and $Q_{low} = 50$, respectively. The choice of $Q_{high} = 100$ (40% of a 256-packet buffer) and $Q_{low} = 50$ (20%) is based on several key factors. First, to prevent bufferbloat, we maintain queue depths below 50% of capacity, as recommended by RFC 7567 [4], thereby avoiding round-trip time (RTT) spikes exceeding 100 ms, which are common in static threshold systems [5]. Additionally, these values align with hardware realities, ensuring compatibility with commercial switches such as Broadcom Tomahawk, which typically have buffer sizes ranging from 128 to 256 packets [6], while allowing for a practical utilization range of 40–80%. This configuration facilitates efficient burst absorption by allocating a 50-packet buffer to manage transient micro-bursts, which generally persist for under 10 ms and constitute less than 30% of the total load. The system dynamically adjusts queue depth thresholds to balance congestion sensitivity and packet loss prevention, ensuring adaptive queue management suited to fluctuating traffic conditions. When $Q > Q_{high}$, congestion is detected and the threshold is reduced to initiate early loss detection (if $Q = 120$, then $\Delta T = -(120 - 100) = -20$); conversely, when $Q < Q_{low}$, the threshold is increased to reduce false positives (if $Q = 30$, then $\Delta T = +(50 - 30) = +20$). If Q is positioned between Q_{low} and Q_{high}, no modifications occur, maintaining stability. This approach continuously adjusts T_n according to

the observed queue depth, effectively balancing sensitivity to congestion and the prevention of packet loss.

B. Selective Packet Recirculation

A key characteristic of programmable networks is selective packet recirculation, enabling the reprocessing of packets in the data plane rather than immediate disposal or forwarding. This process has been investigated in prior studies, illustrating its efficacy in congestion management. Adaptive queue management, which involves dynamically altering packet-handling decisions based on network conditions, is particularly beneficial.

As illustrated in Fig. 1, selective packet recirculation in P4 utilizes a designated recirculating port 511 and (ingress-to-egress) cloning (Clone.I2E) to reinject packets into the ingress pipeline. The diagram depicts how an incoming packet, upon meeting specific recirculation criteria—such as excessive queue depth or high-priority QoS classification—is not immediately forwarded but instead undergoes Clone.I2E cloning, enabling further assessment within the pipeline.

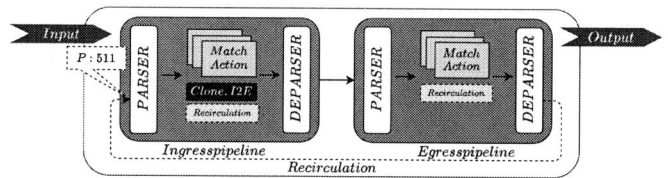

Fig. 1. Selective packet recirculation process in P4 pipelines

The system assesses two main criteria, queue depth and QoS priority, to determine whether a packet requires recirculation. The first criterion relates to queue depth. When the queue depth Q above the threshold Q_{high} (e.g., 100 packets), the packet is marked for recirculation to prevent excessive congestion. This technique reallocates load, reducing sudden packet loss and improving network responsiveness. The second condition addresses QoS priority. Packets of high priority, categorized within QoS classes 6–7, are assigned DSCP values 48–63 in the IPv4 header, as stipulated in RFC 2474 [7]. These parameters pertain to Expedited Forwarding (EF) and Class Selector (CS6, CS7), utilized for latency-sensitive traffic, including VoIP and video conferencing.

To guarantee minimal latency, DSCP values 48–63 are processed via the P4 pipeline, enabling high-priority packets to be recirculated before to arriving to Q_{high}. This emphasizes essential traffic management while ensuring congestion regulation. By including both queue depth and QoS priority, the system averts superfluous recirculation of low-priority packets while dynamically regulating high-priority traffic. Recirculated packets re-enter the ingress pipeline, where metadata such as queue depth and loss thresholds are updated. The subsequent match-action stage ascertains whether packets are forwarded, recirculated, or discarded, therefore minimizing premature packet loss and alleviating congestion. Should congestion continue, packets may experience further recirculation cycles. A recirculation counter is employed to avert infinite cycles. It commences at 0 and increases with each recirculation. If it surpasses three iterations, the packet is deleted to avert buffer congestion illustrated in Fig. 2.

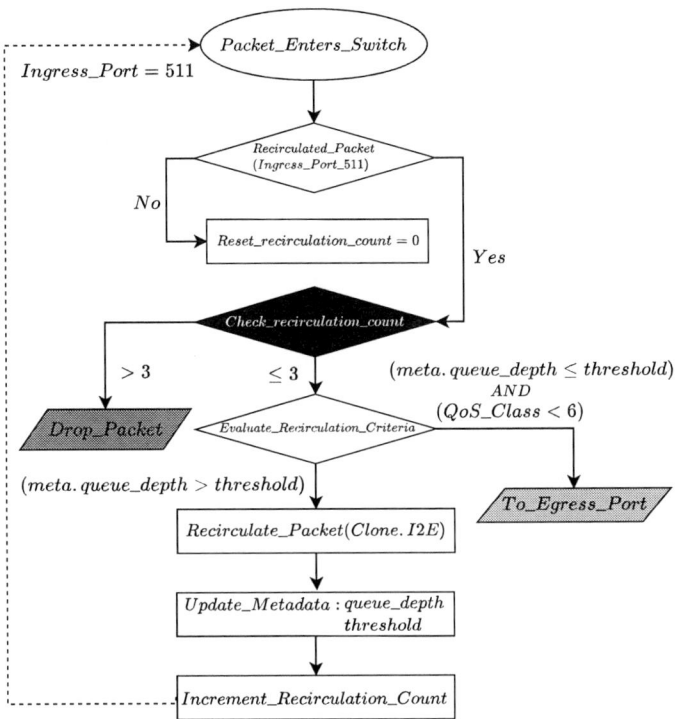

Fig. 2. Loop Prevention Flowchart

Restricting recirculation to three rounds is grounded in practical reasons. One recirculation alone might not be enough to completely adjust the threshold and enhance congestion management. Permitting a maximum of three iterations gives the system sufficient chances to adjust without incurring too much processing burden. If recirculation were not limited, packets might become trapped in endless reprocessing loops, resulting in buffer congestion and diminished switch performance. Therefore, limiting the recirculation count to three achieves a good balance between being responsive and maintaining efficiency.

Including selective recirculation into the P4 pipeline helps to use queues more effectively, leading to a congestion control mechanism that is more responsive and adaptive, which in turn significantly lowers packet loss. This method is especially beneficial when traditional static threshold-based congestion control systems struggle to adjust to changing traffic patterns. Dynamic recirculation offers detailed packet management features, allowing switches to effectively control network congestion and optimize resource allocation in real-time.

C. Digest-Triggered Recovery

In contemporary network monitoring systems, identifying and rectifying packet loss is an essential function, particularly in high-performance and mission-critical networks. The *Digest-Triggered Recovery* technique is essential in this procedure by utilizing acquired information. from *probe* packets to monitor network health and trigger recovery actions when necessary. This technique functions by transmitting a *digest alert* to the controller upon the detection of substantial packet loss or network irregularities.

1) The Digest is sent at a specific time.: A digest is sent when the number of missing packets, computed as the difference between total expected and received packets, exceeds a predefined threshold (`loss_threshold`). This triggers a notification to the controller for further action.

If the number of missing packets exceeds the predefined threshold, the system classifies the digest type based on its severity. Urgent digests are only triggered when packet loss surpasses a critical threshold, ensuring rapid response to severe congestion.

A normal digest (`digest_type = 0`) is sent when packet loss is within the acceptable range, whereas an urgent digest (`digest_type = 1`) is triggered for severe packet loss, ensuring rapid response.

2) Relationship with Adjust Threshold and Recirculation: The *Digest-Triggered Recovery* mechanism is closely linked to the *Adjust Threshold* action, which dynamically modifies the packet loss threshold according to real-time network conditions. This adjustment ensures the system stays responsive to changing network traffic and congestion levels.

The threshold is modified according to the `queue_depth`, which is a parameter kept in the packet metadata. The queue depth indicates how many packets are pending processing by a switch or network device. When the queue depth surpasses a specific limit (for instance, `queue_depth > 100`), the threshold is lowered to enable faster identification and resolution of packet loss. If the queue depth is low, the threshold is raised to prevent unnecessary alerts or false positives.

$$if \ queue\_depth > 100 \quad then \quad reduce\_threshold \quad (3)$$

In addition, the system includes recirculation, a recovery procedure wherein packets deemed lost or incorrectly handled are reintegrated into the network for additional management. In the streamlined iteration of our approach (without QoS checks), recirculation is activated when the network experiences high congestion—specifically, when `queue_depth` exceeds 100 and the number of `missing_packets` surpasses the predefined threshold. Under these conditions, the system redirects the affected packet to a designated `recirculation port 511`. By reprocessing lost packets through this port, the network can quickly recover from disruptions and mitigate the impact of packet loss.

3) Sending Digest Through P4Runtime/gRPC Ports: Digests are sent to the controller via the P4Runtime/gRPC interface (default ports 9090–9092 for BMv2 switches). The older Thrift-based APIs have largely been superseded by P4Runtime and gRPC in recent versions of BMv2. The controller is usually a PC host running BMv2 (Behavioral Model v2) or another P4Runtime client. By using BMv2, the controller can analyze

979-8-3315-1550-8/25 $31.00 © 2025 IEEE

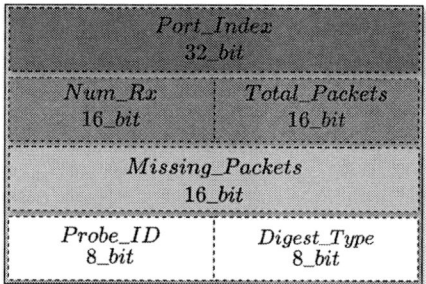

Fig. 3. The Link Loss Digest Packet Architecture.

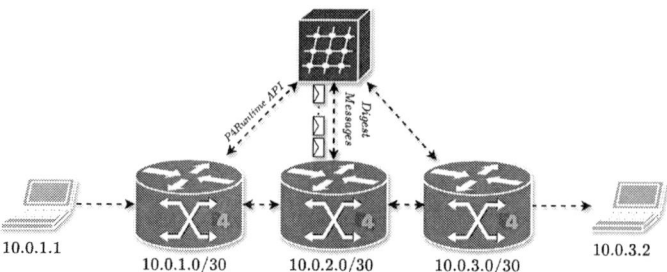

Fig. 4. Experimental Topology.

the received digest and respond by adjusting network configurations or triggering further recirculation when necessary.

4) LinkLossDigest_t Packet Architecture: The Digest-Triggered Recovery mechanism relies on probe packets to monitor packet loss and network health. Inspired by the work in [8], this system introduces a *Probe header* between the IP header and the OSI layer 2 header. Similar to this approach, our probe packets include fields such as `Probe ID`, `Num Tx`, and `Total Tx`, which help track transmission cycles and detect link loss in real-time. However, our system extends this mechanism by integrating an adaptive loss threshold and a digest-based recovery strategy to enhance congestion response and network resilience.

- `port_idx (bit<32>)`: The index of the port where the packet loss occurred.[1]
- `num_rx (bit<16>)`: The number of packets that have been successfully received.
- `total_pkt (bit<16>)`: The total number of packets expected to be received.
- `probe_id (bit<8>)`: The ID of the probe packet that triggered the report.
- `digest_type (bit<8>)`: The type of digest, with type 0 indicating a normal digest and type 1 indicating an urgent digest.
- `missing_packets (bit<16>)`: The number of packets that were lost during transmission.

These fields enable the controller to pinpoint the location of the loss, assess its severity, and correlate the digest with specific probe sessions.

In the data plane, the number of missing packets is computed using a simple arithmetic operation:

$$missing\_packets = total\_pkt - received\_count \quad (4)$$

where `total_pkt` represents the total number of packets expected, and `received_count` is the number actually received. Based on this value, the system determines the digest type as follows:

$$digest\_type = \begin{cases} 0, & if\ missing\_packets \leq threshold \times 2, \\ 1, & if\ missing\_packets > threshold \times 2. \end{cases} \quad (5)$$

[1]The `port_idx` field is cast to 32 bits for compatibility with digest processing systems, though the physical port index uses only 9 bits.

TABLE I. Traffic Profile Descriptions

Traffic Load	Description
50 Mbps	Background monitoring traffic
100 Mbps	Mixed web/streaming traffic
300 Mbps	Stress test with `netem`, introducing 20% packet loss or 5 ms delay

The factor of 2 ensures urgent digests are only triggered for severe congestion, avoiding false alarms caused by transient fluctuations. This prevents unnecessary processing overhead while maintaining responsiveness to critical network issues.

Furthermore, to ensure the integrity and authenticity of the digest message, cryptographic techniques can be integrated into the process. In addition to using CRC16 for error detection, a cryptographic hash (such as SHA-256) or an HMAC can be computed over the digest fields. While cryptographic techniques such as HMAC (SHA-256) provide strong integrity guarantees, it is important to note that P4 does not natively support complex cryptographic operations. Instead, simple integrity mechanisms like CRC16 can be implemented in the data plane, while cryptographic authentication should be handled by the control plane.

III. EXPERIMENTAL RESULTS

A. Simulation Setup

All experiments run on a Linux WSL platform (Intel Core i7-11800H, 8 vCPUs, 8 GB RAM). Due to CPU limitations in the emulation environment, throughput tests are restricted to 50 Mbps, 100 Mbps, and 300 Mbps to prevent saturation. Latency is measured using `ping`, while packet loss and throughput are evaluated using sFlow sampling and `iperf3` UDP streams, respectively. Scapy scripts inject probe packets with QoS markings, and `tc` Network Emulation simulates artificial loss. Our experiments use Mininet with the BMv2 software switch, arranged in a linear topology (Fig. 4) connecting two hosts (h1, h2) via three switches (s1--s2--s3). Each switch runs a P4-based pipeline that dynamically adjusts congestion thresholds ($Q_{low} = 50$, $Q_{high} = 100$), applies QoS-aware packet recirculation via RECIRC_PORT, and triggers digest-based loss recovery to minimize packet drops.

979-8-3315-1550-8/25 $31.00 © 2025 IEEE

TABLE II. Adaptive Threshold Performance

Metric	50 Mbps	100 Mbps	300 Mbps
Queue Depth (pkts)	27 ± 2	64 ± 3	120 ± 5
Threshold (units)	$10 \to 20$	10 (fixed)	$10 \to 5$
Adjustment Freq. (Hz)	0.25	0	0.15
Latency (ms)	1.5 ± 0.4	3.6 ± 0.5	14.5 ± 3.8

B. Metrics Collection and Evaluation

Packet loss is measured via sFlow sampling at switch egress ports, latency via ping probes (0.1s interval), and throughput using iperf3 UDP streams. Each scenario runs for 10s, repeated five times, with mean ± std.dev. reported. To ensure a fair comparison, we evaluate packet loss, latency, and throughput under different congestion levels and compare our adaptive queue management approach against RED (Random Early Detection). RED is configured using Linux tc, with a 1000-packet queue, dynamic drop thresholds, and a 0.02 drop probability. For high-priority traffic (QoS 6–7), DSCP values 48–63 are assigned in the IPv4 DiffServ field, with recirculation applied via ternary matching in the P4 pipeline.

C. Evaluation

This section evaluates our P4-based adaptive system under varying traffic conditions, focusing on (i) threshold adjustment effectiveness, (ii) packet loss mitigation through selective recirculation, and (iii) digest-triggered recovery for rapid fault handling. We also present an overall performance comparison and analyze system behavior during sudden traffic surges.

1) Adaptive Threshold Performance: Table II summarizes how the system dynamically adapts to different traffic loads. At 50 Mbps, the threshold increases (from 10 to 20) to prevent unnecessary congestion alarms. At 100 Mbps, it remains constant at 10, while at 300 Mbps, it decreases to 5 to detect congestion earlier. The adjustment frequency is 0.15 Hz at 300 Mbps, reflecting real-time reactivity. Although latency increases up to 14.5 ms, it remains within an acceptable range. Fig. 5 illustrates these variations.

2) Comparison with RED: Fig. 6a shows the loss comparison among the static-threshold scheme, RED, and our Adaptive approach under a 300 Mbps load. RED reduces packet loss compared to the static scheme (14.8% vs. 20.5%), yet its loss rate still exceeds that of the Adaptive method (8.7%). Meanwhile, Fig. 6b illustrates that while RED's early drop mechanism achieves lower latency (3.2 ms), the Adaptive approach experiences a slightly higher latency (4.5 ms) due to prioritizing recirculation over packet drop, resulting in additional processing delay. Finally, Fig. 6c demonstrates that this extra processing facilitates higher throughput (270 Mbps) in the Adaptive method, ensuring better packet retention and making it more suitable for real-time applications. These results confirm that in-network programmability can effectively respond to real-time queue conditions and prioritize critical flows, outperforming traditional AQM methods.

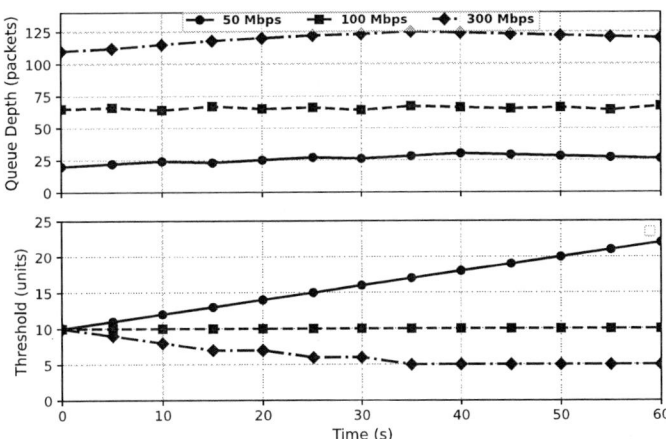

Fig. 5. Adaptive threshold adjustment and queue depth under different traffic loads.

TABLE III. Effectiveness of Packet Recirculation

Scenario	Recirc. Rate	Loss Reduction	Latency
No Recirc.	0	20–22%	3.3 ± 0.5 ms
QoS 6–7 Only	75–90%	$2.0 \pm 0.5\%$	5.0 ± 0.4 ms
All Traffic	45%	$8.0 \pm 2.0\%$	5.1 ± 0.5 ms

TABLE IV. Digest-Triggered Recovery Metrics

Cond.	Lost Pkts	Digest Type	Recovery	Throughput
Normal	$\leq 2 \times T$	Type 0	130 ± 15 ms	200 ± 10 Mbps
Urgent	$> 2 \times T$	Type 1	70 ± 10 ms	240 ± 15 Mbps

3) Packet Recirculation Impact: To assess the impact of recirculation on packet loss mitigation, Table III compares different configurations. Without recirculation, packet loss remains 20–22% under 300 Mbps traffic. Recirculating all traffic, with an estimated recirculation rate of approximately 45%, reduces loss to 8%, while recirculating only QoS 6–7 traffic, with a higher recirculation rate of 75–90%, further reduces loss to 2%, improving packet retention. However, the latter approach increases latency slightly due to additional processing.

4) Digest-Triggered Recovery Performance: For fault handling, our system employs a digest mechanism that differentiates Normal and Urgent packet loss conditions. Table IV shows that under normal conditions, recovery time is 130 ms with 200 Mbps throughput, while urgent conditions trigger faster recovery (70 ms) with 240 Mbps throughput. This 46% faster response confirms the benefit of digest-based recovery.

5) Overall Performance Analysis: Table V presents the performance comparison between Static, RED, and Adaptive methods. While RED reduces packet loss and improves recovery time compared to Static, it is still less effective than the Adaptive approach. Specifically, Adaptive reduces packet loss by approximately 59%, shortens recovery time by 67%, and increases throughput by 27% over Static. However, this improvement comes with a minor latency increase of 1.8 ms over Static and 1.9 ms over RED. Despite this trade-off, the

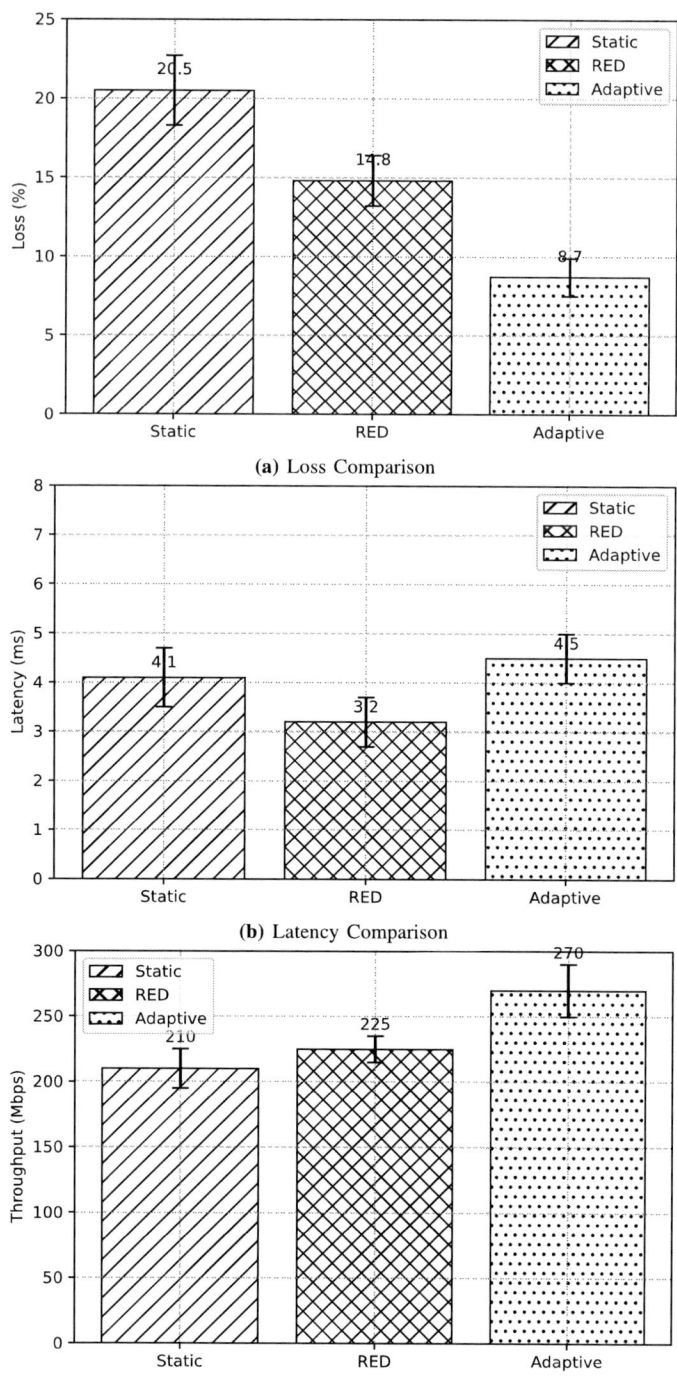

(a) Loss Comparison

(b) Latency Comparison

(c) Throughput Comparison

Fig. 6. Comparison of Static, RED & Adaptive at 300 Mbps

TABLE V. Overall Performance Gains (Mean ± Std.)

Metric	Static	RED	Adaptive
Packet Loss (%)	20.0 ± 2.1	14.8 ± 1.6	8.2 ± 1.0
Recovery Time (ms)	150 ± 10	90 ± 8	50 ± 5
Throughput (Mbps)	220 ± 15	225 ± 10	280 ± 20
Latency (ms)	3.3 ± 0.5	3.2 ± 0.5	5.1 ± 0.6

justment, selective packet recirculation, and digest-triggered recovery. Leveraging programmable data planes, our solution dynamically mitigates congestion, prioritizes high-importance traffic, and significantly reduces packet loss. Experiments in a Mininet/BMv2 environment demonstrated superior performance over static-threshold baselines, achieving lower loss and faster recovery with minimal latency overhead.

While our study was conducted in a virtualized WSL-based BMv2 environment, performance may differ on hardware P4 switches due to CPU and virtualization constraints. Additionally, our evaluation used a linear Mininet topology, and further validation is required for larger, real-world networks. The current QoS model differentiates only two priority classes, limiting fine-grained traffic management.

In summary, in-network programmability enables adaptive and efficient queue management, reducing congestion while prioritizing critical traffic. Future P4-based systems can further minimize packet loss and enhance network resilience, even in more complex and high-throughput environments.

REFERENCES

[1] Open Networking Foundation, "Programming protocol-independentpacket processors (P4)," [Online], accessed: Jan. 10, 2025, available: https://opennetworking.org/p4/

[2] P. Bosshart, D. Daly, G. Gibb, M. Izzard, N. McKeown, J. Rexford, C. Schlesinger, D. Talayco, A. Vahdat, G. Varghese, and D. Walker, "P4: Programming protocol-independent packet processors," ACM SIGCOMM Computer Communication Review, vol. 44, no. 3, pp. 87–95, Jul. 2014. Doi: 10.1145/2656877.2656890.

[3] K. Kumazoe and M. Tsuru, "P4-Based Implementation and Evaluation of Adaptive Early Packet Discarding Scheme," in Proc. Advances in Network-Based Information Systems. Advances in Intelligent Systems and Computing, vol 1263. Springer, Cham, pp. 442-453, Aug. 2020. Doi: 10.1007/978-3-030-57796-4_44.

[4] "Guidelines for Buffer Management and Queueing in Packet-Switched Networks," RFC 7567, 2015, [Online], accessed: Jan. 10, 2025, available: https://www.rfc-editor.org/info/rfc7567.

[5] G. Appenzeller, I. Keslassy, and N. McKeown, "Sizing router buffers," in Proc. 2004 conference on Applications, technologies, architectures, and protocols for computer communications (SIGCOMM). Association for Computing Machinery, New York, NY, USA, pp. 281–292, Oct. 2004. Doi: 10.1145/1030194.1015499.

[6] Intel, "Intel® Tofino™ Programmable Ethernet Switch Architecture", [Online], accessed: Jan. 10, 2025, available: https://www.intel.com/content/www/us/en/products/docs/network-io/ethernet-switches/tofino-architecture-brief.html.

[7] "Definition of the Differentiated Services Field (DS Field) in the IPv4 and IPv6 Headers." RFC 2474, 1998, [Online], accessed: Jan. 10, 2025, available: https://www.rfc-editor.org/info/rfc2474

[8] N. V. Ha, T. N. T. Kiet, B. N. T. Binh, N. M. Tri, T. T. T. Nguyen, and M. Tsuru, "Real-Time In-Band Network Link Loss Detection With Programmable Data Plane," in Proc. 16th International Conference on Knowledge and Smart Technology (KST), Krabi, Thailand, Feb. 2024, pp. 167–172. Doi: 10.1109/KST61284.2024.10499673.

latency remains well within the 50 ms SLA requirement for real-time applications, ensuring Adaptive is the most effective method for reducing loss and optimizing throughput.

IV. CONCLUSION AND FUTURE WORK

In this paper, we introduced a P4-based adaptive queue management framework that integrates adaptive threshold ad-

979-8-3315-1550-8/25 $31.00 © 2025 IEEE

A Data Labeling Method in Deep Learning Model for User Clustering in the NOMA Systems

Ngo Minh Nghia[1,2], Nguyen Thi Xuan Uyen[1,2], Nguyen Dung[1,2], Thai Ngoc Duy Kha[1,2], Dang Le Khoa[1,2]*

[1]VNUHCM - University of Science, Ho Chi Minh City, Vietnam
[2]Vietnam National University, Ho Chi Minh City, Vietnam
*Email: dlkhoa@hcmus.edu.vn

Abstract—**Deep Learning has emerged as a promising candidate for future wireless networks because of its strong adaptability to dynamic environments. In particular, Deep Learning algorithms are capable of learning optimal outcomes and making rapid decisions in complex scenarios. This study addresses the user clustering problem in Non-Orthogonal Multiple Access (NOMA) systems by leveraging Deep Learning techniques to maximize the system's sum rate. We initially employ a brute-force search algorithm to explore all possible clustering strategies and identify the optimal one. This resulting dataset serves as the training input for Deep Neural Network (DNN) models. However, the generated data are often highly imbalanced, as they tend to overrepresent a limited number of high-frequency scenarios. To address this issue, the study proposed a two-step balanced sampling. First, infrequent scenarios are grouped into a single class. Second, under-sampling is applied to balance the dataset across all classes. The simulation results demonstrate that the proposed method significantly improves learning performance.**

Index Terms—**Data Labeling, Deep Learning, NOMA System, User Clustering**

I. INTRODUCTION

Deep Learning (DL) has become a key enabler in wireless communications, offering powerful capabilities to model complex, nonlinear relationships in dynamic and data-rich environments. As wireless networks grow in complexity, with diverse user demands and rapidly changing channel conditions, DL provides adaptive and data-driven solutions that improve system performance, efficiency, and resilience across various tasks. Early applications of DL in wireless communications include signal detection in Orthogonal Frequency Division Multiplexing (OFDM) systems, where Deep Neural Networks (DNNs) were employed to mitigate channel impairments and improve symbol detection accuracy [1]. DL techniques have also been applied to automatic channel estimation and modeling of channel state information (CSI), allowing systems to adapt effectively to fast-varying or unpredictable channel conditions [2], [3]. Furthermore, DL-based solutions have shown promise in addressing practical challenges such as imperfect successive interference cancellation (SIC) through energy-efficient power allocation schemes [4] and Convolutional Neural Network (CNN)-based decoding strategies that alleviate sum-rate degradation [5].

Building on these successes, recent research has extended the use of DL to more complex scenarios such as Non-Orthogonal Multiple Access (NOMA) systems, where efficient resource allocation and interference management are crucial.

NOMA enhances spectral efficiency by enabling multiple users to share the same radio resources through power-domain multiplexing [6]. To maximize system throughput, various user clustering and pairing strategies have been proposed, including grouping based on channel disparity [7], user proximity [8], Signal-to-Interference-plus-Noise Ratio (SINR) differences and power allocation [9]. Adaptive User Clustering (AUC) in [10] further improves flexibility by dynamically forming user groups based on real-time channel conditions. More recently, DL has been employed to enhance user clustering in NOMA, offering better adaptability and performance in complex scenarios [11]. Techniques like Extreme Learning Machine (ELM) [12], Deep Neural Networks (DNNs) [13], and unsupervised K-means++ algorithms [14] have demonstrated superior capabilities in managing interference, optimizing power allocation, and improving spectral efficiency over traditional methods.

One critical challenge in applying DL to user clustering in NOMA systems lies in the severe class imbalance typically present in the training data. This imbalance often arises from the use of Brute-force search methods to generate optimal user cluster configurations based on sum-rate maximization. This leads to two major problems: a highly multi-class classification setting with hundreds of unique labels and severe class imbalance, as some cluster configurations appear far more frequently than others. Additionally, the way NOMA clustering labels are created from different user combinations makes it hard to use regular data balancing methods. Therefore, specialized approaches are needed to address the imbalance in this context. Recent studies [15], [16] have investigated domain-specific balancing strategies, and ensemble learning or augmentation methods [17] have shown promise in enhancing robustness and efficiency.

In addition to domain-specific solutions, general techniques for handling data imbalance fall into three categories: data-level, algorithm-level, and hybrid approaches [18]. Data-level methods adjust sample distributions through undersampling such as random undersampling, Tomek links, and Edited Nearest Neighbors (ENN) or oversampling such as Synthetic Minority Oversampling Technique (SMOTE) [19], [20], though both have trade-offs. Algorithm-level strategies like cost-sensitive learning and specialized loss functions such as Focal Loss, Label-Distribution-Aware Margin (LDAM) [18] focus training on rare or hard examples. Hybrid methods such as SMOTE-ENN and SMOTE-Tomek [19] combine sampling with noise reduction for better generalization, especially in wireless systems where detecting rare but critical events is essential.

Choosing the right way to deal with data imbalance is very important, especially when using DL for user clustering in NOMA systems. To address these challenges, this study proposes a novel approach that jointly handles class imbalance and label complexity. Specifically, we introduce a new user clustering and relabeling strategy that restructures the original training set into a more balanced and learnable form. This approach not only reduces the label space to a manageable size but also mitigates the risk of biased learning caused by dominant classes, based on the imbalance handling techniques discussed earlier.

The key contributions of this study are summarized as follows:

1) We found that there is a significant imbalance in the sample sizes when creating the best user groupings for NOMA, especially when using a brute-force search focused on maximizing the sum rate. This imbalance renders standard DNN training ineffective due to label sparsity and dominance.

2) To address this, we propose a novel imbalance handling approach tailored to the NOMA clustering problem. Our method combines label grouping with under-sampling techniques, followed by a relabeling process that reduces class granularity while preserving key structural distinctions among clusters.

3) The newly relabeled and balanced dataset enables successful DNN-based training for user clustering. Extensive evaluation based on metrics such as F1-score, Precision, Recall demonstrates the effectiveness of the proposed strategy in enabling reliable multi-class learning under severe class imbalance conditions.

The remainder of this paper is structured as follows: Section II presents the system model and details the data generation process. Section III introduces the proposed data labeling method for Deep Learning. In Section IV, we develop a Deep Learning model to evaluate the effectiveness of the proposed data processing approach. Finally, Section V concludes the paper.

II. System Model and Data Generation

A. Downlink NOMA System with User Clustering for Sum-Rate Maximization

We consider a NOMA downlink system with N users, where the users are divided into K clusters, each consisting of two users. In this system, both the transmitter and each receiver are equipped with a single antenna. Within each cluster, the second user (UE2), which is farther from the transmitter than the first user (UE1), is allocated corresponding power levels P_1 and P_2. The signals for UE1 and UE2, denoted by s_1 and s_2, are decoded via successive interference cancellation (SIC).

The overall transmitted signal in the k-th cluster at the base station can be expressed as:

$$s_k = \sqrt{P_k}\left(\sqrt{\alpha_{k1}}s_{k1} + \sqrt{\alpha_{k2}}s_{k2}\right), \quad (1)$$

where P_k is the transmit power of the k-th cluster, and α_{k1}, α_{k2} are the power distribution factors for UE1 and UE2, respectively. We have $E\left[\|s_{k1}\|^2\right] = E\left[\|s_{k2}\|^2\right] = 1$, where $E[\cdot]$ denotes expectation and $\|\cdot\|$ is the second-order norm. In a Rayleigh fading channel, the channel responses for UE1 and UE2 at the k-th cluster are h_{k1} and h_{k2}, with average gains $g_{k1} = E\left[\|h_{k1}\|^2\right]$ and $g_{k2} = E\left[\|h_{k2}\|^2\right]$. The distances from the transmitter to UE1 and UE2 are d_k^{UE1} and d_k^{UE2}, respectively, with $g_{k1} > g_{k2}$ since $d_k^{\mathrm{UE1}} < d_k^{\mathrm{UE2}}$. The received signal at j-th UE at the k-th cluster can be expressed as:

$$
\begin{aligned}
y_{kj} &= h_{kj} * s_k + n_{kj} \\
&= h_{kj} * \sqrt{P_k}(\sqrt{\alpha_{k1}}s_{k1} + \sqrt{\alpha_{k2}}s_{k2}) + n_{kj}, \forall j = \{1,2\}.
\end{aligned}
\quad (2)
$$

Each received signal is affected by complex Gaussian noise with zero-mean and variance, modeled as $n_{ki} \sim \mathcal{CN}(0, \sigma^2)$.

The goal is to allocate power to meet QoS requirements and maximize the sum-rate of user pairs. For a cluster with $\gamma_1 > \gamma_2$, where $\gamma_k = \frac{h_k}{N_0 B}$, users require minimum rates R_1^* and R_2^*, where $R_i^* > 0$. Let B be the bandwidth per user, P_1 and P_2 the transmit powers, and P the total transmission power in the cluster. The optimal power allocation can then be formulated as:

$$\textbf{max: } B\left(\log_2(1 + P_1\gamma_1) + \log_2\left(1 + \frac{P_2\gamma_2}{P_1\gamma_1 + 1}\right)\right), \quad (3)$$

$$\text{subject to: } P - (P_1 + P_2) > 0, \quad (4)$$

$$B\log_2\left(1 + \frac{P_i\gamma_i}{\sum_{j=1}^{i-1} P_j\gamma_j}\right) - R_i^* \geq 0, \quad \forall i = \{1,2\}, \quad (5)$$

$$P_2\gamma_1 - P_1\gamma_1 - P_{SIC} \geq 0; \quad (6)$$

Equation (4) representing the total power constraints of the base transceiver station (BTS). This optimization problem is investigated in [8], where the optimal power allocations for the two users are derived:

$$P_1 = \frac{P}{\varphi_2} - \frac{\varphi_2 - 1}{\varphi_2\gamma_2}, \quad (7)$$

$$P_2 = \frac{P(\varphi_2 - 1)}{\varphi_2} + \frac{\varphi_2 - 1}{\varphi_2\gamma_2}, \quad (8)$$

where $\varphi_i = 2^{\frac{R_i^*}{B}}$.

B. Data Generation based on Brute-force Search

The Brute-force search algorithm is an exhaustive approach that evaluates all possible solutions to identify the optimal one. In this study, the data generation process for user clustering is conducted using the Brute-force search method, as outlined below:

979-8-3315-1550-8/25 $31.00 © 2025 IEEE

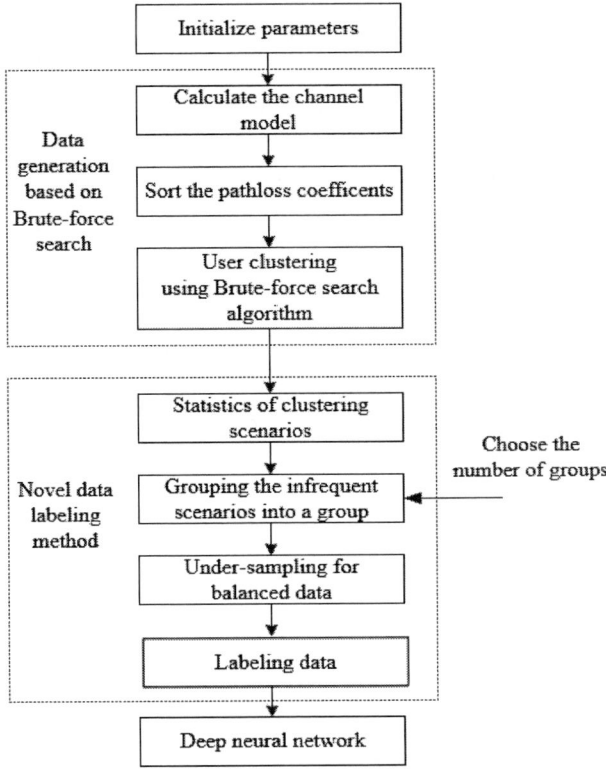

Fig. 1: The workflow of the proposed method.

- All possible user pairing combinations are exhaustively generated based on the predefined number of users and the specified group size.
- For each pairing combination, power allocation is performed for users in a group, followed by the calculation of the corresponding sum-rate.
- The combination that yields the highest sum-rate is then considered the optimal user grouping.

The data generation process based on the Brute-force algorithm poses significant challenges when integrating the data into a DL approach. These challenges can be summarized as the following aspects:

- Labeling each sample with its corresponding optimal user pairing is time-consuming and involves handling a large number of categories, increasing the complexity of the task.
- Without a sufficient amount of simulation-generated data, a multi-class classification problem like this offers limited opportunities for the model to converge effectively.
- The output dimension is much larger than the input dimension, leading to challenges in model training and computational efficiency.
- A significant class imbalance is observed in the dataset, which poses difficulties for effective model training and generalization.

To overcome the aforementioned challenges, this paper proposes an efficient labeling method that facilitates more effective learning for the model. The proposed approach is presented in Section III.

III. A NOVEL DATA LABELING METHOD FOR DEEP LEARNING

A. A Novel Data Labeling Method

During the data generation process using the Brute-force search algorithm, it was observed that only a few clustering patterns frequently appeared. The results indicated an uneven distribution of clustering patterns within the data space, with certain patterns emerging more prominently than others. As a result, rather than labeling all possible cases, the labeling process was refined to focus only on the most frequently occurring patterns. This approach not only optimizes the labeling process by reducing the workload but also ensures the labeled data retains high quality and relevance. The novel data labeling method helps to highlight and prioritize the most significant data patterns, which could be crucial for subsequent analysis and modeling. This approach is clearly illustrated in the block diagram shown in Fig. 1.

Initially, system parameters including the number of users, antenna configuration per base station, and cell radius are defined. These parameters govern the subsequent computation of the wireless channel model, which incorporates path loss attenuation to reflect signal degradation over distance. Following channel model generation, path loss coefficients are sorted in descending order to facilitate user clustering. This sorting prioritizes users with stronger channel conditions, thereby improving the potential for achieving higher sum-rates during the clustering process. User clusters are then exhaustively generated via the Brute-force algorithm, and all resulting scenarios are recorded for analysis.

In the label stage, the statistical distribution of the clustering scenarios is analyzed to identify frequent and infrequent cases. Based on a predefined number of frequent cases, infrequent clustering scenarios are aggregated into a single group to mitigate the data imbalance issue. A balancing procedure is then applied, in which under-sampling is used to reduce the number of samples from overrepresented classes. After balancing, new class labels are assigned following an incremental indexing scheme. The resulting labeled and balanced dataset is finally used to train a DNN model for classification.

B. Deep Learning Model for User Clustering

In this study, a straightforward DL architecture is developed to address the user pairing classification problem. The model comprises four fully-connected layers, each utilizing the Rectified Linear Unit (ReLU) as the activation function to capture nonlinear relationships in the input data, except for the final output layer which utilizes a softmax function to produce class probability distributions across possible user grouping configurations. To stabilize training and accelerate convergence, normalization layers are included after each dense block. The output dimensions of each layer, along with

TABLE I: The structure of DNN model for user clustering classification tasks, where N is the number of users and N_f is the number of frequent cases.

Layer type	Size (output)	Parameters	Activation function
Feature input	N	-	-
Fully-connected	32	352	ReLU
Normalization	32	64	-
Fully-connected	64	2112	ReLU
Normalization	64	128	-
Fully-connected	32	2080	ReLU
Normalization	32	64	-
Fully-connected	$N_f + 1$	$33 \left(N_f + 1 \right)$	Softmax

TABLE II: Simulation parameters.

Simulation cell radius	100 m
Number of users	10
Number of antenna transmissions	8
Number of antenna receiver	1
Path loss exponent	4
Transmission power	40 dBm
Power threshold	10 dBm
Transmission bandwidth	200 kHz
Noise power	-116 dBm
Achievable rate	1.5 Mbps
Number of radio channels	100

the number of parameters and activation functions, are detailed in Table I.

The model is trained using a Supervised Learning framework. The input feature vector consists of path-loss values between BTS and UEs, forming a fixed-length representation of size N, where N is the number of users. This vector captures key spatial channel characteristics that influence user pairing decisions. All input features are normalized to the $[0, 1]$ range prior to training in order to improve learning generalization and reduce training variance. The output corresponds to a user grouping configuration that maximizes the achievable sum-rate performance. For the classification task, the target labels are re-encoded using the one-hot encoding scheme, allowing the model to be optimized using a categorical cross-entropy loss function.

IV. SIMULATION RESULTS

In this section, several DNN with varying numbers of frequent cases are implemented to evaluate the effectiveness of the proposed data processing approach. The simulation considers the downlink NOMA system operating within a cell radius of 100 meters. The base station is equipped with 8 antennas to simultaneously serve 10 single-antenna users. The system operates over a 200 kHz bandwidth, and the total receiver noise power is set to -116 dBm. Each user cluster is allocated a downlink transmit power of 40 dBm from the base station. During the SIC process, the minimum power difference of 10 dBm is required to reliably separate the target signal from the non-decoded interfering signals. Additionally, each user is required to achieve a minimum data rate of 1.5 Mbps to ensure quality of service. Further simulation parameters are summarized in Table II.

The models are initially trained on a dataset comprising 35000 samples generated using the brute-force search algorithm. To enhance the training effectiveness and mitigate

class imbalance issues, a sample balancing process is subsequently applied based on the proposed algorithm described in Section III. The training dataset is randomly split into 80% training, 10% validation, and 10% testing subsets. To promote generalization, the training data is shuffled at the beginning of each epoch. The model is trained for 30 epochs with a batch size of 32 using the Adam optimizer, selected for its adaptive learning rate and stable convergence behavior. The initial learning rate is set to 0.001. Early stopping based on validation loss is employed to prevent overfitting.

A. Handling Class Imbalance via the Proposed Balancing Algorithm

Fig. 2 illustrates the distribution of several pairing cases generated with 10 users and a group size of 2. Under this configuration, a total of 945 distinct pairing combinations are produced. However, the resulting dataset exhibits significant class imbalance, with the majority of samples concentrated in a small subset of frequent cases. To address this issue in the dataset comprising 35,000 samples, the balancing algorithm proposed in Section III is employed. The core idea of the algorithm is to enhance the representation of underrepresented classes while limiting the dominance of overrepresented ones through a combination of sample augmentation and reduction techniques. Initially, all infrequent clustering cases are aggregated into a single group to mitigate data sparsity and simplify the classification task. For dominant classes, redundant samples are selectively reduced to avoid bias during the training process. This balancing strategy aims to ensure a more uniform class distribution, thereby enhancing the generalization capability of the DL model trained on the dataset. A new dataset is generated after applying this algorithm and is subsequently relabeled according to the principle of incremental labeling for the cases.

979-8-3315-1550-8/25 $31.00 © 2025 IEEE

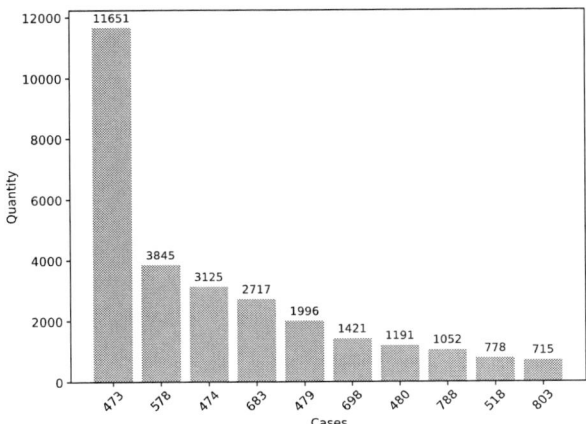

Fig. 2: The chart of statistics of some resulting cases in the dataset using the Brute-force search.

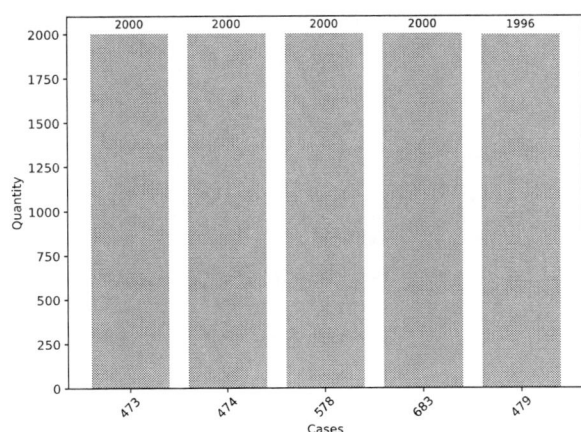

Fig. 3: The chart of statistical of some resulting cases in the dataset after the proposed data processing.

Consider a scenario where the top five most frequent clustering cases are selected. A balancing threshold of 2000 samples is applied uniformly across all classes, and the resulting distribution is illustrated in Fig. 3. As shown, the dataset now exhibits a more uniform class distribution. This uniform distribution enhances the fairness of the model training process, ensuring that no class dominates the learning process and improving the generalization ability of the Deep Learning model on unseen data. Table III illustrates the mapping process from the original clustering labels to the newly assigned labels. The most frequent clustering scenarios are relabeled from 1 to 5, while all other less common cases are grouped under a single label.

B. Effect of Training Data Rebalancing on Model Performance Metrics

Fig. 4 illustrates the effect of varying the number of frequent cases on the training loss of the DNN model. As the number of frequent cases increases, a degradation in model performance is observed. Specifically, models trained with fewer frequent cases (e.g., 3 or 5) demonstrate faster convergence and achieve lower final training loss values. In contrast, when the number of frequent cases increases to 10 or 20, the training process becomes slower and the convergence performance declines. These results indicate that limiting the number of frequent cases simplifies the learning task, enabling the model to fit the training data more effectively and reducing the risk of underfitting. Furthermore, the application of the proposed sample balancing strategy leads to improved model capability in distinguishing between different grouping configurations. The balanced dataset facilitates a more equitable allocation of learning capacity across all classes, thereby mitigating bias toward dominant cases. When evaluated in a test set of 2000 samples, the model trained with three frequent cases achieves a test accuracy of 81%, demonstrating stable generalization performance. These findings underscore the effectiveness of combining a limited frequent-case strategy with data balancing

TABLE III: Mapping of original clustering labels to new labels with five frequent cases.

Original labels	Pairing scheme	New labels
473	[1,6;2,7;3,8;4,9;5,10]	1
474	[1,6;2,7;3,8;4,10;5,9]	2
578	[1,7;2,6;3,8;4,9;5,10]	3
683	[1,8;2,6;3,7;4,9;5,10]	4
479	[1,6;2,7;3,10;4,8;5,9]	5
Others	Otherwise	6

techniques to enhance model robustness and generalization across diverse user grouping scenarios.

Table IV presents the performance metrics of the DNN model under varying sample frequency thresholds. It is evident that reducing the threshold for frequent cases—from 20 to 3—not only leads to significant improvements in evaluation metrics but also reduces the number of classes the model must learn. Specifically, the accuracy increases from 0.69 to 0.88, and the weighted F1-score rises from 0.68 to 0.87. These results indicate that the proposed sample balancing strategy, which limits the dominance of frequent cases during training, enhances the model's generalization ability. Importantly, although the number of classes is reduced, the pairwise matching performance remains stable, indicating that less frequent cases do not substantially contribute to matching accuracy. This demonstrates that the proposed approach effectively streamlines the training process without compromising matching effectiveness, making it more suitable for real-world, imbalanced datasets.

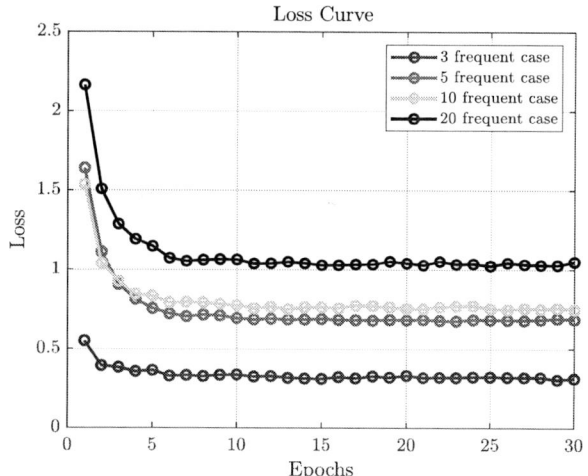

Fig. 4: The effect of the number of frequent cases on the training loss across epochs for the DNN Model.

TABLE IV: Performance metrics for different Top-N_f frequent case models. Top-N_f refers to models trained on the N_f most frequent cases in the dataset.

Model	Top-3	Top-5	Top-10	Top-20
Accuracy	0.88	0.81	0.75	0.69
Weighted Precision	0.88	0.82	0.76	0.69
Weighted Recall	0.88	0.81	0.75	0.69
Weighted F1-score	0.87	0.81	0.75	0.68

V. CONCLUSION

This paper examined Deep Learning-based user clustering methods in NOMA. The drawback of this system is the imbalanced datasets. This paper proposes a novel two-stage method to solve this problem. The effectiveness of this approach is confirmed via a DNN, showing improved loss and clustering accuracy. These findings indicate the potential for Deep Learning to enhance user clustering in NOMA, facilitating more adaptive and scalable wireless communication solutions. The sum rate of this system will be considered in more detail.

ACKNOWLEDGMENT

This research is funded by University of Science, VNU-HCM under grant number ĐT-VT 2023-02.

REFERENCES

[1] H. Ye, G. Y. Li and B. -H. Juang, "Power of Deep Learning for Channel Estimation and Signal Detection in OFDM Systems," in IEEE Wireless Communications Letters, vol. 7, no. 1, pp. 114-117, Feb. 2018

[2] G. Gui, H. Huang, Y. Song and H. Sari, "Deep Learning for an Effective Nonorthogonal Multiple Access Scheme," in IEEE Transactions on Vehicular Technology, vol. 67, no. 9, pp. 8440-8450, Sept. 2018

[3] Lin, C.; Chang, Q.; Li, X. A Deep Learning Approach for MIMO-NOMA Downlink Signal Detection. Sensors 2019

[4] W. Saetan and S. Thipchaksurat, "Application of Deep Learning to Energy-Efficient Power Allocation Scheme for 5G SC-NOMA System with Imperfect SIC," 2019 16th International Conference on Electrical Engineering/Electronics, Computer, Telecommunications and Information Technology (ECTI-CON), Pattaya, Thailand, 2019

[5] Sim, I.; Sun, Y.G.; Lee, D.; Kim, S.H.; Lee, J.; Kim, J.-H.; Shin, Y.; Kim, J.Y. Deep Learning Based Successive Interference Cancellation Scheme in Nonorthogonal Multiple Access Downlink Network. Energies 2020

[6] Y. Li, M. Jiang, Q. Zhang, and J. Qin, "Joint Beamforming Design in Multi-Cluster MISO NOMA Reconfigurable Intelligent Surface-Aided Downlink Communication Networks," IEEE Transactions on Communications, vol. 69, no. 1, pp. 664-674, 2021.

[7] A. Salari, M. Shirvanimoghaddam, M. B. Shahab, R. Arablouei, and S. Johnson, "Design and Analysis of Clustering-Based Joint Channel Estimation and Signal Detection for NOMA," IEEE Transactions on Vehicular Technology, vol. 73, no. 2, pp. 2093-2108, 2024.

[8] M. S. Ali, H. Tabassum, and E. Hossain, "Dynamic User Clustering and Power Allocation for Uplink and Downlink Non-Orthogonal Multiple Access (NOMA) Systems," IEEE Access, vol. 4, pp. 6325-6343, 2016.

[9] N. S. Mouni, A. Kumar, and P. K. Upadhyay, "Adaptive User Pairing for NOMA Systems With Imperfect SIC," IEEE Wireless Communications Letters, vol. 10, no. 7, pp. 1547-1551, 2021.

[10] S. Prabha Kumaresan, T. Chee Keong, L. Ching Kwang, and N. Yin Hoe, "Adaptive user clustering for downlink nonorthogonal multiple access based 5G systems using Brute-force search," Transactions on Emerging Telecommunications Technologies, vol. 31, no. 11, p. e4098, 2020/11/01 2020.

[11] S. M. Hamedoon, J. N. Chattha, and M. Bilal, "Towards intelligent user clustering techniques for non-orthogonal multiple access: a survey," EURASIP Journal on Wireless Communications and Networking, vol. 2024, no. 1, p. 7, 2024/01/24 2024.

[12] S. P. Kumaresan, C. K. Tan, and Y. H. Ng, "Extreme Learning Machine (ELM) for Fast User Clustering in Downlink Non-Orthogonal Multiple Access (NOMA) 5G Networks," IEEE Access, vol. 9, pp. 130884-130894, 2021.

[13] S. P. Kumaresan, C. K. Tan, and Y. H. Ng, "Deep Neural Network (DNN) for Efficient User Clustering and Power Allocation in Downlink Non-Orthogonal Multiple Access (NOMA) 5G Networks," symmetry, vol. 13, no. 8, 2021.

[14] Q. N. Le, V. D. Nguyen, O. A. Dobre, N. P. Nguyen, R. Zhao, and S. Chatzinotas, "Learning-Assisted User Clustering in Cell-Free Massive MIMO-NOMA Networks," IEEE Transactions on Vehicular Technology, vol. 70, no. 12, pp. 12872-12887, 2021.

[15] W. Wang, M. Zhang, L. Zhang, and Q. Bai, "Imbalanced Data Classification for Multi-Source Heterogenous Sensor Networks," IEEE Access, vol. 8, pp. 27406-27413, 2020.

[16] W. Chen, S. Zhao, R. Zhang, H. Chen, and L. Yang, "Generalized User Grouping in NOMA: An Overlapping Perspective," IEEE Transactions on Wireless Communications, vol. 20, no. 5, pp. 2876-2887, 2021.

[17] A. A. Khan, O. Chaudhari, and R. Chandra, "A review of ensemble learning and data augmentation models for class imbalanced problems: Combination, implementation and evaluation," Expert Systems with Applications, vol. 244, p. 122778, 2024/06/15/ 2024.

[18] M. Altalhan, A. Algarni and M. Turki-Hadj Alouane, "Imbalanced Data Problem in Machine Learning: A Review," in IEEE Access, vol. 13, pp. 13686-13699, 2025.

[19] Fadi Thabtah, Suhel Hammoud, Firuz Kamalov, and Amanda Gonsalves. 2020. Data imbalance in classification: Experimental evaluation. Inf. Sci. 513, C (Mar 2020).

[20] Rendón, E.; Alejo, R.; Castorena, C.; Isidro-Ortega, F.J.; Granda-Gutiérrez, E.E. Data Sampling Methods to Deal With the Big Data Multi-Class Imbalance Problem. Appl. Sci. 2020.

RTL Design of Convolution for CNN Using Baugh Wooley and Wallace Tree Multipliers

Truong Quang Vinh
Falcuty of Electrical and Electronics Engineering
Ho Chi Minh City University of Technology
Vietnam National University – Ho Chi Minh
Ho Chi Minh, Vietnam
tqvinh@hcmut.edu.vn

Doan Duy Quan
Falcuty of Electrical and Electronics Engineering
Ho Chi Minh City University of Technology
Vietnam National University – Ho Chi Minh
Ho Chi Minh, Vietnam
quan.doansuryxin4112@hcmut.edu.vn

Nguyen Minh Khang
Falcuty of Electrical and Electronics Engineering
Ho Chi Minh City University of Technology
Vietnam National University – Ho Chi Minh
Ho Chi Minh, Vietnam
khang.nguyenk21@hcmut.edu.vn

Abstract— This paper presents an implementation of convolution module for Convolutional Neural Network (CNN) on FPGA. The proposed hardware architecture utilizes Baugh Wooley and Wallace tree multipliers. The Baugh-Wooley algorithm is leveraged for efficient signed multiplication, while the Wallace Tree structure ensures fast partial product reduction, combining the benefits of low power consumption and reduced latency. The pipeline architecture further enhances the design by breaking down the multiplier operation into multiple stages, enabling concurrent processing and significantly improving throughput. The proposed design has been implemented on Xilinx Zynq UltraScale+ ZCU106 FPGA, achieving a frequency of 769.231 MHz.

Keywords—Convolution, Baugh Wooley Wallace tree multiplier, FPGA, CNN

I. INTRODUCTION

Convolutional neural networks, which are extensively employed in image processing and computer vision tasks, are optimized by specialized hardware components called hardware accelerators. The speed and efficiency of computing are greatly increased by these accelerators, which include GPUs, FPGAs, and ASICs, in comparison to conventional CPU designs[1]. They do this by utilizing parallel computing power and customized data pathways that lower energy and latency. Hardware accelerators are becoming necessary in data centers and edge devices as CNN models get more complicated in order to provide real-time processing and scalable performance[2]. To handle huge amounts of data, CNN convolution procedures sometimes depend on many multipliers running in parallel, which speeds up inference and lowers latency. Researchers have created creative, complex solutions that combine conventional multiplier structures with innovative architectural improvements especially hardware accelerators, motivated by the constantly increasing needs for speed and energy efficiency in modern computing.

Hybrid approaches have also been explored to optimize multipliers for high-performance applications. Beura et al. [3] integrated a 4:2 compressor within the Baugh-Wooley multiplier to improve the addition of partial products. Their method balanced design time overestimation while improving error tolerance and reducing area-delay tradeoffs, making it particularly effective for edge detection applications. However, integrating inexact 4:2 compressors into the Baugh-Wooley multiplier could limit the maximum operating frequency due to the propagation delays introduced by approximate logic components, potentially

impacting the multiplier's suitability for extremely high-frequency digital signal processing applications. Thamizharasan et al. [4] proposed a hybrid compressor-based multiplier implemented on FPGA, modifying the Vedic multiplier for enhanced signal and image processing. Their results showed up to a 35.83% improvement in processing speed compared to conventional multipliers, while also reducing energy consumption. However, the hybrid compressor-based multiplier could encounter constraints on its maximum achievable operating frequency due to increased critical path complexity inherent in the hybrid compressor logic. This frequency limitation might restrict its effectiveness in ultra-high-speed real-time processing applications. Alternative approaches such as Vedic mathematics and floating-point multiplication have also been explored. Al-Nounou et al. [5] proposed a hybrid approach for FPGA-based binary multiplication by integrating customized basic cells to optimize area, power, and delay. Their design utilizes a combination of logic compression techniques and FPGA-specific optimizations to enhance multiplication efficiency. By modifying traditional architectures and employing resource-aware synthesis techniques, the proposed method achieves a balanced trade-off between speed and power consumption. The results demonstrate a significant reduction in processing delay and hardware complexity, making it suitable for high-performance computing applications. However, the method relies on customized basic cells and FPGA-specific optimizations, which may limit its portability and scalability. Nguyen et al. [1] presented FPGA-based multi-level approximate multipliers that leverage novel approximate logic compressors to achieve significant power-delay-area product (PDAP) gains up to 7.1× for 8-bit multipliers. Their designs demonstrate impressive performance in image processing applications, delivering high PSNR and SSIM values along with notable dynamic power savings. However, while the proposed multipliers significantly reduce delay, area, and power consumption, the inherent approximation in the compressor stages can lead to error accumulation especially as the multiplier size increases which may restrict their use in applications demanding ultra-high computational precision. Pathan et al. [7] introduced an efficient ROM size reduction technique for distributed arithmetic-based multipliers, optimizing memory utilization while maintaining computational accuracy. Their approach minimized storage requirements in FPGA implementations, making it suitable for high-performance multiplier designs. By leveraging optimized LUT structures, their method significantly reduced area and power consumption compared

979-8-3315-1550-8/25 $31.00 © 2025 IEEE

```
                              y_{m-1}   ...   y_4    y_3    y_2    y_1    y_0
                              x_{n-1}   ...   x_2    x_1    x_0
                          x_0y_{m-2}    ...  x_0y_4  x_0y_3 x_0y_2 x_0y_1 x_0y_0
                   x_1y_{m-2}     ...  x_1y_4  x_1y_3 x_1y_2 x_1y_1 x_1y_0
              x_2y_{m-2}     ...  x_2y_4  x_2y_3 x_2y_2 x_2y_1 x_2y_0
                         .                        .
                         .                        .
   x_{n-1}y_{m-1}  0   x_{n-2}y_{m-2}  .   .   .  x_{n-2}y_3 x_{n-2}y_2 x_{n-2}y_1 x_{n-2}y_0
   x̄_{n-1}   x_{n-1}ȳ_{m-2}  x_{n-1}ȳ_{m-3}  .       .   x_{n-1}ȳ_2 x_{n-1}ȳ_1 x_{n-1}ȳ_0
   ȳ_{m-1}
 1
P_{n+m-1} P_{n+m-2} r_{n+m-3} r_{m+n-4} ... r_{m-1} ... r_{(n+1)} r_n r_{n-1} r_{n-2} ... r_3 r_2 P_1 P_0
```

Fig. 1. Algorithm with all positive partial product bits

to conventional memory-based multipliers. However, this dual-port ROM-based approach may encounter scalability limitations as increasing operand sizes can lead to an exponential growth in memory requirements, potentially restricting its practicality for large-word-length multiplications.

This research contributes to the RTL (Register Transfer Level) design and implementation of a high-performance pipelined Baugh-Wooley and Wallace Tree Multiplier, specifically optimized for the convolutional stage of CNNs to enhance the efficiency of multiply-accumulate (MAC) operations. By leveraging pipelining techniques, the proposed multiplier minimizes latency, improves throughput, and optimizes power consumption, making it well-suited for real-time and energy-efficient deep learning applications. A key innovation is its ability to achieve high-speed computation without compromising accuracy, ensuring efficient hardware utilization. To validate its real-world applicability, the design is integrated into FPGA platforms (Xilinx Virtex-7 and Xilinx Zynq UltraScale+) and simulated using Vivado, demonstrating its effectiveness in accelerating CNN computations while maintaining hardware efficiency and scalability.

II. BACKGROUND

A. Baugh Wooley Algorithm

This paper proposes a parallel two's complement multiplier where every partial-product bit is given a non-negative weight. With all partial products treated as positive, the entire multiplication can be carried out using a straightforward array-adder network. Conventional two's complement designs, on the other hand, generate both positive and negative partial products and therefore require extra correction circuitry to handle the negative terms [8].

a) Overview

In binary multiplication $n + m$ − bit product $P = (p_{n+m-1}, p_{n+m-2}, ..., p_0)$ formed by multiplying the n − bit multiplicand $Y = (Y_{m-1}, Y_{m-2}, ..., y_0)$. The AND of each multiplier bit and each multiplicand bit is formed to produce the partial product bits. The partial products are then summed to form the product.

The main challenge in two-'s-complement multiplication is dealing with the signs of the operands. Let Y_v denotes the numeric value of the multiplicand Y and X_v denotes the numeric value of the multiplier X. In two's complement representation, the values of X_v and Y_v are expressed by

$$Y_v = -y_{m-1}2^{m-1}\sum_{i=0}^{m-2} y_i 2^i \qquad (1a)$$

$$X_v = -x_{n-1}2^{n-1}\sum_{i=0}^{n-2} x_i 2^i \qquad (1b)$$

The value P_v of the product P

$$P_v = -p_{m+n-1}2^{m+n-1} + \sum_{i=0}^{m+n-2} p_i 2^i = Y_v X_v$$

$$= (-y_{m+n-1}2^{m+n-1} + \sum_{i=0}^{m-2} y_i 2^i)(-x_{n-1}2^{n-1} + \sum_{i=0}^{n-2} x_i 2^i)$$

$$= (x_{n-1}y_{m-1}2^{m+n-2} + \sum_{i=0}^{n-2}\sum_{i=0}^{m-2} x_i y_i 2^{i+j})$$

$$-(\sum_{i=0}^{m-2} x_{n-1}y_i 2^{n-1+i} + \sum_{i=0}^{n-2} y_{m-1}x_i 2^{m-1+i}) \quad (2)$$

When assembling P from its partial products, the sign of the partial product bits must be considered. In particular, the signs of $x_{n-1}y_i$ for $i = 0, ..., m-2$ and $y_{m-1}x_i$ for $i = 0, ..., n-2$ are negatives. By rewriting the partial product bits, all the partial product bits with the negative signs are placed in the last 2 rows. The product is formed by adding the first $n - 2$ partial product rows and subtracting the last two rows.

Instead of subtracting the partial products that have negative signs, the negation of the partial products can be added. The value of the negation of a two's complement number $Z_v = (z_{k-1}, ..., z_0)$ with the value Z_v is

$$-Z_v = 1 - z_{k-1}2^{k-1} + \sum_{i=0}^{k-2} \bar{z}_i 2^i \qquad (3)$$

Where $\underline{z}$ is the complement of z_i. Therefore, the subtraction of

$$2^{n-1}(-0 \cdot 2^m + 0 \cdot 2^{m-1} + \sum_{i=0}^{m-2} x_{n-1}y_i 2^i) \quad (4)$$

can be replaced with the addition of

$$2^{n-1}(-1 \cdot 2^m + 1 \cdot 2^{m-1} + 1 + \sum_{i=0}^{m-2} \overline{x_{n-1}y^i} 2^i) \quad (5)$$

Thus, the partial product row of Fig. 1 contains

$$0\ \ 0\ \ x_{n-1}y_{m-2}\ \ x_{n-1}y_{m-3} ..... x_{n-1}y_0$$

is replaced by

$$1\ \ 1\ \ \overline{x_{n-1}y_{m-2}}\ \ \overline{x_{n-1}y_{m-3}} ..... \overline{x_{n-1}y_0}$$

With a "1" added to the p_{n-1} column. Similarly, the row contains

$$0\ \ 0\ \ x_{n-2}y_{m-1}\ \ x_{n-3}y_{n-1} ..... x_0 y_{m-1}$$

is replaced by

$$1\ \ 1\ \ \overline{x_{n-2}y_{m-1}}\ \ \overline{x_{n-3}y_{m-1}} ..... \overline{x_0 y_{m-1}}$$

with a "1" added in the p_{m-1} column. Following these substitutions, all partial product bits can be treated in the same manner with respect to the sign.

b) Technical Implementation and circuit consideration

The substitution of (5) for (4) in Fig. 1 results in nonuniformity regarding partial product bits since some partial product bits are the NAND of a multiplier bit and multiplicand bit, while others are formed with an AND. To simplify this situation, the following equivalences are used. Note that (5) has the value

$$0 \qquad\qquad , for\ x_{n-1} = 0 \quad (6)$$
$$2^{n-1}(-2^m + 2^{m-1} + 1 + \sum_{i=0}^{m-2} \overline{y_i} 2^i)\ , for\ x_{n-1} = 1$$

From (6) it follows that (5) can be rewritten as
$$2^{n-1}(-2^m + 2^{m-1} + \overline{x_{n-1}} 2^{m-1} + x^{n-1} + \sum_{i=0}^{m-2} x_{n-1}\overline{y_i} 2^i) \qquad\qquad (7)$$

Consequently, the equation (5) can be shown to be algebraically identical to equation (7). Because the penultimate row of partial-product bits in Fig 1 follows the form of (5), it is replaced by the equivalent expression in (7). Applying a similar substitution to the final row, together with the incorporation of the requisite constant terms, yields the uniform partial-product array depicted in Figure 1. The defining characteristic of this arrangement is the consistency of its partial-product bits, which confers two principal benefits:

- Each partial-product bit is produced by ANDing one bit of the multiplier with one bit of the multiplicand.

- All of these partial-product bits are assigned positive weights.

As a result, the product can be generated using only logical ANDs followed by addition. No subtraction is necessary, nor is the NAND function needed to form $\overline{x_i y_j}$. In hardware implementation, separate AND gates are not needed to form the nm partial product bits. Both Pezaris and Hampel et al. have designed circuits that readily incorporate the AND operation into the additional circuits. Also, the five extra partial product bits of the algorithm, x_{n-1}, $\underline{x}_{n-1}, y_{m-1}, \underline{y}_{m-1}$, and "1", and can be included in a circuit implementation without increasing the total propagation delay of the two's complement multiplication.

B. Wallace Tree Algorithm

a) Overview

Multipliers are classified as fully parallel, fully serial, or digit-serial. Digit-serial designs process multi-bit digits to balance throughput and area, allowing pipeline, constant-rate operation. The Wallace tree multiplier, by contrast, is fully parallel: it generates all partial products, compresses them with a logarithmic-depth carry-save adder tree, and finishes with a carry-propagate adder. Although faster than array multipliers, its irregular interconnect and higher power draw complicate layout and impede low-power use.

b) Technical Implementation and circuit consideration

In real-time digital signal processing applications, the principal design objectives for computational building blocks are maximum throughput and numerical accuracy [9]. Because multiplication is fundamental to the operation of microprocessors, process controllers, DSP engines, and graphics units, overall system performance is ultimately governed by the efficiency of the multiplier [10],[11]. Advances in semiconductor technology have therefore driven extensive research into multiplier architectures that deliver higher speed and lower power dissipation, making them well suited for increasingly complex and portable VLSI implementations. Conventional multiplier architecture consists of three stages: (i) generation of partial products, (ii) accumulation (or reduction) of these partial products, and (iii) final carry-propagate addition to produce the product.

Fig. 2 shows a 4×4 Wallace-tree multiplier. 16 partial products are generated in parallel and then reduced by grouping three bits per column into carry-save adders, with any leftover bits passed to the next stage. This compression repeats until only two rows of bits remain. These final two rows are summed by a fast carry-propagate or ripple-carry adder to produce the 8-bit product.

Binary multiplication generates N^2 partial products in parallel using AND gates, arranging them into columns by weight. A Wallace-tree reduction then processes each column by grouping bits in threes: a carry-save adder produces one sum bit at the current weight and one carry to the next, while any unpaired bit simply advances. This three-to-two compression recurs until only two rows of bits remain, which are finally merged with a high-speed carry-propagate adder. By performing reductions in a logarithmic-depth, fully parallel fashion, the Wallace-tree multiplier achieves minimal critical-path delay.

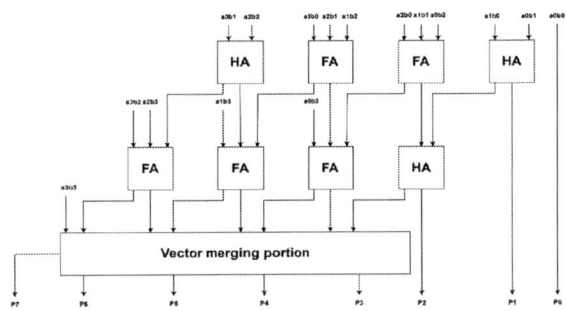

Fig. 2. Circuit diagram of 4x4 Wallace Tree multiplier architecture

III. PROPOSED ARCHITECTURE

A. Proposed Multiplier using Baugh Wooley algorithm and Wallace Tree structure

Creating the partial product by using Baugh Wooley, we multiply each bit of one operand (multiplicand, a) with each bit of the other operand (multiplier, b). For an n-bit signed multiplier, the partial products are formed as follows:

- From row 1 to row 7, the partial products (pp) are calculated as $pp_{ij} = a_i\ \&\ b_j$, where i and j denote the bit positions, for example: $pp_{45} = a_4\ \&\ b_5$. For the MSB of each row, the partial product is calculated as $!pp_{ij} = !(a_i\ \&\ b_j)$, for example: $!pp_{85} = !(a_8\ \&\ b_5)$.

								a7	a6	a5	a4	a3	a2	a1	a0
								b7	b6	b5	b4	b3	b2	b1	b0
							1	!pp70	pp60	pp50	pp40	pp30	pp20	pp10	pp00
							!pp71	pp61	pp51	pp41	pp31	pp21	pp11	pp01	
					!pp72	pp62	pp52	pp42	pp32	pp22	pp12	pp02			
				!pp73	pp63	pp53	pp43	pp33	pp23	pp13	pp03				
			!pp74	pp64	pp54	pp44	pp34	pp24	pp14	pp04					
		!pp75	pp65	pp55	pp45	pp35	pp25	pp15	pp05						
	!pp76	pp66	pp56	pp46	pp36	pp26	pp16	pp06							
1	pp77	!pp67	!pp57	!pp47	!pp37	!pp27	!pp17	!pp07							
						s14	s12	s10	s8	s6	s4	s3	s2	s1	pp00
						c12	s13	s11	s9	s7	s5	c2	c1		
			c16	s16	s15	c13	c10	c8	c6	c4	c3	pp03			
			!pp75	c15	c14	!pp72	c11	c9	c7	c5					
		!pp76	pp66	pp56	pp46	pp36	pp26	pp16	pp06						
1	pp77	!pp67	!pp57	!pp47	!pp37	!pp27	!pp17	!pp07							
					s15	s27	s25	s23	s21	s20	s19	s18	s17	s1	pp00
				c16	s16	s29	s28	s26	s24	s22	c19	c18	c17		
			pp77	s32	s31	s30	c27	c25	c23	c21	c20	c5			
1	c32	c31	c30	c29	c28	c26	c24	c22							
		s32	s42	s41	s40	s39	s38	s37	s36	s35	s34	s33	s17	s1	pp00
		pp77	c31	c41	c40	c39	c38	c37	c36	c35	c34	c33			
1	c32	c42	c30	c29	c28	c26	c24	c22							
1	s53	s52	s51	s50	s49	s48	s47	s46	s45	s44	s43	s33	s17	s1	pp00
	c53	c52	c51	c50	c49	c48	c47	c46	c45	c44	c43				
s64	s63	s62	s61	s60	s59	s58	s57	s56	s55	s54	s43	s33	s17	s1	pp00

(Legend: dotted-box groupings marked "Using Half Adder" and "Using Full Adder".)

Fig 3. Diagram of Baugh Wooley Wallace Tree 8-bit multiplier

- In row 8, the partial products (pp) are calculated as $!pp_{ij} = !(a_i\ \&\ b_j)$, for example, $!pp_{18} = !(a_1\ \&\ b_8)$. For the MSB of each row, the partial product is calculated as $pp_{ij} = a_i\ \&\ b_j$, for example: $pp_{88} = a_8\ \&\ b_8$.

By using Wallace Tree structure, we can reduce the large number of partial products to just two rows (a sum and carry) for final addition. This is done in stages using Half Adders (HA) and Full Adders (FA).

- Step 1: Grouping of Bits: Bits in the same column (corresponding to the same weight) are grouped together for reduction. Adders are applied to reduce the number of bits in each column to at most two.

- Step 2: Use of Adders

  - Full Adders (FA): Used when three bits are available in a column. Outputs a sum bit (placed in the current column) and a carry bit (propagated to the next higher column).

  - Half Adders (HA): Used when only two bits remain in a column. Outputs a sum and a carry.

For instance, in the 3rd column, we use Full adder to add pp_{20}, pp_{11} and pp_{02} and the output is sum s_1 and carry out is c_1.

As shown in Fig. 3, in Stage 0, the first step is to generate partial products that account for signed multiplication. This involves creating bitwise products of the multiplier and multiplicand, while also handling sign bits in a way that allows for correct two's complement arithmetic. Once the partial products are generated, Stage 1 begins to reduce their number using basic adders. The goal is to organize and combine the partial products in a way that simplifies downstream logic.

From Stage 2 through Stage 4, the focus is on reducing the number of partial products to simplify the final addition. The Wallace tree structure in Stages 2 and 3 further compresses them into two main groups of signals. Finally, Stage 4 uses a conventional adder to sum these groups and produce the final multiplication result.

B. Proposed convolution

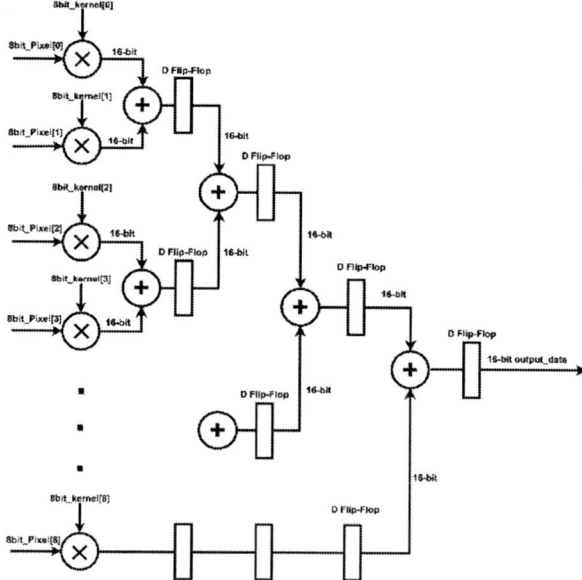

Fig 4. Block diagram of convolution layer using proposed high-speed multiplier

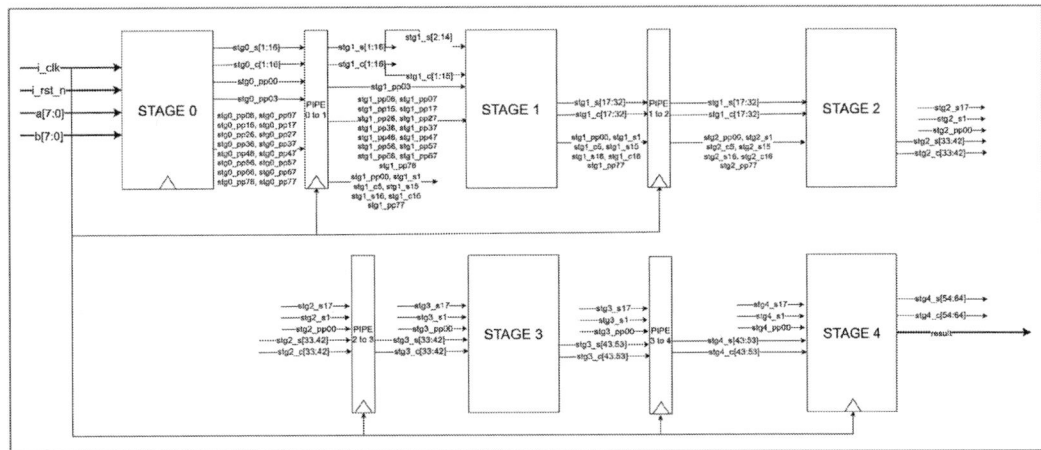

Fig. 5. Block diagram of Pipeline Baugh Wooley Wallace Tree 8-bit multiplier

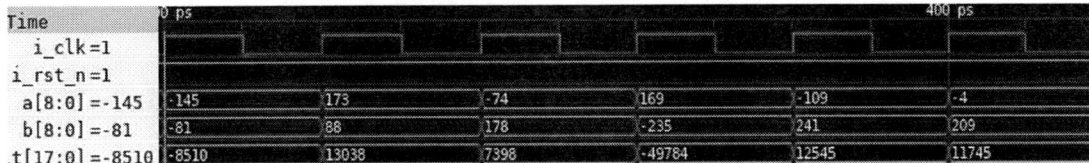

Fig. 6. Waveform of Baugh Wooley Wallace Tree Multiplier

Figure 4 illustrates a hardware-efficient convolutional layer design that utilizes multiple multipliers for convolution operations, with the number of multipliers determined by the selected kernel size. In this 3×3 example, nine multipliers are employed—one for each kernel element—to perform parallel multiplications between the kernel values and corresponding image pixels. The individual results are then summed to form the final output pixel, which is passed on for subsequent image reconstruction. This design can be scaled for larger kernels (e.g., 5×5 or 7×7) simply by increasing the number of multipliers to match the added kernel elements. Moreover, as depicted in Figure 5, each multiplier is pipelined to improve throughput and reduce the critical path, allowing partial computations to be processed in stages and enabling higher operating frequencies. The proposed multiplier is optimized for efficient hardware implementation, potentially leveraging approximate computing, reduced-bit multiplication, or custom logic designs to minimize power consumption and area usage while maintaining computational accuracy.

IV. EXPERIMENTAL RESULT

A. Experimental setup

In this proposed design, the RTL code is written in SystemVerilog, offering advanced constructions for modeling and describing complex digital systems. For simulation, we use GTKWave software to verify the design's functionality and behavior, while Vivado 2018.3 is utilized for synthesis to implement the design on target hardware. In the convolution layer, we use an image with a size of 256 x 256 and a 3x3 sharpening kernel. The multiplier handles multiplying pixel values with the values in the kernel. This setup aims to enhance the image clarity by emphasizing edges and fine details, thus testing the performance of the Wallace Tree combined with the Baugh-

Wooley multiplier in executing intensive computational tasks.

B. Simulation result

To verify the functionality of the Baugh Wooley Multiplier, we use a testbench to randomly choose the values of a and b in Fig. 6. However, this multiplier runs on a 4-stage pipeline, so the output results are delayed by 4 clock cycles for the first input data. For example, at 300ps, the values of a and b are -145 and -81 respectively, and 4 cycles later (at 400ps), the result is 11745.

Fig. 7. Output grayscale image size 254x254 after applying Sharpening kernel

In Fig. 7, the simulation output shows the successful application of a 3x3 Sharpening kernel to the 256x256 grayscale input picture, which results in a final size of 254x254. The sharpened image visually shows sharper edges and greater information, proving that the filter logic in the design is working as it expected. This outcome not only demonstrates that the hardware implementation was right, but it also demonstrates how well the sharpening technique highlighted important aspects of the image.

C. FPGA synthesis

The proposed design has been synthesized on Xilinx Zynq UltraScale+ FPGA and Xilinx Virtex – 7 FPGA by Vivado software. Fig. 8 and Fig. 9 provide a summary of the resource usage on the Xilinx FPGA. The results indicate that only a small portion of the FPGA's capabilities are utilized, as seen by the generally low percentages of resource usage across most categories.

Resource	Estimation	Available	Utilization %
LUT	120	230400	0.05
LUTRAM	3	101760	0.01
FF	207	460800	0.04
IO	38	360	10.56
BUFG	1	544	0.18

Fig. 8. Analysis & Synthesis Resource Usage Summary of proposed design on Xilinx Zynq UltraScale+ FPGA

Resource	Estimation	Available	Utilization %
LUT	171	303600	0.06
LUTRAM	3	130800	0.01
FF	207	607200	0.03
IO	38	700	5.43
BUFG	1	32	3.13

Fig. 9. Analysis & Synthesis Resource Usage Summary of proposed design on Xilinx Virtex – 7 FPGA

D. Performances comparison

TABLE I. PROPOSED POWER CONSUMPTION AND FREQUENCY COMPARED TO RELATED WORKS

Para-meters	Proposed	Proposed	[4]	[5]	[6]	[7]
Power (mW)	349	705	0.05825	65	0.228	-
Frequency (MHz)	540.5	769.231	72.73	92.52	270.27	122.2
Slice of LUTs	171	120	90	146	38	61
Latency (ns)	1.85	1.3	13.75	10.81	3.7	8.183
Technology (FPGA)	Xilinx Virtex-7	Xilinx Ultra-Scale+	Xilinx Spartan-6	Xilinx Spartan-3E	Xilinx Sparta-6	Xilinx Virtex-7

Table I compares the proposed designs with related works, highlighting differences in power consumption, frequency, LUT usage, and latency for various FPGA implementations. The design in [4] using a hybrid compressor achieves the lowest power consumption at 0.05825 mW, but it has the highest latency at 13.75 ns, indicating a focus on energy efficiency over speed. The fast binary multiplication based on customized basic cells in [5] consumes 65 mW of power with a latency of 10.81 ns, striking a balance between power consumption and latency, offering decent performance with relatively low power usage. In contrast, the multi-level approximate multipliers in [6] have the power consumption at 0.228 mW and the latency of 3.7 ns, making it highly efficient both in terms of energy and speed. The study on FPGA's dual-port ROM-based 8x8 multiplier for area optimization in [7] does not specify the power consumption, but it operates at a frequency of 122.2 MHz with a latency of 8.183 ns, indicating moderate latency. The proposed design which has been implemented on Xilinx Virtex-7 and Xilinx Ultra-Scale+ demonstrates a significant enhancement in operating frequency compared to other studies. This improvement is thanks to the use of the efficient Baugh-Wooley algorithm and the Wallace Tree structure. The pipeline technique

allows the proposed design to process one multiplication operation in each clock cycle. Therefore, the throughput of the proposed design is equal to the operating frequency with 540.5 and 769.231 million operations per second on Xilinx Virtex-7 and Xilinx Ultra-Scale+, respectively.

V. CONCLUSION

The implementation of the pipelined Baugh-Wooley Wallace Tree Multiplier demonstrates significant improvements in performance metrics such as latency, operating frequency, and the throughput, when integrated into the convolutional stage of CNNs. By utilizing the pipelining technique on the Xilinx Virtex – 7 and Xilinx Zynq UltraScale+ and verifying the design through GTKWave simulations, this work highlights the multiplier's potential to address the computational challenges of deep learning workloads. The results validate the superiority of the pipelined approach over the single-stage design, particularly in real-time processing scenarios, while maintaining the accuracy of CNN computations. This research serves as a step forward in developing hardware-efficient accelerators for modern deep learning applications.

ACKNOWLEDGMENT

We acknowledge the support of time and facilities from Ho Chi Minh City University of Technology (HCMUT), VNU-HCM for this study.

REFERENCES

[1] Soulef Bouaafia, Seifeddine Messaoud, Randa Khemiri, Fatma Ezahra Sayadi., "An FPGA-SoC based Hardware Acceleration of Convolutional Neural Networks", IEEE, 2022.

[2] Truong Quang Vinh, Dinh Viet Hai, "Optimizing Convolutional Neural Network Accelerator on Low-Cost FPGA," 2021.

[3] Beura SK, Devi BP, Saha PK, Meher PK, "Design of a Novel Inexact 4: 2 compressor and its placement in the partial product array for area, delay, and power-efficient approximate multipliers". Circuits Syst Signal Process 43(6):3748–3774, 2024

[4] Thamizharasan V, Parthipan V, "Design of efficient binary multiplier architecture using hybrid compressor with FPGA implementation". Sci Rep 14(1):8492, 2024

[5] Al-Nounou A.A., Al-Khaleel O., Obeidat F., Al-Khaleel M, "FPGA Implementation of Fast Binary Multiplication Based on Customized Basic Cells". Journal of Universal Computer Science, vol. 28, no. 10, 2022, pp. 1030-1057.

[6] N. V. Toan; J.-G. Lee, "FPGA-based Multi-Level Approximate Multipliers for High-Performance Error-Resilient Applications," IEEE Access, vol. 4, pp. 1-17, 2020.

[7] Pathan A, Memon T, "FPGA's Dual-Port ROM-Based 8x8 Multiplier for Area Optimized Implementation of DSP Systems". Iranian Journal of Electrical and Electronic Engineering, Vol. 17, No. 4, pp. 2011, 2021.

[8] C. R. Baugh and B. A. Wooley, "A two's complement parallel array multiplication algorithm", IEEE Transactions on Computers, vol. C-22, no. 12, pp. 1045–1047, Dec. 1973.

[9] G Wallace, C. S., "A Suggestion for a Fast Multiplier", IEEE Transactions on Computers, vol. 13, pp. 14-17, 1964.

[10] Y. Harata; Y. Nakamura; H. Nagase; M. Takigawa; N. Takagi, "A high speed multiplier using a redundant binary adder tree", IEEE Journal of Solid-State Circuits, vol. 22, pp. 28- 34,1987.

[11] P. G. McCrea; W. S. Matheson, "Design of high-speed fully serial tree multiplier", IEEE Proceedings E - Computers and Digital Techniques, vol. 128, pp. 13-20,1981.

High-Efficiency 4:2 Compressor Designs: A Comparative Study on Hardware Cost and Error Trade-Offs

Vishnu Padmakumar
Electronics and Communication
Engineering Department
National Institute of Technology
Silchar
Silchar, India
vishnu.padmakumar00@gmail.com

Adhiraj Nandy
Electronics and Communication
Engineering Department
National Institute of Technology
Silchar
Silchar, India
adhirajnandy9@gmail.com

Sourav Nath
Electronics and Communication
Engineering Department
National Institute of Technology
Silchar
Silchar, India
souravnath_rs@ece.nits.ac.in

Koushik Guha
Electronics and Communication
Engineering Department
National Institute of Technology
Silchar
Silchar, India
koushik@ece.nits.ac.in

Krishna Lal Baishnab
Electronics and Communication
Engineering Department
National Institute of Technology
Silchar
Silchar, India
klb@ece.nits.ac.in

Saroj Kr Biswas
Computer Science and Engineering
Department
National Institute of Technology
Silchar
Silchar, India
saroj@nits.ac.in

Abstract—**Approximate computing enhances energy, area, and performance in error-resilient applications like image processing and ML. This paper proposes two novel 4:2 compressors optimized for hardware efficiency while preserving accuracy. Integrated into an 8×8 approx. multiplier with a PPR structure, our compressors achieve significant gains in hardware and error metrics. We benchmark them against state-of-the-art designs, evaluating area, power, delay, PDP, and PDAP, along with error metrics (NMED, MRED, PRED). Experimental results show Compressor I reduces power by 8.2%, PDP by 11.7%, and PDAP by 9.5% over conventional designs. Trade-off analysis highlights compressors near the origin in error-efficiency plots as optimal in cost and accuracy. Our designs outperform existing architectures, making them ideal for power-efficient AI/ML accelerators and high-performance applications. FPGA validation (Vivado), ASIC synthesis (Genus, TSMC 65nm), and Python-based error analysis confirm accuracy and reliability.**

Index Terms—**Approximate computing, approximate multipliers, 4:2 compressor, partial product reduction, error analysis, FPGA implementation, ASIC synthesis.**

I. INTRODUCTION

ENERGY efficiency is a major challenge in modern system-on-chips (SoCs), where exact circuits face performance and power limitations [1]. With increasing multimedia applications in portable devices, traditional components lead to higher power dissipation and costs [2]. Approximate computing offers a solution for error-tolerant applications like image processing and machine learning, where exact precision is unnecessary [3], [4].

Multiplication, a key yet energy-intensive operation, has driven research into approximate multipliers, categorized by approximation at partial product generation, reduction, or final addition. The reduction stage, being the most power- and area-intensive, is a prime target for approximation [10]. Truncation further minimizes hardware by stabilizing least significant bits (LSBs) [9].

Building on preliminary work [12], this paper proposes two novel 4:2 compressors designed to balance hardware efficiency and computational accuracy. The key contributions are:

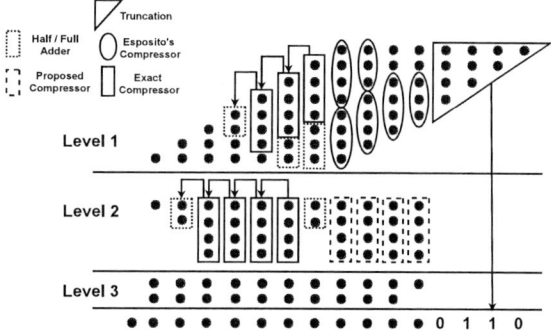

Fig. 1: Partial product reduction structure [6].

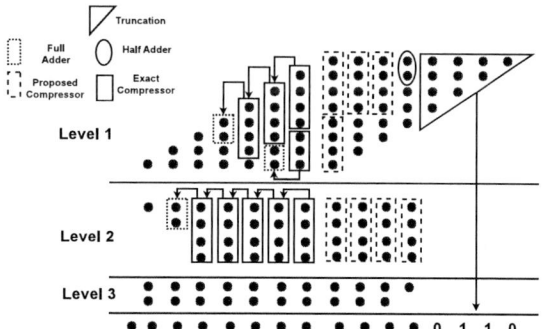

Fig. 2: Partial product reduction structure [2].

1) Two new 4:2 compressors optimized for power, area, and delay while maintaining superior accuracy.

2) Comprehensive hardware and error analysis using FPGA (Vivado) and ASIC synthesis (Cadence Genus, TSMC 65nm), evaluating power, area, delay, ER, NMED, and PRED.

3) A detailed comparative study of state-of-the-art compressors, analyzing hardware efficiency and error characteristics across multiple benchmarks.

II. RELATED WORK

A. In-Literature PPR Structure

We consider two partial product reduction (PPR) structures in Fig. 1 and Fig. 2, based on [6] and [2].

In Fig. 1, a hybrid 8-bit approximate multiplier integrates Esposito's compressor [9] with the proposed design. Level 1 combines exact and Esposito's compressors, while Level 2 primarily uses the proposed compressors. An RCA sums the partial products. Fig. 2 employs exact compressors in the least significant region and proposed compressors elsewhere. Both structures replace the least significant region with a constant

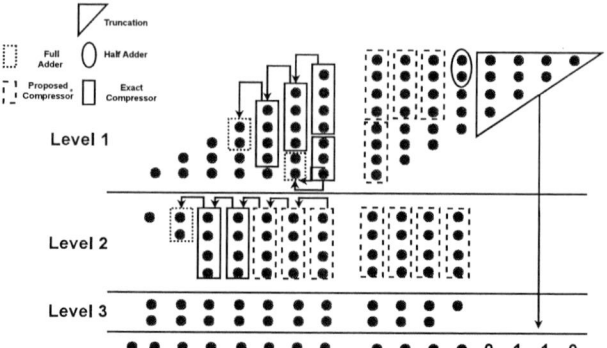

Fig. 3: Partial Product Reduction Structure [12].

approximation for hardware savings. The PPR structure in [2] partitions partial products into three regions: least, middle, and accurate—using constant correction logic, half adders, approximate 4:2 compressors, and exact adders finalized by an RCA.

Error correction modules are omitted for broader applicability across compressor architectures.

B. Previously Proposed Approximate 4:2 Compressors

To reduce area and power, most approximate 4:2 compressors use only S and C outputs, simplifying design at the cost of accuracy. The Lin compressor [1] employs logic gates and a multiplexer for high accuracy but increases complexity. The Kumar compressor [2] also uses a multiplexer, introducing four errors while maintaining accuracy. The Strollo compressor [3] integrates a full adder, causing two errors but increasing area and power. The Momeni compressor [7] reduces gate usage, leading to four errors but improving efficiency.

These designs illustrate the trade-off between efficiency and accuracy.

C. Preliminary Work Presented at MNCDCS

Our preliminary investigations [12] introduced a novel PPR structure as shown in Fig. 3.

III. PROPOSED 4:2 COMPRESSORS

The proposed 4:2 compressors demonstrate a simpler architecture than existing designs (Fig. 4). Unlike conventional approaches using multiplexers or full adders, both designs rely only on basic logic gates, reducing complexity while maintaining competitive error characteristics. Design I achieves hardware efficiency with a minimal gate combination, limiting errors to 5 while maintaining a maximum error magnitude of ± 1.

TABLE I: Truth Table for both In Literature and Proposed Compressors

A3...A0	Prob	Lin [1] CS	E	Kumar [2] CS	E	Strollo [3] CS	E	Momeni [7] CS	E	Proposed compressor 1 CS	E	Proposed compressor 2 CS	E
0 0 0 0	81/256	00		00		00		01	+1	00		00	
0 0 0 1	27/256	01		01		01		01		10	+1	01	
0 0 1 0	27/256	01		01		01		01		00	-1	01	
0 0 1 1	9/256	10		01	-1	10		01	-1	10		10	
0 1 0 0	27/256	01		01		01		01		10	+1	01	
0 1 0 1	9/256	10		10		10		10		10		10	
0 1 1 0	9/256	10		10		10		10		10		10	
0 1 1 1	3/256	11		10	-1	10	-1	11		11		11	
1 0 0 0	27/256	01		01		01		01		01		01	
1 0 0 1	9/256	10		10		10		10		11	+1	01	-1
1 0 1 0	9/256	10		10		10		10		10		01	-1
1 0 1 1	3/256	11		10	-1	11		11		11		11	
1 1 0 0	9/256	10		10		10		01	-1	10		11	+1
1 1 0 1	3/256	11		11		11		11		11		11	
1 1 1 0	3/256	11		11		11		11		11		11	
1 1 1 1	1/256	10	-2	11	-1	11	-1	11	-1	11	-1	11	-1

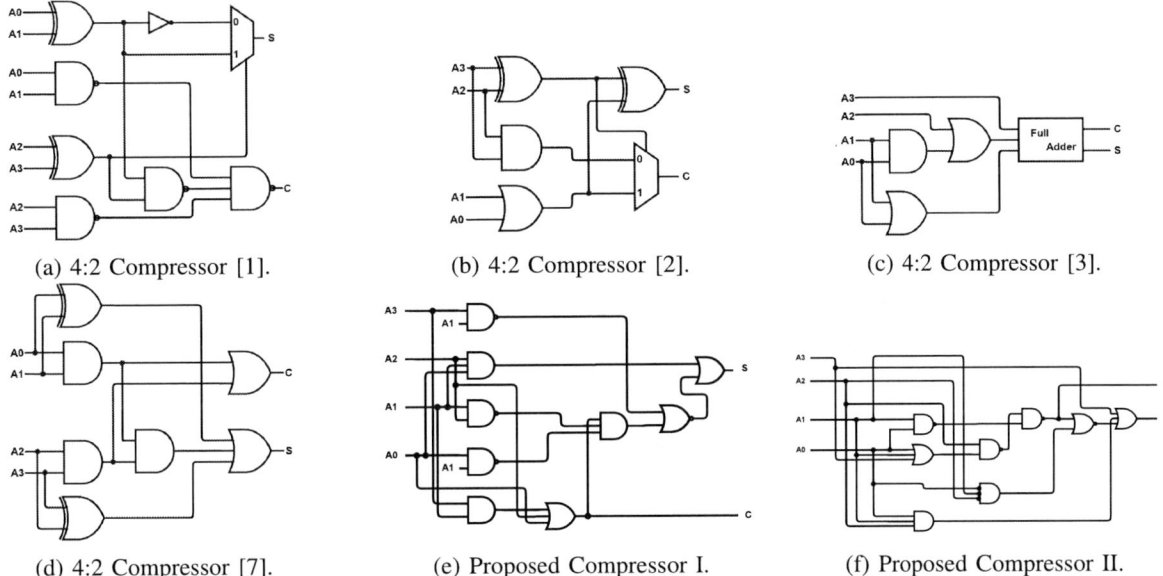

(a) 4:2 Compressor [1].   (b) 4:2 Compressor [2].   (c) 4:2 Compressor [3].

(d) 4:2 Compressor [7].   (e) Proposed Compressor I.   (f) Proposed Compressor II.

Fig. 4: Structure of existing and proposed 4:2 compressors.

Design II improves accuracy with just 4 errors, qualifying as a High-Accuracy Compressor per [11], while retaining the same error bound. This combination of simplified gate-level implementation and constrained error magnitude enables both designs to balance hardware efficiency and computational accuracy, making them suitable for error-resilient applications requiring moderate precision.

IV. METHODOLOGY

A. Hardware Implementation and Analysis

The 4:2 compressors were first designed in Verilog and validated using Xilinx Vivado. Insights from simulations guided refinements for better area, power, and performance trade-offs while ensuring accuracy. For hardware analysis, designs were synthesized in Cadence Genus using TSMC 65nm technology to extract area, power, and delay metrics. This process was repeated for various 4:2 compressor combinations across PPR architectures, ensuring consistent com-

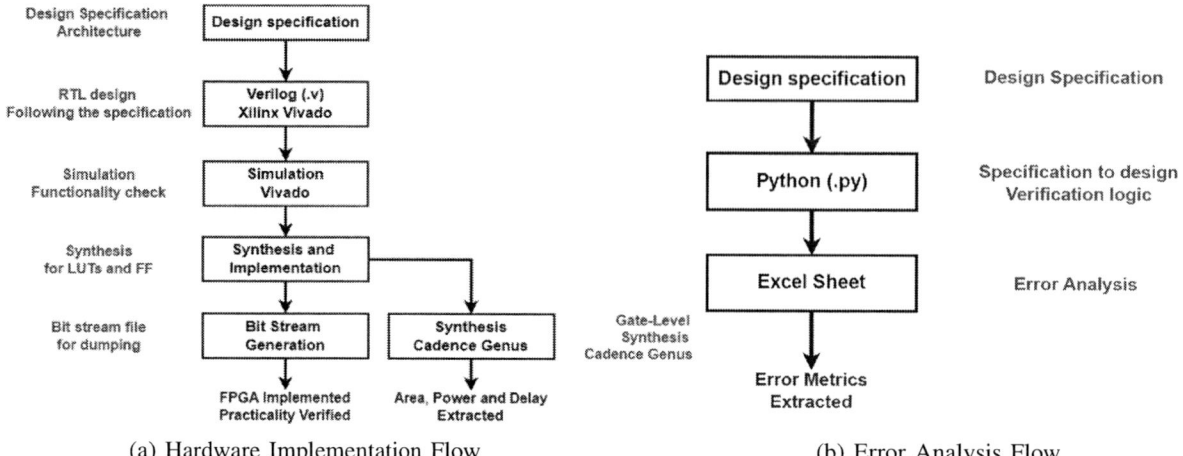

(a) Hardware Implementation Flow   (b) Error Analysis Flow

Fig. 5: Hardware and Error Analysis Flow for All Results Obtained.

TABLE II: Error and Hardware Analysis of Different 4:2 Compressors in PPR Structure [6]

Compressor Type	Error Analysis				Hardware Analysis				
	ER (%)	NMED ($\times 10^{-3}$)	MRED ($\times 10^{-2}$)	PRED ($\times 10^{-1}$)	Area (μm^2)	Power (μW)	Delay (ps)	PDP (fJ)	PDAP (fJ-μm^2)
Compressor I	99.76	29.44	19.30	6.83	613.92	30.71	891	27.36	16798.45
Compressor II	99.34	29.18	17.57	6.30	621.60	31.48	1011	31.83	19783.22
Lin [1]	99.90	29.73	24.73	7.23	625.92	30.97	960	29.73	18609.35
Kumar [2]	98.87	28.69	17.11	6.09	600.96	31.07	990	30.76	18485.11
Strollo [3]	98.75	28.96	17.22	6.04	629.76	30.92	978	30.24	19043.79
Momeni [7]	99.84	29.13	23.98	6.76	599.52	28.62	950	27.19	16300.35

TABLE III: Error and Hardware Analysis of Different 4:2 Compressors in PPR Structure [2]

Compressor Type	Error Analysis				Hardware Analysis				
	ER (%)	NMED ($\times 10^{-3}$)	MRED ($\times 10^{-2}$)	PRED ($\times 10^{-1}$)	Area (μm^2)	Power (μW)	Delay (ps)	PDP (fJ)	PDAP (fJ-μm^2)
Compressor I	99.88	4.09	6.39	4.84	596.64	25.57	916	23.42	13974.57
Compressor II	99.94	4.11	5.30	4.55	585.60	27.58	925	25.51	14939.53
Lin [1]	99.92	6.01	27.60	5.94	608.16	28.24	960	27.11	16487.46
Kumar [2]	99.53	4.21	5.32	4.65	572.16	30.31	941	28.52	16318.98
Strollo [3]	99.09	3.92	4.99	4.34	585.12	30.83	1061	32.71	19139.64
Momeni [7]	99.84	4.00	19.85	4.53	528.96	24.55	958	23.51	12440.55

parisons (Fig. 5a). The practical implementation was conducted on the Artix-7 Nexys 4 Board (XC7A100T-1CSG324C) using Xilinx Vivado, selected for its balance of performance, power efficiency, and cost-effectiveness.

B. Error Evaluation

Various metrics evaluate approximate arithmetic circuits [3], [9]. Given exact and approximate outputs A and A', the Error Distance (ED) is defined as

$$ED = |A - A'|. \qquad (1)$$

The Relative Error Distance (RED) is given by

$$RED = \begin{cases} \dfrac{ED}{|A|}, & A \neq 0, \\ 0, & A = 0. \end{cases} \qquad (2)$$

For an n-bit multiplier, the maximum possible output (MaxOut) is

$$MaxOut = 2^{2n-2}. \qquad (3)$$

Key performance metrics include Error Rate (ER)(percentage of operations with ED $>$ 0), Normalized Mean Error Distance (NMED) (mean ED normalized by MaxOut), Mean Relative Error Distance (MRED) (mean RED), and Probability

TABLE IV: Error and Hardware Analysis of Different 4:2 Compressors in Preliminarily Proposed PPR Structure

Compressor Type	Error Analysis				Hardware Analysis				
	ER (%)	NMED ($\times 10^{-3}$)	MRED ($\times 10^{-2}$)	PRED ($\times 10^{-1}$)	Area (μm^2)	Power (μW)	Delay (ps)	PDP (fJ)	PDAP (fJ-μm^2)
Compressor I	99.94	9.69	13.60	7.09	564.96	21.51	986	21.21	11982.16
Compressor II	99.94	5.81	6.13	5.40	561.60	25.48	891	22.70	12749.83
Lin [1]	99.97	16.88	91.77	8.38	589.92	26.12	935	24.42	14407.14
Kumar [2]	99.54	6.07	6.10	5.68	544.32	27.19	918	24.96	13586.46
Strollo [3]	99.09	4.17	5.10	4.51	563.04	28.87	1020	29.45	16580.06
Momeni [7]	99.91	9.94	90.87	6.41	486.24	22.90	896	20.52	9976.87

TABLE V: Average Error and Hardware Analysis of Different 4:2 Compressors Across Three Designs

Compressor Type	Error Analysis				Hardware Analysis				
	ER (%)	NMED ($\times 10^{-3}$)	MRED ($\times 10^{-2}$)	PRED ($\times 10^{-1}$)	Area (μm^2)	Power (μW)	Delay (ps)	PDP (fJ)	PDAP (fJ-μm^2)
Compressor I	99.86	14.41	13.10	6.25	591.84	25.93	931.00	24.00	14251.73
Compressor II	99.74	13.60	9.94	5.70	589.60	28.18	942.33	26.68	15824.19
Lin [1]	99.93	17.54	48.03	7.18	608.00	28.44	951.67	27.09	16501.32
Kumar [2]	99.31	12.99	9.51	5.47	572.48	29.52	949.67	28.08	16130.18
Strollo [3]	98.98	12.35	9.10	4.96	592.64	30.21	1019.67	30.80	18254.50
Momeni [7]	99.91	16.34	68.57	6.53	524.00	24.81	914.00	22.74	12084.70

of RED exceeding 2% (PRED). These metrics collectively assess accuracy and efficiency.

All designs were translated from Verilog to Python, and using a Jupyter Notebook we exhaustively simulated all $2^8 \times 2^8 = 65,536$ input combinations. The resulting outputs were recorded to compute the error metrics shown in Fig. 5b.

V. RESULTS AND DISCUSSION

A. Hardware Analysis

Our experiments demonstrate that the proposed designs, Compressor I and Compressor II, deliver notable hardware gains over traditional implementations. As shown in Tables II, III, and IV, both compressors consistently reduce area, power, delay, PDP, and PDAP across various PPR structures. From the averages in Table V, Compressor I lowers power consumption to 25.93μW versus 28.24μW for conventional designs—a reduction of approximately 8.2%. Its delay drops by 2.9% (931.00ps vs. 958.75ps), PDP falls by 11.7% (24.00fJ vs. 27.18fJ), and PDAP is trimmed by 9.5% (14251.73fJ-μm² vs. 15742.68fJ-μm²). Compressor II achieves more modest benefits, with only slight improvements in delay and PDP, and a marginal uptick in PDAP. These findings underscore Compressor I's suitability for high-performance, low-power applications.

B. Error Analysis

Beyond hardware improvements, our compressors maintain competitive accuracy relative to traditional designs. Although error metrics (NMED and MRED)

vary by PPR topology, both Compressor I and Compressor II uphold acceptable error bounds. Table V confirms a balanced compromise between efficiency and accuracy. While some conventional methods (e.g., Lin [1], Strollo [3]) may slightly outperform in raw error metrics, they incur substantially higher hardware costs.

C. Trade-Off Analysis

The design closest to the origin indicates the most efficient balance between accuracy and hardware cost. Lower PDAP and lower error metrics are both critical for an efficient compressor in error-tolerant applications. Figure 6 plots PDAP against NMED and MRED. In Fig. 6a, although Momeni achieves the lowest NMED and Strollo and Kumar exhibit superior PDAP, Proposed Compressor I and Compressor II lie nearest to the origin, signifying the best overall trade-off. Similarly, in Fig. 6b, despite Momeni having the smallest MRED, Proposed Compressor I and II provide the optimal balance of accuracy and efficiency. Notably, Proposed Compressor I outperforms all other designs, validating its position as the premier compressor architecture.

VI. CONCLUSION

This paper has introduced two novel 4:2 compressors Compressor I and Compressor II featuring gate-level simplicity and tightly bounded error behaviour. Comprehensive synthesis in TSMC 65nm and exhaustive error simulation demonstrate that Compressor I achieves

979-8-3315-1550-8/25 $31.00 © 2025 IEEE

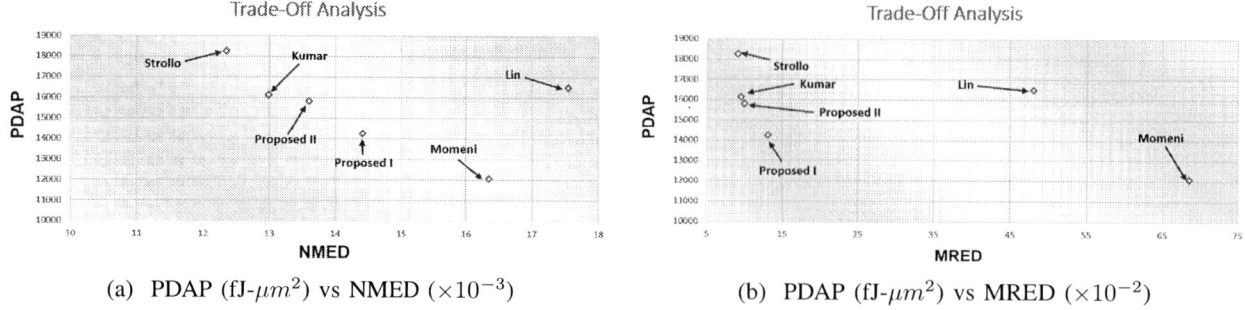

(a) PDAP (fJ-μm^2) vs NMED ($\times 10^{-3}$)   (b) PDAP (fJ-μm^2) vs MRED ($\times 10^{-2}$)

Fig. 6: Trade-off analysis between error metrics and PDAP.

an 8.2% reduction in power, an 11.7% reduction in PDP, and a 9.5% reduction in PDAP relative to conventional designs, while maintaining error metrics comparable to the state of the art. Trade-off analysis shows that designs nearest the origin in PDAP–error plots deliver the most balanced performance; both proposed compressors lie closer to this ideal point than Lin, Kumar, Strollo, and Momeni, with Compressor I exhibiting the best overall balance of accuracy and hardware efficiency. These results validate the suitability of Compressor I for high-performance, low-power, error-tolerant applications.

VII. FUTURE WORK

In future work, we will develop a clear method for designing approximate compressors and use it to create higher order versions (for example, 5:2 compressors). We will add the newest approximate compressor and multiplier techniques to our comparison tests for thorough validation. Finally, we will apply our compressors to real image processing tasks, such as filtering and sharpening, to measure their effect on accuracy and resource usage.

REFERENCES

[1] C. -H. Lin and I. -C. Lin, "High accuracy approximate multiplier with error correction," 2013 IEEE 31st International Conference on Computer Design (ICCD), Asheville, NC, USA, 2013, pp. 33-38, doi: 10.1109/ICCD.2013.6657022.

[2] U. A. Kumar, S. K. Chatterjee and S. E. Ahmed, "Low-Power Compressor-Based Approximate Multipliers With Error Correcting Module," in IEEE Embedded Systems Letters, vol. 14, no. 2, pp. 59-62, June 2022, doi: 10.1109/LES.2021.3113005.

[3] A. G. M. Strollo, E. Napoli, D. De Caro, N. Petra and G. D. Meo, "Comparison and Extension of Approximate 4-2 Compressors for Low-Power Approximate Multipliers," in IEEE Transactions on Circuits and Systems I: Regular Papers, vol. 67, no. 9, pp. 3021-3034, Sept. 2020, doi: 10.1109/TCSI.2020.2988353.

[4] S. Venkatachalam and S. -B. Ko, "Design of Power and Area Efficient Approximate Multipliers," in IEEE Transactions on Very Large Scale Integration (VLSI) Systems, vol. 25, no. 5, pp. 1782-1786, May 2017, doi: 10.1109/TVLSI.2016.2643639.

[5] V. Padmakumar, T. M. Ignatius, T. B. Singha and R. P. Palathinkal, "A Serial-Parallel-Based 4-Bit Novel Multiplier: Design, Implementation, and Performance Analysis," 2023 IEEE Silchar Subsection Conference (SILCON), Silchar, India, 2023, pp. 1-6, doi: 10.1109/SILCON59133.2023.10404909.

[6] M. Zhang, S. Nishizawa and S. Kimura, "Area Efficient Approximate 4–2 Compressor and Probability-Based Error Adjustment for Approximate Multiplier," in IEEE Transactions on Circuits and Systems II: Express Briefs, vol. 70, no. 5, pp. 1714-1718, May 2023, doi: 10.1109/TCSII.2023.3257852.

[7] A. Momeni, J. Han, P. Montuschi and F. Lombardi, "Design and Analysis of Approximate Compressors for Multiplication," in IEEE Transactions on Computers, vol. 64, no. 4, pp. 984-994, April 2015, doi: 10.1109/TC.2014.2308214.

[8] O. Akbari, M. Kamal, A. Afzali-Kusha and M. Pedram, "Dual-Quality 4:2 Compressors for Utilizing in Dynamic Accuracy Configurable Multipliers," in IEEE Transactions on Very Large Scale Integration (VLSI) Systems, vol. 25, no. 4, pp. 1352-1361, April 2017, doi: 10.1109/TVLSI.2016.2643003.

[9] D. Esposito, A. G. M. Strollo, E. Napoli, D. De Caro and N. Petra, "Approximate Multipliers Based on New Approximate Compressors," in IEEE Transactions on Circuits and Systems I: Regular Papers, vol. 65, no. 12, pp. 4169-4182, Dec. 2018, doi: 10.1109/TCSI.2018.2839266.

[10] L. Sayadi, S. Timarchi and A. Sheikh-Akbari, "Two Efficient Approximate Unsigned Multipliers by Developing New Configuration for Approximate 4:2 Compressors," in IEEE Transactions on Circuits and Systems I: Regular Papers, vol. 70, no. 4, pp. 1649-1659, April 2023, doi: 10.1109/TCSI.2023.3242558.

[11] T. Kong and S. Li, "Design and Analysis of Approximate 4–2 Compressors for High-Accuracy Multipliers," in IEEE Transactions on Very Large Scale Integration (VLSI) Systems, vol. 29, no. 10, pp. 1771-1781, Oct. 2021, doi: 10.1109/TVLSI.2021.3104145.

[12] V.Padmakumar, A.Nandy, S.Nath, K.Guha and KL Baishnab, "Design and Synthesis of Area and Power-Efficient Approximate Multipliers with Novel 4:2 Compressors" has been presented in MNDCS 2025 to be published in Springer Lecture Notes.

2025 10th IEEE International Conference on Integrated Circuits, Design, and Verification (ICDV)

SDR Implemented Algorithm for Real-Time Intra-pulse Modulated Radar Signal Analysis

[1]Van Minh Duong, [2]Thi Phuong Nguyen, Nhat Giang Phan, Duy Cong Nguyen, Quang Hoa
Nguyen,Manh Long Nguyen, Tan Phat Huynh
Le Quy Don Technical University.
[1]minhdv_k31@lqdtu.edu.vn, [2]phuongnt@lqdtu.edu.vn

Abstract— **This paper proposes an algorithm for detecting and recognizing intra-pulse modulated radar signals based on software-defined radio (SDR). The proposed algorithm is divided into two sub-processes. The first sub-process solves the problem of detecting signals under noise conditions. The second is used for recognition. First, the functionality of the proposed method was tested with simulated signals in MATLAB. Then, the algorithm's performance was implemented and verified with real-time signals using SDR. The simulated and experimental results showed that the proposed algorithm could recognize intra-pulse modulated signals such as linear frequency modulation (LFM), Barker, and Costas signals under strong noise conditions. Also, the proposed method can be implemented in an actual system.**

Keywords—intra-pulse modulated radar signals, software defined radio, technical analysis, recognition.

I. INTRODUCTION

Intra-pulse radar-modulated signals are crucial and widely used in modern electronic warfare systems [1], [12], and [13]. The advantages of these signals include ensuring the essential functions of radar systems such as surveillance range, target resolution in range and speed, and maintaining confidentiality due to their broad spectrum and sub-noise level presence [2]. As a result, the analysis and identification of intra-pulse modulated signals have recently received significant attention. These methods can be broadly categorized into two groups: using transformations and using Artificial Intelligence

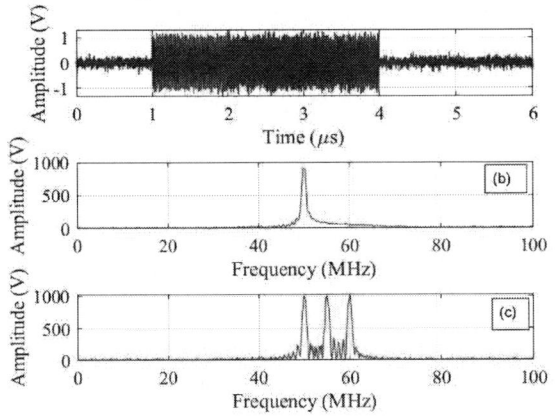

Fig. 1. Example of analyzing Costas signal.

One of the initial methods used to analyze pulse modulation signals is the Fast Fourier Transform (FFT). The advantage of this approach is its rapid processing time. The major disadvantage is the poor performance with signals in low Signal-to-Noise Ratio (SNR ≤ 10 dB) conditions and its accuracy dependence on the number of DFT points. For example, computing FFT with noise and varying FFT points is illustrated in Fig. 1. Fig. 1(b) shows that with 2048 points, only one frequency can be determined, while 8912 points are required to determine the three frequencies of a Costas signal (see Fig. 1(c)) [3]. Other groups for analyzing the signals are time-frequency transformations, such as Short-Time Fourier Transform (STFT) [4], Wigner-Ville Distribution (WVD) [5], Choi-Williams Distribution (CWD) [6], and Wavelet transform (WT) [7]. The drawback of these methods is the substantial computational requirement, rendering them impractical for real-world applications.

In addition, with the development of AI, numerous network architectures have been proposed to address the problem of recognizing pulse modulation signals, such as SqueezeNet, AlexNet, and DarkNet [8]. The primary use of these network structures relies on time-frequency analysis results for recognition. As a result, the recognition accuracy depends on the quality of the input images. Also, the effective operation of these networks requires a large number of sample trials. Moreover, the mentioned network structures have only been tested in simulations and not yet in real-world devices. To overcome these challenges, this paper proposes and conducts empirical validation of a new recognition algorithm for pulse modulation signals using SDR. The paper is organized into three main sections: Section 2 presents the theoretical foundation of the proposed algorithm. Section 3 details the simulation results. The summarized results are presented in Section 4

II. PROPOSED METHOD

The block diagram of the proposed method is shown in Fig. 2. The proposed method includes three modules. The ANTSDR E310 module is used to capture signals from radar. Inside the ANTSDR E310, the received signals are converted into a baseband and passed to the ADC block; then the I-Q output signals from the SDR will be entered into a computer (PC). The detection and recognition process will be performed on the PC, as described below.

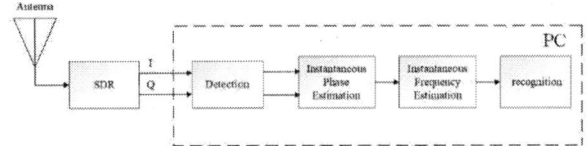

Fig. 2. Block diagram of the proposed method.

979-8-3315-1550-8/25 $31.00 © 2025 IEEE

A. Amplitude detector

This paper uses the cell-average constant false alarm rate (CA-CFAR) concept to detect signals [9]. The block diagram of CA-CFAR is shown in Fig. 3.

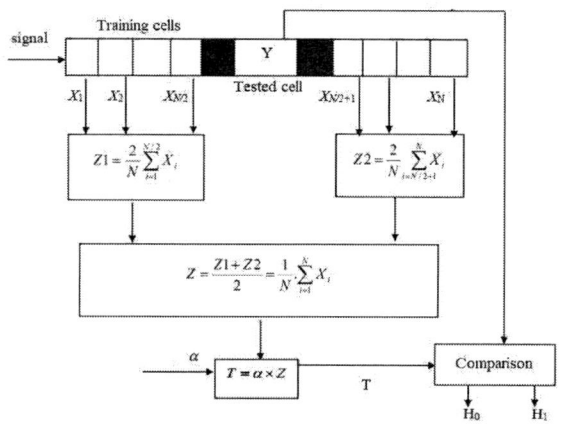

Fig. 3. Block diagram of CA-CFAR [9].

This paper uses the cell-average constant false alarm rate (CA-CFAR) concept to detect signals. The block diagram of CA-CFAR is shown in Fig. 3. The structure of the CA-CFAR system includes training memories, a test cell, and guard cells. The CA-CFAR operates on the principle that the received signals are stored in the CA-CFAR's memory buffer, where it measures the average values on the left branch Z_1, right branch Z_2, and the combined average of both branches Z. This average Z is then multiplied by a fixed constant α. The result of that multiplication is called the detection threshold. Particularly, the detection threshold is calculated according to the following formula (1):

$$T = \frac{\alpha}{N} \sum_{i=1}^{N} X_i \tag{1}$$

In the formula, T is the detection threshold, N is the number of data points, and X_i is the data point in the input series. The constant α is calculated by (2):

$$\alpha = N \times \left(P_{fa}^{-\frac{1}{N}} - 1 \right) \tag{2}$$

where P_{fa} is the probability of false alarm.

The detection threshold T is then compared to the tested cell Y; in this case, it is the received signal. At the beginning of the comparison, if Y > T, the decision is that there is target if $Y \leq T$, the decision is that there is no target. The most significant advantage of the CA-CFAR is that it adapts to different noise environments or non-homogeneous noise features, which may lead to false alarm phenomena. Alternatively, other methods might not be able to achieve the exact initial detection without encountering the same issue.

B. Recognition process

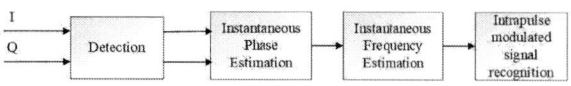

Fig. 4. Block diagram of the recognition process.

The block diagram of the recognition part is shown Fig. 4. The principle of the algorithm is written below. The principle of the algorithm is described below. The detected signals pass through the phase estimator, where their amplitude $A(t)$ and instantaneous phase $\phi(t)$ are calculated by (3) and (4):

$$\begin{cases} A = \sqrt{I^2 + Q^2} \\ \phi = \arctan\left(\dfrac{Q}{I}\right) \end{cases} \tag{3}$$

$$\Delta\phi = \arctan\left(\frac{Q_2}{I_2}\right) - \arctan\left(\frac{Q_1}{I_1}\right) \tag{4}$$

Frequency is the rate of phase change; the average frequency over a sample interval is the phase change over the sampling interval divided by the sampling interval. Since the phase is defined over a range of 360 degrees, it is necessary to add 360 degrees when the phase change from one sample to another is too significant. This can be avoided by using the expression for the difference between two arctangent values (4). The instantaneous frequency $f(t)$ is calculated by (5).

$$f(t) = \frac{d\phi(t)}{dt} = \frac{\Delta\phi}{dt} \tag{5}$$

III. SIMULATION RESULTS

This section tests the performance of the proposed method using simulated radar signals such as radar pulse, linear frequency modulation (LFM), Barker code, Costas frequency and Frank coded signals in MATLAB environments [10]. The general parameters for the simulation setup, related to noise, common parameters of simulated signals, and the CFAR detector, are listed in TABLE I. The parameters for simulated signals are listed in TABLE II.

TABLE I. GENERAL SIMULATION PARAMETERS

Names	Parameters	Value
Sample rate	$f_s\,(MHz)$	1000
Carrier frequency	$f_c\,(MHz)$	100
Pulse width	$\tau\,(\mu s)$	1
Pulse repetition interval	PRI (μs)	10
Type of noise	White Gaussian noise	
Signal to noise ratio	SNR (dB)	0
Probability of detection	$P_d\,(-)$	0.9
Probability of false alarm	$P_{fa}\,(-)$	1e-6
Training cell	N (-)	6

TABLE II. PARAMETERS OF SIMULATED SIGNALS

Signals	Parameters	Value
LFM	$BW\,(MHz)$	100
Costas frequency coded	Frequency step $\Delta f\,(MHz)$	20
Barker coded	Code length	7, 11, 13

A. Radar pulse

Fig. 5 shows the transmitted and received signals with SNR = 0 dB. The output of the amplitude detector is shown in Fig. 6(a). The pulse width and estimated error can be calculated by (6). The instantaneous frequency of the received signal is given by (7) and Fig. 6(b) shows that the IF does not change during the pulse width. It concludes that

979-8-3315-1550-8/25 $31.00 © 2025 IEEE

the type of signal is a radar pulse. (6) and (7) prove that the proposed method can be used for estimating parameters (pulse width and carrier frequency) and type of received signal. The proposed method will be applied to analyze the LFM signal in the following subsection.

$$\begin{cases} \tau_e = t_2 - t_1 = 1.997 - 1.002 = 0.995 \, (\mu s) \\ \Delta \tau = |\tau_e - \tau| = |0.995 - 1.0| = 5 \, (ns) \end{cases} \quad (6)$$

$$\begin{cases} f_{ce} = 95.18 \, (MHz) \\ \Delta f_c = |f_{ce} - f_c| = |95.18 - 100| = 4.82 \, (MHz) \end{cases} \quad (7)$$

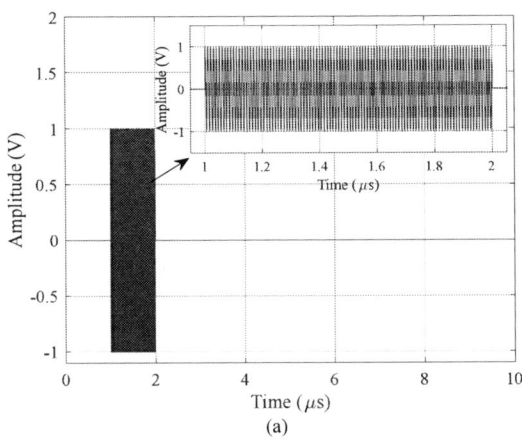

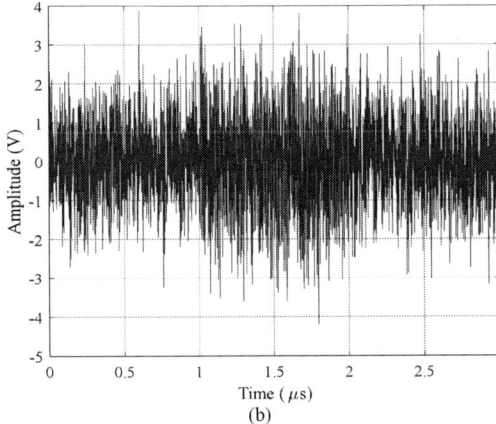

Fig. 5 Radar pulse: (a) transmitted; (b) received with SNR = 0 dB.

B. Linear frequency modulation

This subsection will analyze the LFM signal with SNR = 0 dB. The output of the amplitude detector and the instantaneous frequency are shown in Fig. 7. Fig. 7(a) shows the envelope of the received signal from the amplitude detector (red line). Also, the detected signal passes through pulse width and repetition interval measurements. Instantaneous (blue line) and estimated frequencies (red line) are given in Fig. 7(b).

It can be seen that the start $f_d = 95.28 \, (MHz)$ and stop frequencies $f_h = 194.63 \, (MHz)$ and bandwidth (BW) and pulse width of LFM are calculated by (8). The simulation result showed that the proposed method could detect and estimate the parameters of the LFM signal with SNR = 0 dB ($\Delta BW = 0.65 MHz$ and $\Delta \tau = 3 \, (ns)$).

$$\begin{cases} BW = 194.63 - 95.28 = 99.35 \, (MHz) \\ \Delta BW = |100 - 99.35| = 0.65 \, (MHz) \\ \tau_e = t_2 - t_1 = 2.0 - 0.997 = 1.003 \, (\mu s) \\ \Delta \tau = |\tau_e - \tau| = |1.003 - 1.0| = 3 \, (ns) \end{cases} \quad (8)$$

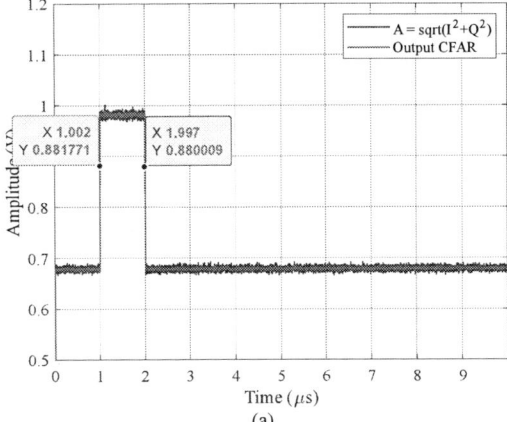

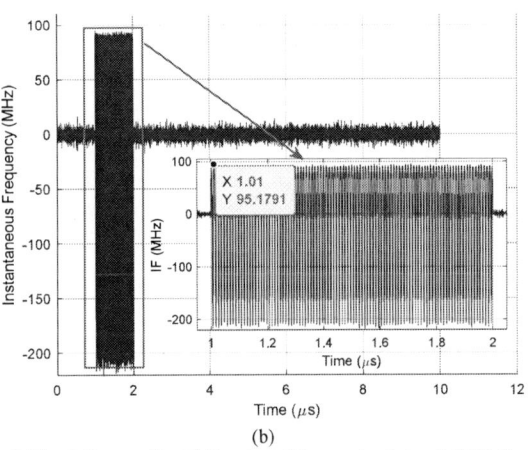

Fig. 6. Simulation results: (a) Envelop of the received signal; (b) Estimated frequency

C. Polyphase and Costas coded signals

In the same step, we tested the algorithm's functionality with different radar signals, such as polyphase coded (Barker, Frank) and Costas 3. The estimated parameters of these signals are shown in Fig. 8. Fig. 8(a) shows a frequency change of the received signal at times t = 4, 6, 7 (us) (zoom-in figure). Based on the results, we can identify it as a phase-coded signal (specifically, Barker 7).

Fig. 8(b) shows that the estimated frequencies of the received signal are $f_1 = 99.98 \, (MHz)$, $f_2 = 137.43 \, (MHz)$ and $f_3 = 119.19 \, (MHz)$. The frequency step Δf of Costas is calculated by (9).

$$\Delta f = \frac{|f_2 - f_1|}{2} = \frac{|137.43 - 99.98|}{2} = 18.75 \, (MHz) \quad (9)$$

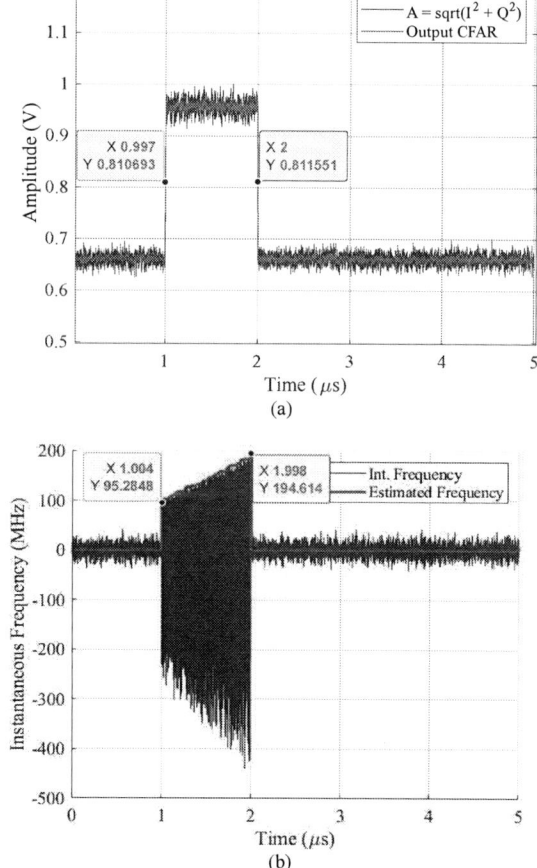

Fig. 7. Simulation results: (a) Envelop of the received signal; (b) Estimated frequency

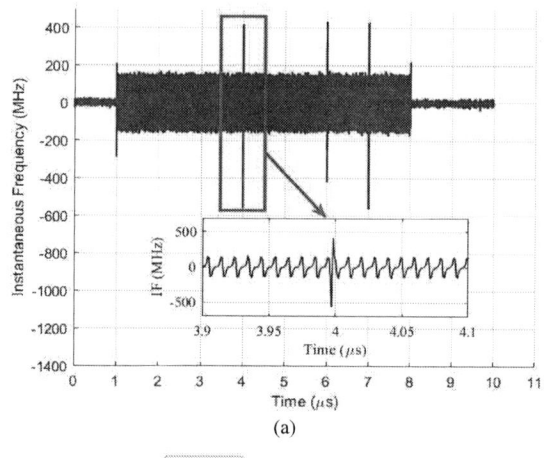

(a)

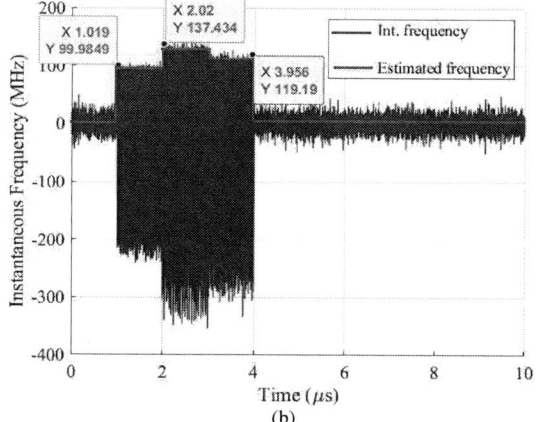

(b)

Fig. 8. Simulation results: (a) Barker; (b) Costas

Fig. 9 illustrates the instantaneous frequency over time of the received signal. It can be seen that the frequency of the signal changes five times (as shown in the zoomed-in figure) over the pulse lifetime. As a result, we can conclude that the form of the signal is a polyphase coded signal.

TABLE III.shows the processing time and required SNR value for analyzing 2000 samples per signal modulation type. It is seen on the side of SNR that AI provided the best performance (SNR = -8 dB). However, it requires the longest processing time (t=500 s). The fastest method is FFT, which only analyzes signals with high SNR (SNR = 6 dB). Also, it is seen that the proposed method can process signals with SNR = 0 dB (better than FFT) and requires a shorter time (t=1.64 s) than AI (t=500 s). From the results, the proposed method has a great opportunity to be implemented in a real electronic system. In the next step, the functionality of the proposed method will be verified using real-time signals.

TABLE III. Performance comparison at P_d = 90 %

Methods	Time processing (s)	SNR (dB)
FFT [1]	0.56	6
STFT [11]	8.29	-2
WVD [11]	236.58	-4
AI [8]	500	-8
Proposed method	1.64	0

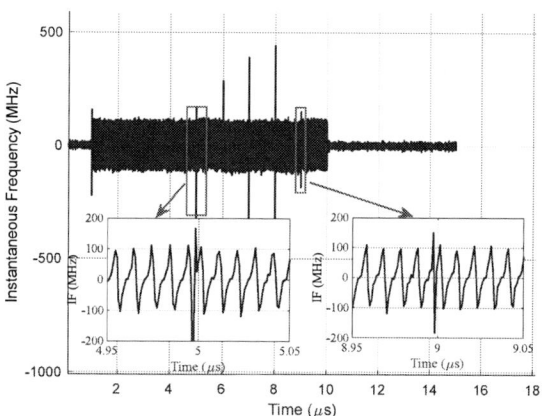

Fig. 9. Instantaneous frequency of polyphase coded signal

IV. EXPERIMENTAL RESULTS

Based on the simulation results, this section verifies the functionality of the proposed method using real-time signals. The experimental measurement setup to evaluate the performance of the proposed method is shown in Fig. 10 and Fig. 11. The parameters of the generated signals are listed in TABLE IV. The LFM signal generated in the time and frequency domain is shown in Fig. 12. It is seen that the FFT cannot determine the type of signal (Fig. 12(b)). The detected signal and its instantaneous frequency are shown on Fig. 13.

979-8-3315-1550-8/25 $31.00 © 2025 IEEE   118

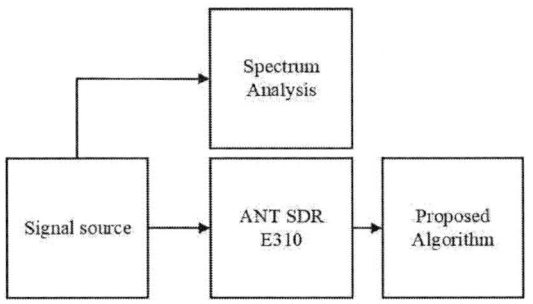

Fig. 10. Block diagram for experimental setup.

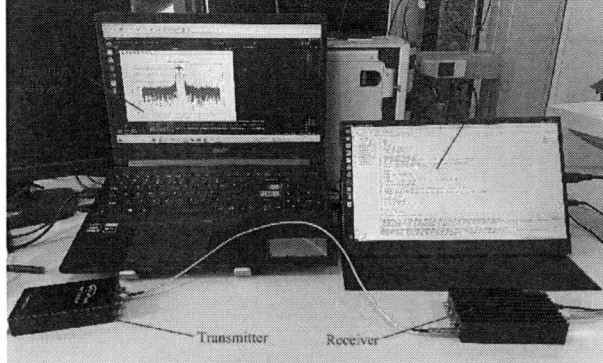

Fig. 11. Experimental measurement.

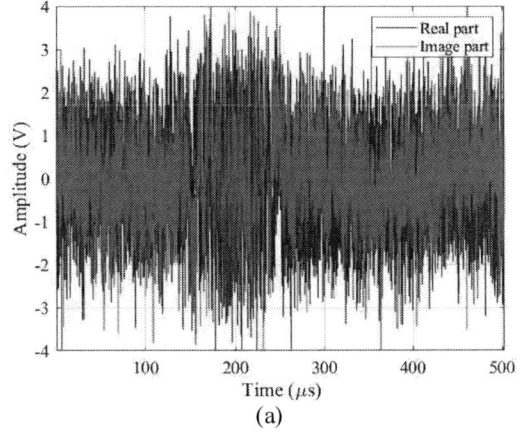

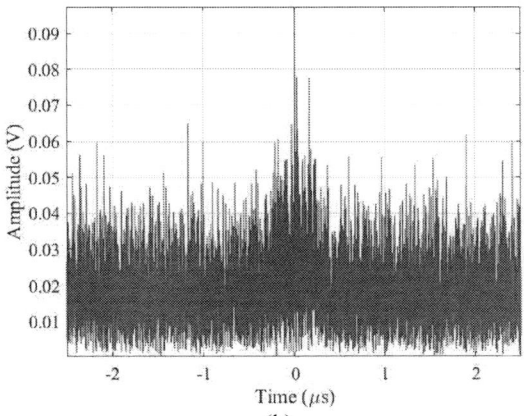

Fig. 12. Generated signal: (a) time domain; (b) frequency domain.

TABLE IV. PARAMETERS OF GENERATED SIGNALS

Signals	Parameters	Value
LFM	$f_c \, (MHz)$	2000
	BW (MHz)	1
	$\tau \, (\mu s)$	100
	Power (dBm)	-60

Fig. 13(a) shows the amplitude of the received signal. From this, the pulse width τ and the error ε are given by (10). The frequency change of the received signal is shown in Fig. 13(b). We can estimate the type of received signal and strongly confirm that it is the LFM.

$$\begin{cases} \tau_e = t_2 - t_1 = 250.9 - 149.7 = 101.2 \, (\mu s) \\ \Delta\tau = |\tau_e - \tau| = |101.2 - 100| = 1.2 \, (\mu s) \end{cases} \quad (10)$$

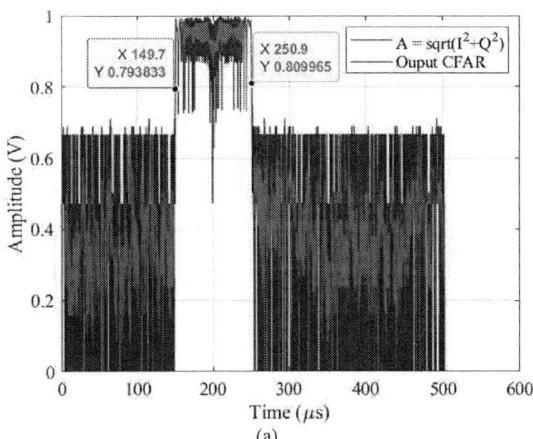

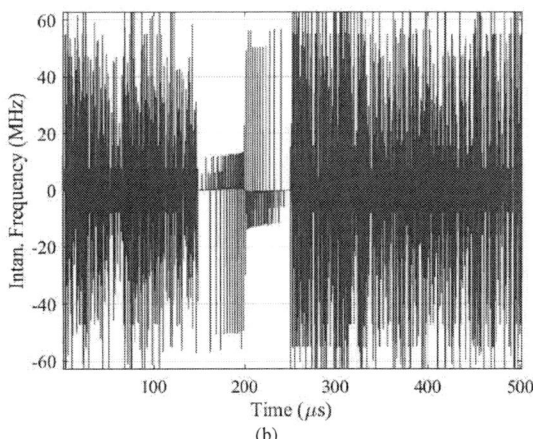

Fig. 13. Detected signal: (a) amplitude; (b) instantaneous frequency

V. CONCLUSION

This paper presented and implemented an algorithm for analyzing intra-pulse modulated radar signals in SDR. First, the performance of the proposed method was tested with simulated radar signals in a MATLAB environment. The simulation results showed that the proposed method could analyze radar signals such as LFM, Barker, Costas, and Frank-code with a low SNR value (SNR = 0 dB). Later, the functionality of the proposed method was verified using real-time radar signals. The experimental results showed that the algorithm can analyze weak radar signals (P = -60 dBm). On the other hand, although AI-based methods can recognize intra-pulse radar signals with SNR = -8 dB, the current AI-based methods are challenging to deploy on real electronic warfare systems because they require high-configuration computers, large databases, and long processing times, while our method does not.

REFERENCES

[1] R. G. Wiley, ELINT: The Interception and Analysis of Radar Signals. Norwood, MA, USA: Artech House, 2006.

[2] N. Levanon and E. Mozeson, Radar Signals. Wiley, New Jersey, 2004.

[3] J. R. Pennington, "Radar signal characteristic extraction with FFT-based techniques," M.S. thesis, Dept. Electr. Eng., Air Force Inst. Technol., Wright-Patterson AFB, OH, USA, 2007.

[4] F. Hlawatsch and G. F. Boudreaux-Bartels, Time-Frequency Analysis: Concepts and Methods. IEEE Press, New York, USA, 1992.

[5] P. Flandrin, Time-Frequency/Time-Scale Analysis. Academic Press, San Diego, CA, USA, 1998.

[6] L. Cohen, Time-Frequency Analysis. Prentice Hall, Upper Saddle River, NJ, USA, 1995.

[7] S. Mallat, A Wavelet Tour of Signal Processing: The Sparse Way. Academic Press, Burlington, MA, USA, 2008.

[8] D. Li, R. Yang, X. Li, and S. Zhu, "Radar signal modulation recognition based on deep joint learning," IEEE Access, vol. 8, pp. 48515–48528, Mar. 2020, doi: 10.1109/ACCESS.2020.2978875.

[9] M. A. Richards, Fundamentals of Radar Signal Processing, 2nd ed. New York, NY, USA: McGraw-Hill, 2014, ch. 8.

[10] B. R. Mahafza, Radar Signal Analysis and Processing Using MATLAB, 2nd ed. CRC Press, 2016.

[11] B. Boashash, Time-Frequency Signal Analysis and Processing: A Comprehensive Reference, 2nd ed. Oxford, UK: Academic Press, 2015.

[12] N. Levanon and E. Mozeson, Radar Signals. Hoboken, NJ, USA: Wiley, 2004.

DDoS Attack Detection for Software-Defined Network Architecture Based on Artificial Intelligence

Bao Pham-Thai[1,2], My Nguyen-Le-Ha[1,2], Luan Van-Thien[1,2], Thuat Nguyen-Khanh[1,2] and Quan Le-Trung[1,2]

[1]*Faculty of Computer Networks and Communications, University of Information Technology, Ho Chi Minh City, Viet Nam*
[2]*Vietnam National University, Ho Chi Minh City, Vietnam*
21520156@gm.uit.edu.vn, 23520964@gm.uit.edu.vn, luanvt@uit.edu.vn, thuatnk@uit.edu.vn and quanlt@uit.edu.vn

Abstract—The Software-Defined Networking (SDN) model provides significant advantages for network administrators by facilitating traffic initialization, control, and management. SDN optimizes bandwidth and resource utilization through efficient traffic routing and network management automation. Distributed Denial-of-Service (DDoS) attacks target websites and servers by disrupting network services and depleting application resources. In SDN, control functions are centralized in the control plane, while network devices handle packet forwarding. DDoS attacks in SDN impact all network layers, exploiting bandwidth and limiting infrastructure scalability. To mitigate DDoS attacks, Artificial Intelligence (AI)-based approaches, specifically Machine Learning (ML) and Deep Learning (DL), are integrated into Intrusion Detection and Prevention Systems for enhanced detection and response. Our proposed approach involves building an SDN network architecture, collecting traffic, and utilizing ML and DL models for DDoS detection. In ML models, feature extraction is performed before classification, while DL models preprocess network traffic by normalizing and converting it into images for classification. Among the evaluated models, Random Forest (RF) and DenseNet demonstrated the best performance, achieving a high accuracy of 99.87% and 99.4% while excelling in training time, inference speed, and model size.

Index Terms—Software-Defined Network, DDoS Attack Detection, Artificial Intelligence, Intrusion Detection

I. INTRODUCTION

SDN is an effective solution for facilitating the management and control of networks within modern architectures. SDN offers superior flexibility, making network management more streamlined while optimizing network resources and enhancing methods for network control. The SDN architecture consists of three primary layers: the Application Layer, which includes network applications communicating their requirements to the SDN controller (Ryu) via the northbound interface; the Control Layer, a logically centralized entity responsible for processing requests from the application layer and managing the underlying network infrastructure via the southbound interface while providing an abstract view of the network; and the Data Layer processing and forwarding data packets, consisting of physical or software-based network devices, such as Open vSwitch, Ryu. Unlike traditional networking architectures, where devices such as routers or switches can only recognize the states of neighboring devices, SDN addresses several inherent drawbacks. These limitations include scalability, security concerns, high operational costs, and the complexity of troubleshooting due to the distributed nature of the traditional system architecture. Conversely, the state of the entire network can be monitored and managed centrally through a centralized SDN controller. This process removes the need to handle individual devices within the system, improving management efficiency and reducing operational costs.

There are various general attack methods targeting network models, with DDoS attacks utilizing botnet models to overwhelm targets. The primary goal of these attacks is to disrupt network operations, leading to congestion and compromising availability, particularly for service-providing network systems. While many countermeasures have been deployed to mitigate such attacks, adversaries continuously refine their techniques. Furthermore, misclassifications between normal and DDoS-induced traffic can negatively impact users. Consequently, defensive mechanisms require continuous improvement, and the application of AI for DDoS detection in SDN networks has gained significant attention. ML approaches have also been explored in this context. AI-based approaches, particularly in SDN, have attracted interest in DDoS detection. Traditional ML methods are effective in feature extraction but require expert selection [1]. ML techniques, including Support Vector Machine (SVM) and Decision Tree (DT), were employed to classify network traffic based on extracted features conducted by Sudar et al. [2]. DL methods have also shown promise in the working of utilizing SDN controller-inspected traffic data, including IP addresses, protocols, and ports to process through a Gated Recurrent Unit (GRU) model [3]. Both approaches demonstrated high classification accuracy while DL models have proven effective, further research is needed to assess the practical applicability of both ML and DL techniques for real-world DDoS detection. Existing studies share a common approach of extracting network traffic features and applying ML and DL models. However, most focus solely on accuracy, while aspects such as training, inference time, or model size remain underexplored.

Hence, the main contributions of our study are summarized as follows:

- **Development of an SDN network model:** We develop an SDN environment using Miniedit, enabling the generation of both normal and attack traffic for capturing and analysis.
- **Application of ML and DL for intrusion detection via network traffic imaging:** We evaluate the effectiveness of multiple models' performance in network traffic classification, including ML (RF, DT, LightGBM, Dummy) and DL (CNN, DenseNet, ResNet, EfficientNet, Inception, VGG). By transforming raw network traffic data into RGB images, DL models can learn complex patterns directly, reducing manual feature processing and enhancing accuracy.
- **Comprehensive performance analysis and insights:** We assess the models using metrics like accuracy, precision, recall, F1-score, and Matthews correlation coefficient (MCC), offering insights into ML/DL for SDN intrusion detection. Furthermore, performance is accessed comprehensively via training, inference time and model size.

The remainder of this study is organized as follows. **Section II** presents methods for model development, including AI algorithms and feature extraction techniques for data processing, providing an overview of the experimental steps conducted in our study. **Section III** details the build of the SDN network model and the proposed AI-based approach covering data preprocessing, feature extraction for ML, image transformation for DL models, and model application for DDoS detection in SDN environments. **Section IV** evaluates the performance of developed models based on predefined criteria. Finally, **Section V** summarizes the findings and outlines future directions for improving the study.

II. RELATED WORKS

There are many methods to detect and mitigate DDoS attacks in SDN. The most commonly used approach involves packet capturing, traffic analysis, and mitigation. Network analysis can be either packet-based or flow-based. Each packet containing fundamental network information is treated as an instance in packet-based analysis, while flow-based analysis aggregates packets by certain extracted features. These flows are then analyzed to classify network behavior and detect potential DDoS attacks. In our model, flow-based analysis is applied for ML as it reduces data volume, summarizing flos into CSV rows before being fed into ML models. Conversely, packet-based analysis is used for DL to handle detailed and complex data. Packets are processed directly from PCAP files and converted into images, serving as inputs for DL models to classify network traffic.

Numerous ML and DL defense mechanisms have been implemented for DDoS detection. Models like SVM, k-Nearest Neighbors (KNN), and Naïve Bayes (NB) are widely used and can be combined to enhance classification effectiveness. Hu et al. [4] propose FADM. This lightweight SDN framework collects network data from the SDN controller and agents and then analyzes it using an SVM classifier to identify network anomalies. Experimental results demonstrate that this model could detect and effectively mitigate DDoS attacks through a whitelist mechanism, ensuring rapid network recovery. Deepa et al. [5], by integrating ML algorithms like KNN, NB, SVM, and Self-Organizing Map, the model can determine anomalies among the incoming network. Since the SDN controller does not directly process packets, switches handle packets from connected devices. Their adoption of a hybrid model improves accuracy, detection rate, and false alarm reduction over single-algorithm models. The application of ML and DL in DDoS attack detection and mitigation has been increasingly researched and deployed, particularly in SDN environments.

In addition to ML models, DL models have proven highly effective, outperforming ML by directly extracting features from network data. Janabi et al. [6] introduce a model named DL-EWPS, which consists of three distinct modules. This model collects flow statistics, extracts IP-based features, and converts them into RGB images for classification, demonstrating near-perfect accuracy with low latency in large-scale networks. In another study, Alashhab et al. [7] propose LD-DoS, a DL model leveraging a Recurrent Neural Network (RNN) with Long Short-Term Memory (LSTM) activation functions for detecting low-rate DDoS attacks in SDN-enabled IoT networks. This model effectively analyzes characteristic values of different LDDoS attack types alongside legitimate traffic and achieves high accuracy. Clinton et al. [8] proposes a CNN model that transforms network traffic data into images, which achieves superior results over existing works. While DL models offer high accuracy and resilience, their computational demands may impact real-time applicability in resource-constrained environments.

III. PROPOSED SYSTEM

This study comprises two main models. The SDN network model is designed to simulate a network environment, ensuring that the collected traffic aligns with real-world scenarios. Regular traffic consists of Web and video services, while DDoS traffic is generated using Hping3. The network topology and traffic collection process details are illustrated in **Fig. 1**.

The second model focuses on traffic classification, as illustrated in **Fig. 2**. This architecture integrates both ML and DL techniques for comprehensive network traffic analysis. In the DL pipeline, raw PCAP files are first divided into 1MB chunks, from which byte-level values are extracted and formatted into 3D arrays. These arrays are then converted into RGB images and resized before being passed through deep learning models for classification. Conversely, the ML pipeline involves extracting essential features from PCAP files, converting them into CSV format, and applying conventional machine learning algorithms for classification. Both approaches aim to distinguish between normal and attack traffic effectively.

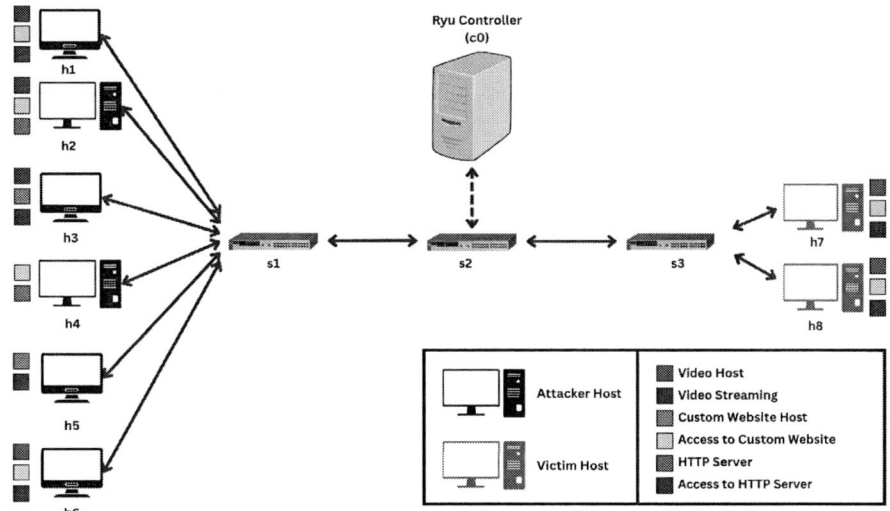

Fig. 1. SDN Testbed Model for Traffic Generation.

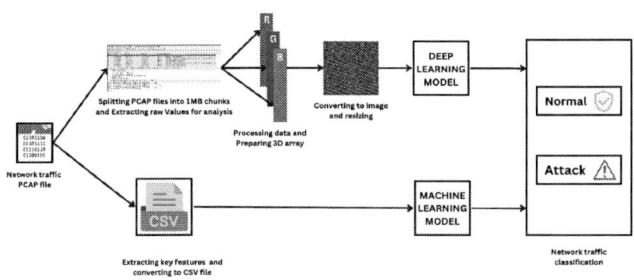

Fig. 2. Training Model.

A. Network setup

In our study, we built an SDN testbed to generate attack and normal traffic flows. Our model was constructed with an SDN controller, 3 OpenFlow switches, and 8 hosts, developed with Miniedit as a graphical interface integrated into the Mininet network emulator. This setup enabled intuitive and efficient network design.

DDoS attacks were simulated in an SDN environment using Hping3, with six attacker hosts targeting two victims via randomized source IPs. Packet capture was conducted at the central switch interface using TCPDUMP on the SDN controller (c_0) for real-time analysis. Normal traffic included video streaming and web services. VLC streamed videos over HTTP, UDP, and RTP, with hosts 6–8 as clients and hosts 1–4 as streamers. Web services were hosted on hosts 3 and 5, accessible to all other hosts. Additionally, hosts 2 and 4 provided HTTP services to the network.

B. Pre-processing

For DL models, network traffic data is converted into images. Following [8], traffic files are split into 1MB segments with IP, UDP, and ETHER values normalized (0-255). These values are then mapped to an RGB image's R, G, and B channels, forming a three-dimensional data structure. OpenCV2[1] is used to convert this structure into an image, while PILLOW[2] adjusts the dimensions as needed.

For ML models, relevant features are extracted and stored in a CSV file instead of image conversion, serving as input for DDoS detection.

C. Model classification

AI techniques, including ML and DL models, are employed to address the problem of DDoS detection in SDN environments based on contextual requirements. Both approaches are applied as detection tools in this study to validate the suitability of AI-based approaches.

CNNs [9] have proven effective in image-based DDoS detection due to their capability for efficient feature extraction from visual presentations of network data. Several well-established CNN architectures are employed in this domain. VGG [10], known for its deep structure, is widely used in image classification. ResNet [11] employs residual connections and enables deeper networks to mitigate vanishing gradients flow. DenseNet [12] enhances gradient flow and feature reuse by connecting all layers, reducing parameters while improving learning. EfficientNet [13] applies compound scaling to balance depth, width, and resolution for optimal performance. Inception [14] leverages parallel convolutions for multi-scale feature extraction, boosting accuracy and training speed. Each

[1] https://github.com/opencv/opencv
[2] https://github.com/python-pillow/Pillow

of these models has variants tailored for improved performance. Furthermore, the CNN model from [8] is integrated into SDN frameworks to enhance DDoS detection capabilities.

Beyond DL models, ML approaches like RF, DT, Light-GBM, and Dummy classifiers are explored to assess AI in SDN-based DDoS detection comprehensively. These models are widely recognized for their accessibility, effectiveness, and adaptability, demonstrating the versatility of ML in this context.

IV. PERFORMANCE EVALUATION

A. Experimental Settings

The SDN model is deployed on a system with 16GB RAM running Ubuntu 20.04. The network design includes eight hosts, with two scenarios constructed for traffic collection. Specifically, for regular traffic, services such as web server and VLC streaming are utilized to reflect real-world interactions between clients and servers. DDoS attack traffic is simulated by Hping3, with Wireshark capturing packets. In the attack scenario, hosts 1 to 6 act as attackers, targeting hosts 7 and 8.

Classification models are deployed on the Kaggle using a P100 GPU for predictive modeling. The CNN-based models are implemented using pre-trained architectures, including VGG-19, Inception-V3, ResNet-50, DenseNet-121, and EfficientNet-B0. In addition to DL approaches, ML models are employed with standard parameter configurations to serve as predictive baselines. An image size of 200×200 is optimal for classification accuracy in DL models. The dataset is split into Training, validation, and testing with a ratio of 8:1:1. Training runs for 10 epochs with a batch size 32.

B. Dataset

TABLE I
DATASET DISTRIBUTIONS

Dataset	Normal	Attack	Feature
CTU-13	4,500	4,500	58
InSDN	3,000	3,000	79
Simulation	1,500	1,500	19

Relevant datasets for DDoS detection in the SDN network model are utilized for evaluation, including CTU-13[3] and InSDN[4], which are appropriate and widely used for this context. These datasets have been adopted in similar research. Additionally, a custom-built dataset is incorporated into the study. Detailed information regarding the number of samples and features used for evaluation is provided in the **Table I**.

C. Evaluation Performance

To highlight the performance of the models on the datasets, the evaluation metrics and their significance in the DDoS detection task are detailed in **Table II**.

In real-time network environments, DDoS detection and traffic-blocking mechanisms can sometimes impact legitimate

[3]https://github.com/imfaisalmalik/CTU13-CSV-Dataset
[4]https://www.kaggle.com/datasets/badcodebuilder/insdn-dataset

users due to false alerts. To evaluate model performance, we record True Positive (TP), True Negative (TN), False Positive (FP), and False Negative (FN) values in **Table III**. Among ML models, RF demonstrates strong performance, while DenseNet represents DL models. Both achieve high accuracy while minimizing false alerts.

Experimental results on three datasets are summarized in **Table IV**. RF consistently outperformed other ML models in accuracy and stability. Among DL models, EfficientNet performed best on the CTU-13 and InSDN datasets, while InceptionV3 led on the Simulation dataset. The performance gap between ML and DL models is likely due to differences in input features: ML models use structured features extracted from network traffic (CSV), preserving discriminative patterns, whereas DL models operate on RGB images generated from PCAP files, which may obscure relevant information or introduce noise. This highlights the importance of effective feature representation in detection performance.

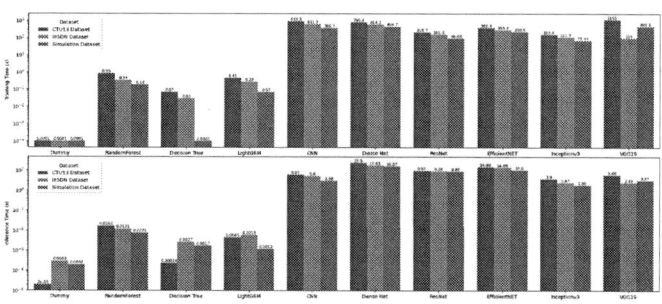

Fig. 3. Training and Inference Time Comparision.

Effective network monitoring requires continuous real-time analysis due to the vast traffic volume. **Fig. 3** presents a comparison of training and inference times, highlighting the efficiency of different models. ML models demonstrate rapid processing, completing training and inference in under one second. In contrast, DL models demand more computational resources due to image-based inputs and complex multi-layer architectures.

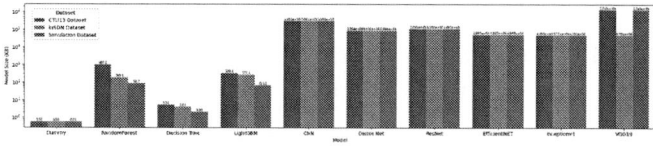

Fig. 4. Model Size Comparision.

Model size is a crucial factor when deploying AI-based monitoring systems. **Fig. 4** presents detailed model size comparisons across all three datasets. ML models scale with dataset volume, while DL models remain consistent, except for VGG19, which is significantly smaller for the InSDN dataset.

Various AI-based approaches, including ML and DL models, have been proposed in the context of DDoS detection in SDN networks. Existing methods are compared with the

TABLE II
METRIC PERFORMANCE

Metric	Formula	Definition
Accuracy	$\frac{TP+TN}{TP+TN+FP+FN}$	Proportion of correctly classified traffic, considering both normal and attack traffic.
Precision	$\frac{TP}{TP+FP}$	Proportion of predicted normal traffic that are actually normal, measuring model reliability for normal predictions.
Recall	$\frac{TP}{TP+FN}$	Proportion of actual normal traffic correctly identified, also known as sensitivity or true positive rate.
F1-score	$2 \times \frac{Precision \times Recall}{Precision + Recall}$	Harmonic mean of precision and recall, providing a balance between them.
MCC	$\frac{TP \times TN - FP \times FN}{\sqrt{(TP+FP)(TP+FN)(TN+FP)(TN+FN)}}$	Balanced metric that considers all four confusion matrix elements, robust for imbalanced datasets.

TABLE III
CONFUSION MATRIX VALUES

Dataset	Value	Machine Learning Models				Deep Learning Models					
		RF	DT	LightGBM	Dummy	CNN	DenseNet	ResNet	EfficientNet	InceptionV3	VGG19
CTU-13	TP	732	727	728	**733**	688	762	763	**770**	759	0
	TN	**766**	762	**766**	0	682	**729**	496	727	726	655
	FP	**1**	5	**1**	767	48	1	178	3	4	**0**
	FN	1	6	5	**0**	82	8	63	**0**	11	695
InSDN	TP	439	**440**	439	**440**	439	439	425	439	438	**440**
	TN	460	460	460	0	447	459	407	**460**	459	457
	FP	**0**	**0**	**0**	460	13	1	53	**0**	1	3
	FN	1	**0**	1	**0**	1	1	15	1	2	**0**
Simulation	TP	224	224	224	224	223	223	106	107	**224**	224
	TN	226	226	226	0	**226**	226	226	226	226	0
	FP	**0**	**0**	**0**	226	**0**	**0**	**0**	**0**	**0**	226
	FN	**0**	**0**	**0**	**0**	1	1	118	117	**0**	**0**

TABLE IV
METRIC PERFORMANCE

Dataset	Metric	Machine Learning Models				Deep Learning Models					
		RF	DT	LightGBM	Dummy	CNN	DenseNet	ResNet	EfficientNet	InceptionV3	VGG19
CTU-13	Accuracy	**99.87**	99.27	99.6	48.87	91.33	99.4	83.93	**99.8**	99	48.52
	Precision	**99.86**	99.27	**99.86**	23.88	93.48	**99.87**	81.08	99.61	99.48	23.54
	Recall	**99.86**	99.27	99.32	48.87	89.35	98.86	92.37	100	98.57	48.52
	F1-Score	**99.86**	99.27	99.59	32.08	91.37	99.41	86.36	**99.81**	99.02	31.7
	MCC	**99.86**	99.27	**99.86**	0	83	99	68	**99.6**	98	0
InSDN	Accuracy	99.89	**100**	99.89	48.89	98.44	99.78	92.44	**99.89**	99.67	99.67
	Precision	**100**	**100**	**100**	48.89	97.12	99.77	88.91	**100**	99.77	99.32
	Recall	99.77	**100**	99.77	**100**	99.77	99.77	96.59	99.77	99.55	**100**
	F1-Score	99.89	**100**	99.89	65.67	98.43	99.77	92.59	**99.89**	99.66	99.66
	MCC	**100**	**100**	**100**	0	97	**100**	85	**100**	99	99
Simulation	Accuracy	**100**	**100**	**100**	49.78	99.78	99.78	73.87	74	**100**	49.78
	Precision	**100**	**100**	**100**	24.78	99.78	99.78	82.77	82.87	**100**	24.78
	Recall	**100**	**100**	**100**	49.78	99.78	99.78	73.78	74	**100**	49.78
	F1-Score	**100**	**100**	**100**	33.09	99.78	99.78	71.8	72.08	**100**	33.09
	MCC	**100**	**100**	**100**	0	**100**	**100**	56	56	**100**	0

proposed models to provide a comprehensive evaluation. As presented by Clinton [8], studies using the CTU-13 and InSDN datasets highlight different ML and DL techniques. For the CTU-13 dataset, ML approaches include REPTree + SVM [15] and entropy-based ML methods [16]. In contrast, DL methods involve RNN [17] and Deep Neural Networks (DNN) [18], all leveraging feature extraction. **Fig. 5** demonstrates the superior performance of the proposed RF and DenseNet models with 99.3% and 99.87%.

Similarly, for the InSDN dataset, DL methods such as CNN

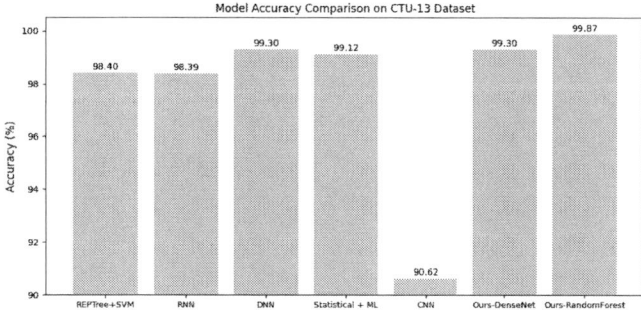

Fig. 5. Comparision in CTU-13 Dataset.

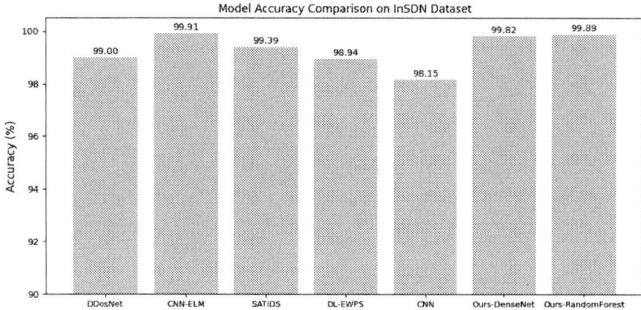

Fig. 6. Comparision in InSDN Dataset.

[6], LSTM [19], RNN with autoencoder [1], and CNN-ELM [20] have been explored. These approaches also rely on feature extraction. As shown in **Fig. 6**, RF and DenseNet achieve 99.82% and 99.89% accuracy, respectively, demonstrating that ML models remain competitive in terms of accuracy while being more efficient in model size and training/inference time.

V. CONCLUSION AND FUTURE WORKS

To detect DDoS attacks, the SDN network designed for traffic analysis made excellent use of AI-driven approaches, incorporating both ML and DL. Packet data were processed by extracting features, normalizing values, representing them as RGB pixels, and converting them into images for DL models. For ML models, selected features were extracted directly from packets.

CNN-based DL models were applied alongside ML models for comparison. Beyond classification performance, the study also highlights the practical applicability of these models. DenseNet (for DL) and RF (for ML) excelled in accuracy, efficiency, and model size on the evaluated datasets. Future work aims to optimize performance further and integrate XAI for better interpretability and transparency.

ACKNOWLEDGMENT

This research is funded by University of Information Technology-Vietnam National University HoChiMinh City under grant number D1-2025-12

REFERENCES

[1] M. S. Elsayed, N.-A. Le-Khac, S. Dev, and A. D. Jurcut, "Ddosnet: A deep-learning model for detecting network attacks," in *2020 IEEE 21st International Symposium on" A World of Wireless, Mobile and Multimedia Networks"(WoWMoM)*, pp. 391–396, IEEE, 2020.

[2] K. M. Sudar, M. Beulah, P. Deepalakshmi, P. Nagaraj, and P. Chinnasamy, "Detection of distributed denial of service attacks in sdn using machine learning techniques," in *2021 international conference on Computer Communication and Informatics (ICCCI)*, pp. 1–5, IEEE, 2021.

[3] M. V. Assis, L. F. Carvalho, J. Lloret, and M. L. Proença Jr, "A gru deep learning system against attacks in software defined networks," *Journal of Network and Computer Applications*, vol. 177, p. 102942, 2021.

[4] D. Hu, P. Hong, and Y. Chen, "Fadm: Ddos flooding attack detection and mitigation system in software-defined networking," in *GLOBECOM 2017-2017 IEEE Global Communications Conference*, pp. 1–7, IEEE, 2017.

[5] V. Deepa, K. M. Sudar, and P. Deepalakshmi, "Design of ensemble learning methods for ddos detection in sdn environment," in *2019 International Conference on Vision Towards Emerging Trends in Communication and Networking (ViTECoN)*, pp. 1–6, IEEE, 2019.

[6] A. H. Janabi, T. Kanakis, and M. Johnson, "Convolutional neural network based algorithm for early warning proactive system security in software defined networks," *IEEE Access*, vol. 10, pp. 14301–14310, 2022.

[7] A. Alashhab, M. S. Zahid, A. Muneer, and M. Abdullahi, "Low-rate ddos attack detection using deep learning for sdn-enabled iot networks," *Authorea Preprints*, 2022.

[8] U. B. Clinton, N. Hoque, and K. Robindro Singh, "Classification of ddos attack traffic on sdn network environment using deep learning," *Cybersecurity*, vol. 7, no. 1, p. 23, 2024.

[9] Y. Lecun, L. Bottou, Y. Bengio, and P. Haffner, "Gradient-based learning applied to document recognition," *Proceedings of the IEEE*, vol. 86, no. 11, pp. 2278–2324, 1998.

[10] L. Yang and H. Zhao, "Ddos attack identification and defense using sdn based on machine learning method," in *2018 15th International Symposium on Pervasive Systems, Algorithms and Networks (I-SPAN)*, pp. 174–178, 2018.

[11] K. He, X. Zhang, S. Ren, and J. Sun, "Deep residual learning for image recognition," in *Proceedings of the IEEE conference on computer vision and pattern recognition*, pp. 770–778, 2016.

[12] G. Huang, Z. Liu, L. Van Der Maaten, and K. Q. Weinberger, "Densely connected convolutional networks," in *Proceedings of the IEEE conference on computer vision and pattern recognition*, pp. 4700–4708, 2017.

[13] M. Tan and Q. Le, "Efficientnet: Rethinking model scaling for convolutional neural networks," in *International conference on machine learning*, pp. 6105–6114, PMLR, 2019.

[14] C. Szegedy, W. Liu, Y. Jia, P. Sermanet, S. Reed, D. Anguelov, D. Erhan, V. Vanhoucke, and A. Rabinovich, "Going deeper with convolutions," in *Proceedings of the IEEE conference on computer vision and pattern recognition*, pp. 1–9, 2015.

[15] P. Kalaivani and M. Vijaya, "Mining based detection of botnet traffic in network flow," *Int. J. Comput. Sci. Inf. Technol. Secur*, vol. 6, pp. 535–540, 2016.

[16] A. Banitalebi Dehkordi, M. Soltanaghaei, and F. Z. Boroujeni, "The ddos attacks detection through machine learning and statistical methods in sdn," *The Journal of Supercomputing*, vol. 77, no. 3, pp. 2383–2415, 2021.

[17] A. Bansal and S. Mahapatra, "A comparative analysis of machine learning techniques for botnet detection," in *Proceedings of the 10th international conference on security of information and networks*, pp. 91–98, 2017.

[18] A. Pektaş and T. Acarman, "Deep learning to detect botnet via network flow summaries," *Neural Computing and Applications*, vol. 31, no. 11, pp. 8021–8033, 2019.

[19] R. A. Elsayed, R. A. Hamada, M. I. Abdalla, and S. A. Elsaid, "Securing iot and sdn systems using deep-learning based automatic intrusion detection," *Ain Shams Engineering Journal*, vol. 14, no. 10, p. 102211, 2023.

[20] J. Wang and L. Wang, "Sdn-defend: a lightweight online attack detection and mitigation system for ddos attacks in sdn," *Sensors*, vol. 22, no. 21, p. 8287, 2022.

A Solution for Built-in On-chip Hardware Integrity Protection Adopting Resource-optimized RO-PUF

Quang-Hoa Nguyen, Hoang-Long Nguyen, Tri-Hieu Le,
Van-Toan Tran, Duy-Cong Nguyen, and Quang-Kien Trinh[*]

Le Quy Don Technical University

[*]corresponding author: kien.trinh@lqdtu.edu.vn

Abstract—**Possessing a robust anti-counterfeiting structure, the Physically Unclonable Functions (PUF), particularly Ring Oscillator (RO) PUF, which are inherently non-reproducible, have recently been proposed as a promising tool for hardware authentication and integrity protection against device design attacks. This research aims to optimize the design of RO-PUF for implementation on the Field-programmable Gate Array (FPGA) platform and to automate the computation and evaluation of design integrity, thereby enabling wider applications for RO-PUF. Furthermore, the research conducts experiments to evaluate the uniqueness and reproducibility of the designs, ensuring the detection of unauthorized hardware design modifications. By improving reliability and resource efficiency, this work contributes to the development of more reliable and efficient PUF-based security systems.**

Index Terms—**Hardware security, PUF, FPGA, RO-PUF, physical layer security**

I. INTRODUCTION

In today's rapidly evolving technological landscape, ensuring the authenticity and security of hardware has become increasingly critical, especially with the growing threats of sophisticated and complex information attacks. Against this backdrop, the concept of a Physically Unclonable Functions (PUF) has emerged as a promising solution for authenticating and protecting hardware as well as data against potential threats [1], [2]. PUFs have been suggested [3], [4] as a promising solution against counterfeit electronic devices. This is particularly meaningful in today's widespread development of IoT devices. PUF exploits inevitable mismatches in solid-state structures arising from the manufacturing process to obtain specific statistical data used as the device's IDs. However, due to dependence on specific technology nodes, the implementation of PUFs structure as well as its data extraction scheme faces certain technical challenges. The FPGA-based Ring Oscillator PUF (RO-PUF) belongs to the subclass of PUF designs that exhibits key characteristics such as reproducibility and reliability, while being relatively easy to implement in popular hardware platforms. Although several RO-PUF designs have been proposed [7]–[9], they commonly suffer from drawbacks including the requirement for massive response data, complex statistical calculations, and unoptimized hardware cost. In this paper, we proposed an FPGA-based RO-PUF design that almost halves the RO components resource while still ensuring security and good performance metrics compared to the prior art. Furthermore, by discovering the high sensitivity of RO-PUF responses to the changes in hardware components and design, we proposed a complete embedded solution for hardware design integrity protection using the FPGA SoC platform. In our approach, the Kullback-Leibler (D_{KL}) Divergence is calculated directly using the available embedded ARM core. This approach eliminates the need for off-chip processing, which could lead to another potential security breach, and ensures strong security protection at the physical level.

The remainder of the paper is organized as follows: Section II describes an RO-PUF design with enhanced reliability and compactness. Experiments on FPGA are presented in Section III. Section IV describes the RO-PUF application for design integrity verification. Conclusions are drawn in Section V.

II. IMPLEMENTATION OF RO-PUF DESIGN ON FPGA PLATFORM

A. The PUF performance metrics

The conventional RO-PUF structure was proposed by Suh [3] (Fig. 1). Its primary concept is to leverage

979-8-3315-1550-8/25 $31.00 © 2025 IEEE

the differences between two propagation delays that come from inevitable imperfections of the semiconductor manufacturing process. Based on that basic scheme, many RO-PUF designs have been developed to investigate RO-PUF characteristics and apply them in practical applications such as ID extraction, key generation, and physical level protections. Most of works on RO-PUF focus on improving fundamental performance metrics of reproducibility and uniqueness.

Uniqueness is the ability to distinguish between one device and others or between the different RO-PUFs on the same device. Ideally, all PUF outputs from different designs should be uniformly distributed and statistically independent. If the set of measurements is statistically independent, their Hamming Distance (HD) would be 50%. This capability is quantified by the Kullback-Leibler (KL) Divergence value [11]. If the one distribution is very different from the other, the KL divergence will be high, which could be used to quantify the stability and uniqueness of RO-PUF responses.

Reproducibility is the ability to produce the same output response under different conditions. The ideal reproducibility is 100%, when the circuit would give the same response under varying conditions. This is relatively challenging because the PUF response is generated due to uncontrollable and unpredictable random factors during the chip manufacturing process [5], [12]. Reproducibility can also be quantified by KL divergence, with values approaching zero indicating good reproducibility.

B. RO-PUF design on FPGA platform

The proposed RO-PUF is designed based on the conventional RO-PUF presented by Suh [3] with several modifications for implementation on FPGA. The functional diagram of the RO-PUF is depicted in Fig. 1. A basic RO consists of N INVERTERs and a NAND gate. These components must be manually planned and routed to maintain a consistent and symmetrical layout. The RO frequency is measured by a counter triggered by the generated oscillating signal from RO. The entire design is kept compact to ensure stability and reliability.

The complete core design of RO-PUF based on the structure in prior works [8], [13], is illustrated in Fig. 2. The design employs an array of parallel counters to convert frequency values to a sufficiently large integer number in a fixed interval. Specifically, the RO response, recorded by the counters within a certain counting period (25 ms used in this work), represented

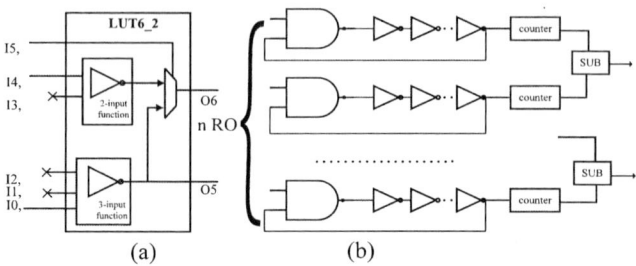

Fig. 1.   a, Structure of LUT6_2; b, RO-PUF diagram in [6]

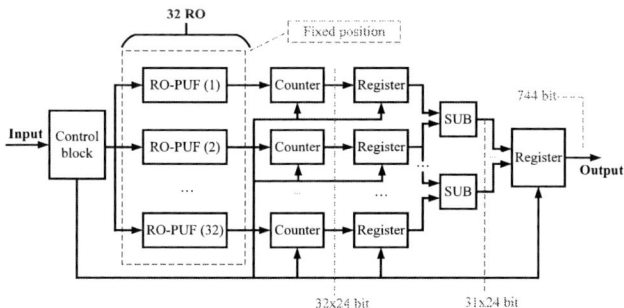

Fig. 2.   RO – PUF structure used

by a 24-bit output, is proportional to the actual ROs' frequency. Thus, the total output response consists of 32×24 bits [6], [13].

1) The circuit design: In this article, we implement the proposed design and carry out experiments on several Artix®-7 Xilinx FPGAs, a series of cost-performance balanced hardware for moderate embedded applications.

2) Design process: Based on the previous design [8], [13], our research group utilized LUT6_2 to optimize resource usage. As usual, all the locations of the LUTs used for RO components locations are fixed at the physical design stage. This ensures the design's overall RO structure and location are not affected by the automatic synthesis and implementation process and guarantees the reproducibility of the RO-PUF responses. To represent each logic gate, we used LUT6_2 (2-output LUT) to optimize resource usage. The 2-output lookup table (LUT), apart from standard functionalities, can realize any two independent 3-input and 2-input logic functions. The previous designs typically utilize one stage INVERTER using one LUT. In this work, we proposed to implement two independent INVERTERS

TABLE I
COMPARE THE RESOURCES OF THE TWO DESIGNS

Resource	Utilization		Available	Utilization %	
	[13]	This work		[13]	This work
LUT	563	320	20800	2.71	1.69
LUTRAM	66	83	41600	0.16	0.29
FF	9	9	250	3.6	4
IO	2	2	32	6.25	6.25

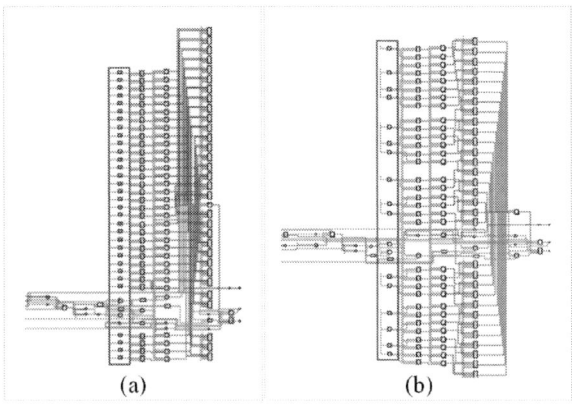

(a)   (b)

Fig. 3. Schematics designs of
a) the design in [10] and (b) the proposed RO-PUF in this work.

on a single LUT. Thus, the RO's core resource usage was almost halved compared to the design in [13], [10].

C. Resource utilization

The main functional block of the RO-PUF, responsible for generating final bit strings, consists of 32 ROs (each containing 16 INVERTER elements and 1 NAND gate). In this section, we compare the resource utilization of two designs implemented on the same FPGA device with the conventional RO-PUF design in previous work [10].

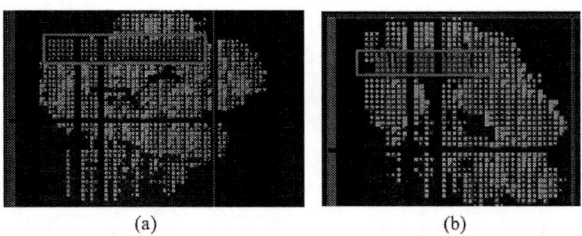

(a)   (b)

Fig. 4. Implementation layout of the RO-PUF (using Xilinx Vivado 2019.2 and targeted for Artix7 XC7A35T FPGA for (a) 32 RO-PUF designed in [10], (b) 32 RO-PUF in this work.

Fig. 3, 4 illustrate the differences between the two designs, where the area bordered in red in Fig. 3 is the core of the RO-PUF. Figs. 3(a), (b) illustrate the RO-PUF schematic in previous work [10] and this work, respectively. The orange area in Fig. 4 represents the occupied LUTs used to form corresponding RO-PUFs. The reduction in RO-PUF area between the proposed design and the previous one could be observed from the figure, and the detailed resource comparison between these two designs is presented in Table I. The results indicated that our proposed design's area is almost half the number of LUTs reported in the previous design [13] (that is, from 563 to 320, by a factor of 1.76). This reduction in resource utilization demonstrates the effectiveness of the proposed design, which can help free up resources for additional functional blocks or for scaling the design to accommodate more RO-PUF instances. In the next Section, we evaluate the fundamental metrics of the proposed PUF design by practical measurements

III. EVALUATION OF THE PROPOSED RO-PUF DESIGN

To evaluate the proposed design, it is essential to consider critical factors that affect the design of the RO-PUF and its metrics, such as the sensitivity of the RO to operating conditions, variations in supply voltage, local and global variations, and the influence of the surrounding components. These factors lead to unstable RO-PUF responses. The frequency of RO is generalized through the following model [13]:

$$
\begin{aligned}
f_{RO} = f_{nominal} &+ \Delta f_{(proc,local)} \\
&+ \Delta f_{(proc,global)} + \Delta f_{OP}
\end{aligned}
\tag{1}
$$

where, $f_{nominal}$ is the nominal RO frequency (i.e., the frequency measured under nominal conditions, in this case, 25°C; 1.0 V core supply voltage). This value theoretically remains constant across FPGA devices manufactured using the same technology. The remaining components in the model are random variables; $\Delta f_{(proc,global)}$ and $\Delta f_{(proc,local)}$ represent frequency fluctuation due to global and local process variations, respectively; Δf_{OP} is the frequency deviation due to operating conditions.

To evaluate the reproducibility and uniqueness of the RO-PUF, we quantified the impacts on the proposed design's output response using the Kullback-Leibler (KL) Divergence. This quantity measures the difference

979-8-3315-1550-8/25 $31.00 © 2025 IEEE   129

between two probability distributions P and Q by the following formula

$$D_{KL}(P \parallel Q) = \frac{1}{2}\left(\frac{\sigma_1^2}{\sigma_2^2} + \frac{\sigma_2^2}{\sigma_1^2} - 2\right)$$
$$+ \frac{1}{2}(\mu_1 - \mu_2)^2\left(\frac{1}{\sigma_2^2} + \frac{1}{\sigma_1^2}\right) \tag{2}$$

where P is a Gaussian distribution with mean μ_1, variance σ_1^2, and Q is a Gaussian distribution with mean μ_2, variance σ_2^2.

A. Evaluating the RO-PUF stability

We collected the output responses of the RO-PUF to check the stability level of the output frequency during operation. In this experiment, we implemented a fixed RTL design and then performed multiple sampling, each time 128 frequency samples are collected and this process was repeated for four different physical FPGAs. After the sampling process, we used the first distribution sample P_i (i=1,2,3,4) of each FPGA as the reference one to calculate the D_{KL} value concerning all subsequent samples $Q_{i,j}$ (j=1,2,...128) on the same FPGA chip. After repeating this process for 2,000 distributions, we got the calculated D_{KL} value ranged from 2.0×10^{-3} to 2.4×10^{-2}, i.e., very close to the theoretical value of zero. These results confirm a good level of stability of the proposed RO-PUF based on LUT6_2 described in Section II.B.

B. Evaluating the impact of RO-PUF position

With the same experiment setup, we now manually change the physical position of RO-PUF to examine the impact of its RO-PUF position changes on the output responses for four different locations on the same FPGA chip. D_{KL} distances between the distributions from RO-PUF at positions 2, 3, 4 with respect to position 1 have been calculated and presented in Table II.

TABLE II
CALCULATED D_{KL} FROM DATA EXTRACTED FROM 5 RO-PUF POSITIONS IN THE SAME FPGA

	pos2 vs pos1	pos3 vs pos1	pos4 vs pos1
[13]	1086.78	56.34	217.12
This work	217.96	646.22	1073.19

The results indicate that the difference between RO-PUF responses from different design positions is quite clear. Specifically, at the worst case of D_{KL} (56.34), this value is still 10^4 times larger than the case of no change applied as reported in the previous sub-section. Thus, the uniqueness of RO-PUF response can be guaranteed.

TABLE III
CALCULATED D_{KL} FROM DATA EXTRACTED FOUR MODIFIED SURROUNDING DESIGNS WHILE FIXING RO-PUF POSITION ON THE SAME FPGA

	p2 vs p1	p3 vs p1	p4 vs p1
[13]	138.77	568.46	370.2
This work	955.42	32.14	595.9

C. Evaluating the impact of surrounding design variation

Furthermore, we examine the impact of the surrounding design on the RO-PUF responses. We conducted measurements of RO-PUF responses on a single chip when applying local changes to the design (apart from the RO-PUF). In particular, within the same design, we shift the position of a dedicated register across four different locations. Then, similarly to the above analysis, we calculate the D_{KL} distances between the distributions from design at positions 2, 3, 4 with respect to the original design position (i.e., position 1). The major results are presented in Table III, including our data from previous work.

From the table, it can be seen that the minor changes from the surrounding design could heavily affect the RO-PUF responses since the D_{KL} are in the scale of 10^3-10^4 larger than those presented in Section III.A for the cases of the same designs.

IV. ON-CHIP DESIGN INTEGRITY PROTECTION USING RO-PUF IN FPGA SoC

In the previous sections, we examined the reliability and security level of the RO-PUF for design integrity protection. However, the processing was performed on an external computer. In order to automate the computation process, we proposed a novel scheme based on the FPGA System-on-a-chip (SoC) platform, where the embedded processor can determine the potential threats and trigger the countermeasure and block any harmful actions if necessary by itself.

We have deployed our design on Zynq 7000 board, a family of FPGA SoC devices from Xilinx that integrates a dual-core ARM Cortex-A9 processor with programmable logic (PL) part sharing the same manufacturing process as Artix 7 devices. The RO-PUF is designed on PL while the data processing are performed in the processing system (PS), as depicted in Fig.5.

A. Design IP RO-PUF on PL

The RO-PUF design on the FPGA will be packaged as an IP core and implemented on PL with the same

979-8-3315-1550-8/25 $31.00 © 2025 IEEE

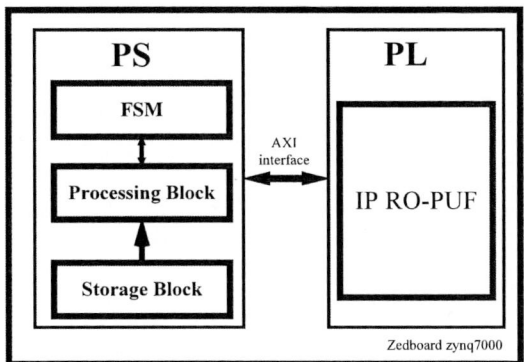

Fig. 5. Deloying RO-PUF on embedded FPGA SoC (Zynq 7000)

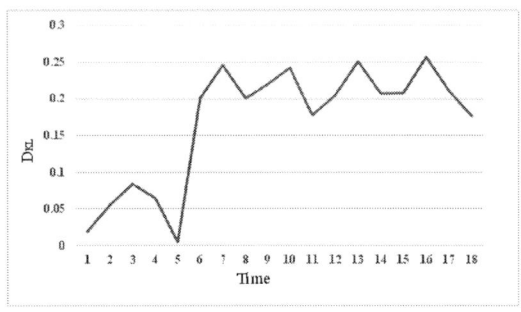

Fig. 6. The D_{KL} values changed under multiple measurement on FPGA SoC

TABLE IV
SUMMARY OF D_{KL} DISTANCES FOR SIX RO-PUF DESIGNS ON FPGA SoC

	Original	With additional IP	Change RO's position (4 variants)
Original	$(4.23 - 256)$ $\times 10^{-3}$	0.15×10^3	$(0.15 - 2.33)$ $\times 10^3$

design structure as described on Section II. The RO-PUF IP communicates with PS through the system AXI4 bus. The key design steps include:

- **RO-PUF core design:** Design ring oscillators (RO) in the PL of the Zynq 7000 PL. These ROs will generate oscillating signals with frequencies unique to each FPGA device, based on intrinsic physical variations of each chip.
- **IP Packaging & mapping:** Package the RO-PUF design into an IP core, facilitating easy integration into other FPGA projects without redesigning from scratch. Now the PS will see the RO-PUF as a system component and can control and access its responses using memory instructions.

B. Processing and analyzing RO-PUF response on PS

Since the RO-PUF responses are mapped to the system address space, PS can access and perform the post-processing on-chip. This process includes:

- **RO-PUF control and data acquisition module:** Develop an embedded software to control and compute the RO-PUF response using the ARM processor. This software will perform measurements and store the PUF bit-string on-chip without the need for an external computer.
- **Module for calculating D_{KL} values:** Use the D_{kl} formula to compare the current RO-PUF values with previously stored reference values. The reference mean and variance of the PUF are stored for comparison.

C. Evaluation of RO-PUF responses on the embedded system

We conducted experiments to evaluate the stability of the system. However, unlike the previous case, this

process is performed in two phases. The first phase will collect and calculate μ_1 and σ_1^2. These two values will be stored as reference values for the second phase. In phase 2, the circuit will automatically collect the RO-PUF response and then calculate the D_{KL} distance with the reference value from phase 1.

- **Stability testing:** Stability testing was conducted over multiple (18) RO-PUF measurements. The results shown in Fig.6 revealed D_{KL} values ranging from 4.23×10^{-3} to 2.56×10^{-1}. It can be seen that this value is relatively low and this suggests that the design has not been altered.
- **Design integrity protection/hardware identification capability testing:** In this experiment, we added an external IP to the design to mimic a hardware Trojan block (here we simply created an IP MAC to perform the calculation $Y = A \times X + B$, which is the block circled in red in Fig. 7). Also, we manually changed the RO-PUF to four different physical locations. In total, we have six different designs: the original without MAC IP, the design with MAC IP, and four others with the RO-PUF position changed.

The calculated D_{KL} with respect to the original design were collected (via PS serial terminal for observation purposes) and summarized on Table IV. From the table, it can be seen that when no change in design, D_{KL} value is relatively small at $(10^{-1}\text{-}10^{-3})$. i.e., in line with

979-8-3315-1550-8/25 $31.00 © 2025 IEEE

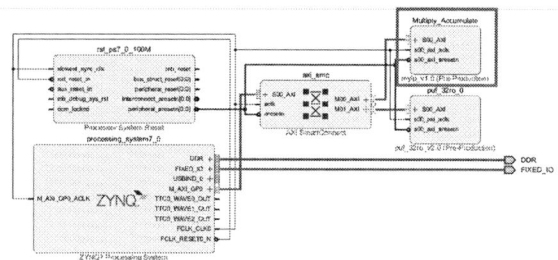

Fig. 7. The D_{KL} divergence varies with changes in the design.

what has been reported in Section III.A. For other cases, when RO position changed or when additional IPs are added, the D_{KL} values range from 153.08 to 2335.16, significantly exceeding the internal D_{KL}. In a practical scenario, the PS can self-define a safe ratio threshold of D_{KL} (e.g., 100 times higher than the case of no design changes) to determine whether the design has been modified. This experiment demonstrates that RO-PUF properties could be deployed on devices and used for chip ID identification and design integrity verification. The advantage of this approach is that the device can self-detect if there is any minor change in its hardware structure, and thus reduce the potential security breaches due to external storage and/or communication.

V. CONCLUSIONS

This research proposed a resource-optimized RO-PUF implemented on FPGA platform that could almost reduce the core of RO-PUF by half while still ensuring its reliability and uniqueness. The experimental results on chosen FPGA families show that under fixed operating conditions, the RO-PUF responses are quite stable, and a minor change in either RO-PUF or surrounding design could lead to a huge difference in the measured data. Furthermore, we have implemented and practically evaluated a complete embedded system on FPGA SoC that is capable of extracting and analyzing the RO-PUF response on-chip. This approach not only simplifies the system design but enhances the security level by eliminating the need for off-chip storage and communication. This could be one of the effective solutions for hardware design integrity protection at the physical level.

ACKNOWLEDGMENTS

This work was supported by the Ministry of Science and Technology project, grant KC-4.0-48/19-25.

REFERENCES

[1] A. Maiti, and P. Schaumont, "Improved ring oscillator PUF: An FPGA-friendly secure primitive," *Journal of Cryptology*, vol. 4, no. 2, pp. 375-397, 2011.

[2] R. Maes. *Physically unclonable functions: Constructions, properties, and applications.* PhD thesis, Katholieke Universiteit Leuven, Belgium, 2012.

[3] G. E. Suh and S. Devadas. "Physical Unclonable Functions for Device Authentication and Secret Key Generation," in *Proceedings of the 44th ACM/IEEE DAC*, 2007.

[4] J. Guajardo, S. S. Kumar, G.-J. Schrijen, and P. Tuyls, "FPGA intrinsic PUFs and their use for IP protection," in *Proc. 9th Int. Workshop on Cryptographic Hardware and Embedded Systems (CHES)*, Sep. 2007, pp. 63–80.

[5] Zhang, Ji-Liang; Qu, Gang. "A Survey on Silicon PUFs and Recent Advances in Ring Oscillator PUFs," *Journal of Computer Science and Technology*, vol. 29, pp. 664-678, 2014.

[6] V. -T. Tran, Q. -K. Trinh, and V. -P. Hoang. "Stabilizing On-chip Secure Key Generation Using RO – PUF," in *2021 ICTC*, Jeju Island, Korea, Republic of, 2021.

[7] S. K. Ram, S. R. Sahoo, B. B. Das, K. Mahapatra, and S. P. Mohanty, "sThing: A Novel Configurable Ring Oscillator Based PUF for Hardware-Assisted Security and Recycled IC Detection," *IEEE Access*, vol. 13, pp. 2994–3013, 2025.

[8] A. A. Zayed, H. H. Issa, and K. A. Shehata, "FinFET Based Low Power Ring Oscillator Physical Unclonable Functions," in *Proc. 2019 ICM*, 2019, pp. 227–230.

[9] Y. Cui, J. Li, Y. Chen, C. Wang, C. Gu, M. O'Neill, and W. Liu, "An Efficient Ring Oscillator PUF Using Programmable Delay Units on FPGA," *ACM Trans. Des. Autom. Electron. Syst.*, vol. 29, no. 1, p. 1, 2024.

[10] V. -D Nguyen,Q. -K. Trinh,V.-P. Hoang, Q.-H. Nguyen, V.-N. Pham. *Research on the Stability of Parameters Extracted from RO-PUF Implemented on FPGA* in 2023 *The 26nd REV ECIT*, vol. 65, pp.482-486,2023.

[11] S. Theodoridis, K. Koutroumbas. *Pattern Recognition*, Academic Press, pp. 279, 2009.

[12] R. Maes. *Physically unclonable functions: Constructions, properties, and applications.* Katholieke Universiteit Leuven, Belgium, 2012.

[13] Van-Toan Tran, Quang-Kien Trinh, Van-Phuc Hoang. "A robust Euclidean metric based ID extraction method using RO – PUFs in FPGA," *Elsevier Integration*, vol. 82, pp. 37-47, 2022.

Correlation Power Analysis of Pipelined and Multi-Threaded Coarse-Grained Reconfigurable Cryptographic Accelerator

Van-Tuan Luu[1], Hoai Luan Pham[2], Van-Tinh Nguyen[1], Van-Phuc Hoang[1], Van-Trung Nguyen[1],
Vu Trung Duong Le[2], and Yasuhiko Nakashima[2]

[1] Le Quy Don Technical University, Hanoi, Vietnam; [2] Nara Institute of Science and Technology, Nara, Japan.
Email: tuanlv.isi@lqdtu.edu.vn, pham.luan@is.naist.jp

Abstract—LiCryptor, our state-of-the-art coarse-grained re-configurable cryptographic accelerator, is designed for emerging Internet of Things (IoT) systems that demand high-speed and energy-efficient hardware capable of supporting multiple cryptographic algorithms. In this paper, we evaluate the vulnerability of LiCryptor to Correlation Power Analysis (CPA) side-channel attacks (SCA), proposing two primary attack models based on Hamming Weight (HW) and Hamming Distance (HD) to target intermediate values stored in registers during encryption. We introduce 8 power model schemes for CPA attacks, covering both single-threaded pipelined operation and dual-threaded pipelined operation on LiCryptor. Experimental results show that, under the pipelined operation, the proposed HD-based energy model can reveal all 16 key bytes with 82,100 traces, achieving a 1.4x improvement over single-threaded methods. In dual-threaded operation, the number of traces required for key recovery is reduced to 54,000 traces. Furthermore, we identify CPA_HD_R10 as the most effective attack model, outperforming other models in both single-threaded and dual-threaded configurations.

Index Terms—CGRA, Cryptography, AES, Side-channel attack (SCA), Correlation Power Analysis (CPA).

I. INTRODUCTION

Cryptography is the science of securing information by transforming plaintext into an unreadable format using mathematical algorithms, ensuring data confidentiality, integrity, and authenticity. It encompasses various cryptographic methods, including block ciphers, stream ciphers, hash functions, and authenticated encryption. TThese techniques secure networks, protect data, and verify authenticity, with applications in SSL/TLS, VPNs, and IoT transactions [1].

Since cryptographic algorithms require efficient implementation to meet the stringent resource constraints of IoT systems, the development of dedicated cryptographic hardware is essential to optimize area, power consumption, and processing speed. Several typical hardware architectures, such as Application-Specific Integrated Circuits (ASICs), Application-Specific Instruction Set Processors (ASIPs), and Coarse-Grained Reconfigurable Arrays (CGRAs), have been proposed to address the challenges related to speed, area-power efficiency, and flexibility in cryptographic tasks. ASIC-based cryptography hardware in [2], [3] provides the highest performance and energy efficiency for specific tasks but lacks flexi-

bility and is expensive to reconfigure for different algorithms. On the other hand, ASIP-based cryptography hardware in [4], [5] offers more flexibility than ASICs by supporting custom instructions but still has higher latency compared to ASICs. Among these, CGRA in [6]–[11] has proven to be the most efficient architecture to support multiple cryptographic algorithms at high speed, offering both flexibility and efficiency. Notably, our CGRA-based cryptographic architecture named *LiCryptor* in [11] is the latest state-of-the-art design, offering superior speed and energy efficiency compared to existing CGRA architectures in [6]–[10].

In cryptographic algorithms, Advanced Encryption Standard (AES), a block cipher, is widely used and essential for securing private data due to its robust security and efficient performance. However, AES remains vulnerable to Side-Channel Attacks (SCA), which exploit the physical characteristics of the system to extract secret keys. Specifically, SCA uncovers cryptographic keys by analyzing the relationship between intermediate data and the device's physical emissions, such as power consumption, electromagnetic radiation, temperature changes, timing differences, and acoustic signals. This paper focuses on Power Analysis Attacks (PPA), which include Simple Power Analysis (SPA), Differential Power Analysis (DPA), and Correlation Power Analysis (CPA). Among these, CPA is the most effective method for revealing AES keys by correlating the power consumption with the cryptographic operations performed. CPA has primarily been analyzed for ASIC-based standard AES architectures. In contrast, CGRA architectures are generally considered resistant to SCA, especially CPA, but remain susceptible for three primary reasons. First, energy fluctuations during execution are a physical characteristic of hardware that can be exploited in side-channel attacks. Second, because CGRA architectures are often optimized for AES, their post-processing configuration becomes similar to ASICs, which makes them more prone to SCA. Third, the absence of sufficient countermeasures against such attacks further exacerbates this vulnerability.

Our *LiCryptor* architecture incorporates both pipelining and multi-threading, setting it apart from other CGRA designs,

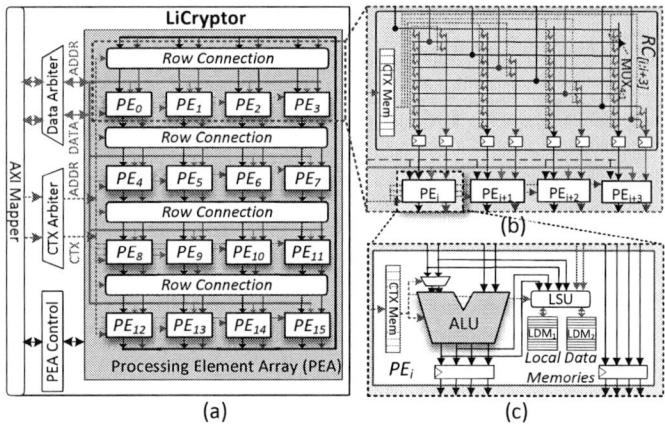

Fig. 1: Architecture of LiCryptor: (a) overall structure, (b) row connection (RC), and (c) processing element (PE).

which typically only utilize pipelining. However, *LiCryptor* does not include built-in SCA countermeasures for three main reasons. First, the low Signal-to-Noise Ratio (SNR) caused by both pipelining and multi-threading makes it difficult to detect side-channel leakage effectively. Second, implementing countermeasures to defend against SCA would significantly increase the area of the design, potentially undermining its efficiency. Third, because LiCryptor supports run-time reconfiguration, each new configuration generates distinct energy characteristics, further complicating the implementation of consistent countermeasures.

Therefore, in this paper, we analyze the security vulnerability and susceptibility of LiCryptor to SCA. To effectively conduct CPA attacks, we employ two main models: Hamming Distance (HD) and Hamming Weight (HW). Our experimental results show that with a 128-bit key, the system can be attacked and the secret key can be revealed after 82,100 traces. To the best of our knowledge, this is the first paper to analyze SCA for a CGRA architecture with the characteristics of both pipelining and multi-threading.

The remainder of this paper is organized as follows. Section II provides the background. Section III details the proposed power model. In Section IV, we present the experimental setup and evaluation results. Finally, Section V concludes the paper.

II. BACKGROUND

A. LiCryptor Overview Architecture

The LiCryptor architecture consists of a Processing Element Array (PEA), which is composed of 4×4 row connections (RCs) and processing elements (PEs), as shown in Fig. 1 (a). The RCs are responsible for routing data between the rows of PEs, ensuring that data flows from one row to the next, and directing data into the PEs and their buffer lanes, as depicted in Fig. 1 (b). Each PE, as illustrated in Fig. 1 1(c), contains two main components: an Arithmetic Logic Unit (ALU) and two Local Data Memories (LDMs), each with

TABLE I: List of algorithms supported by LiCryptor

Types	Algorithm
AEAD	ASCON, Elephant, GIFT-COFB, Grain-128, ISAP, PHOTON, Romulus, SPARKLE, TinyJambu
Block Cipher	AES, SM4, SPECK, SIMON, CLEFIA
Stream Cipher	Chacha20, Salsa20, Trivium
Hashing	SHA-256, SM3, SipHash, Chaskey

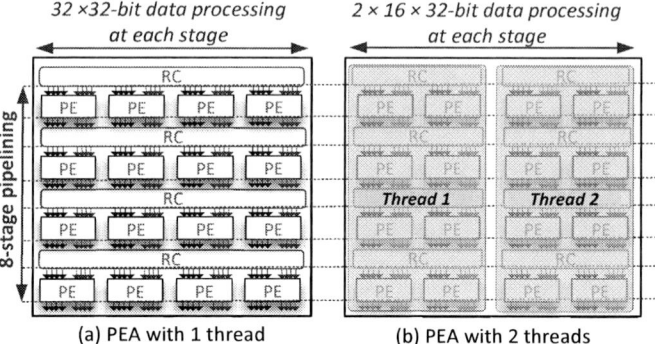

Fig. 2: Pipelined and multi-threaded operations of LiCryptor

a capacity of 1024×32 bits. The ALU allows the PE to perform computations on 8/32/64-bit data, which is crucial for executing various computational tasks. Thanks to the flexibility of the RC and PE in the PEA, LiCryptor can be configured to support various types of cryptographic algorithms, such as Authenticated Encryption with Associated Data (AEAD), block ciphers, stream ciphers, and hashing, with multiple algorithms as shown in Table I.

The LiCryptor architecture features a 4-row structure with alternating RC and PE, each equipped with flip-flops to reduce the critical path and improve processing efficiency. This design results in a total of 8 stages in each PEA (Processing Element Array) loop, where data is processed through the pipeline in 8 sequential stages. By utilizing an 8-stage pipeline, LiCryptor optimizes throughput and efficiently handles multiple data streams in parallel. As shown in Fig. 2 (a), the PEA processes 32×32-bit data with a single thread, while in Fig. 2 (b), dual-threading allows it to handle $2 \times 16 \times 32$-bit data at each stage, doubling the throughput and enhancing parallel processing efficiency. According to our findings, LiCryptor is the first architecture capable of being configured for both pipelined and multi-threaded computation.

B. Correlation Power Analysis

CPA exploits the relationship between a cryptographic device's internal data and power consumption by analyzing power traces, using hypothetical power models like Hamming Distance (HD) and Hamming Weight (HW) to compare with actual power dissipation [12], [13].

The HD model assumes power consumption is influenced by bit flips between two binary values, represented by $S(X, Y)$, where S is the number of differing bits. It is expressed as (1),

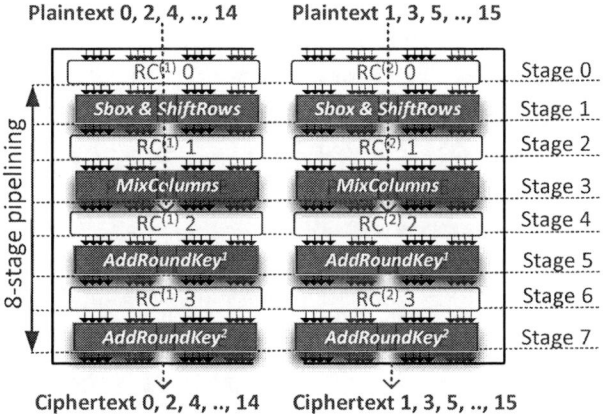

Fig. 3: Mapping of AES-128 with an 8-stage pipeline and dual-threading operation on LiCryptor.

with λ as a scaling constant and δ accounting for offset, noise, or time-dependent variations [14].

$$HD(X,Y) = \lambda \cdot S(X,Y) + \delta \qquad (1)$$

Similarly, the HW model operates on the assumption that the power consumption is proportional to the number of 1's in the data after a certain transformation. It can be simplified by treating one of the intermediate states as zero. It is mathematically expressed as (2).

$$HW(X) = \lambda \cdot S(X,0) + \delta \qquad (2)$$

Once these power models are constructed, the secret key of the cryptographic device can be recovered by comparing the correlation between the hypothetical power model and the actual observed power. Let k denote the potential candidates for the cryptographic key, and let t be the time at which the correlation is measured. The correlation coefficient $\rho_{k,t}$ for each key candidate k at time t is computed as (3), where N is the total number of power traces, M represents the hypothetical power model, and P is the actual power dissipation [14].

$$\rho_{k,t} = \frac{\sum_{i=1}^{N}(M_i - \bar{M}_k) \cdot (P_i, t - \bar{P})}{\sqrt{\sum_{i=1}^{N}(M_i - \bar{M}_k)^2 \cdot \sum_{i=1}^{N}(P_i, t - \bar{P})^2}} \qquad (3)$$

III. PROPOSED POWER MODEL

A. AES-128 Mapping on LiCryptor

SCA primarily targets block cipher algorithms, with AES being one of the most widely analyzed due to its widespread use in various security applications. In this paper, we focus specifically on analyzing CPA-based SCA attacks on AES-128 mapped onto the LiCryptor architecture. AES-128 is a symmetric key encryption algorithm that operates on a 128-bit key and blocks of data. It performs 10 rounds of cryptographic transformations, including SubBytes (Sbox), ShiftRows, Mix-Columns, and AddRoundKey. Each round, except the final

TABLE II: HD and HW power models based on different intermediate values used for CPA attacks on LiCryptor.

Scheme	Power Models	Attack
CPA_HW_R0 (using 1 thread)	$HW(Sbox(PT_0^i \oplus K_0^i))$	Logic state pattern of Sbox and ShiftRows Output
CPA_HW_R0 (using 2 threads)	$HW(Sbox(PT_0^i \oplus K_0^i)) +$ $HW(Sbox(PT_1^i \oplus K_1^i))$	Logic state pattern of Sbox and ShiftRows Output
CPA_HD_R0 (using 1 thread)	$HD(Sbox(PT_0^i \oplus K_0^i),$ $Sbox(PT_2^i \oplus K_2^i))$	Switching pattern between consecutive Outputs of Sbox and ShiftRows
CPA_HD_R0 (using 2 threads)	$HD(Sbox(PT_0^i \oplus K_0^i),$ $Sbox(PT_2^i \oplus K_2^i)) +$ $HD(Sbox(PT_1^i \oplus K_1^i),$ $Sbox(PT_3^i \oplus K_3^i))$	Switching pattern between consecutive Outputs of Sbox and ShiftRows
CPA_HW_R10 (using 1 thread)	$HW(ISbox(CT_0^i \oplus K_0^i))$	Logic state pattern of Sbox Input
CPA_HW_R10 (using 2 threads)	$HW(ISbox(CT_0^i \oplus K_0^i)) +$ $HW(ISbox(CT_1^i \oplus K_1^i))$	Logic state pattern of Sbox Input
CPA_HD_R10 (using 1 thread)	$HD(ISbox(CT_0^i \oplus K_0^i),$ $ISbox(CT_2^i \oplus K_2^i))$	Switching pattern between consecutive Inputs of SBox
CPA_HD_R10 (using 2 threads)	$HD(ISbox(CT_0^i \oplus K_0^i),$ $ISbox(CT_2^i \oplus K_2^i)) +$ $HD(ISbox(CT_1^i \oplus K_1^i),$ $ISbox(CT_3^i \oplus K_3^i))$	Switching pattern between consecutive Inputs of SBox

one, includes all these transformations, while the last round omits MixColumns. In Fig. 3, the AES-128 encryption process is mapped onto the LiCryptor architecture with an 8-stage pipeline and dual-threading operation. Sbox and ShiftRows are handled in the first PE row, MixColumns are executed in the second PE row, and AddRoundKey is configured in the third and final PE rows. AddRoundKey is executed in 2 stages because each PE can only process 32-bit data. In AES-128 computation, the *LiCryptor* is developed to process a large number of plaintext blocks (N), where each block consists of 16 plaintexts. The *PEA* processes each block using two threads: in the first thread, eight plaintexts (denoted as $PT_{j'}^i$, where i represents the ith block and $j' \in \{0, 2, 4, \ldots, 14\}$) are processed, and in the second thread, the remaining eight plaintexts (denoted as $PT_{j''}^i$, where $j'' \in \{1, 3, 5, \ldots, 15\}$) are processed. Each plaintext PT_j^i, associated with its corresponding key K_j^i, is processed using a pipelined and multi-threaded approach, enabling the parallel computation of 16 different plaintexts per block. This results in 16 ciphertexts CT_j^i for each block of plaintext, where i represents the ith block and $0 \leq j \leq 15$.

B. Proposed HD and HW Models for Attacking LiCryptor

In this section, we propose using HD and HW models to conduct CPA-based attacks on LiCryptor, targeting different intermediate values. For the AES-128 algorithm, conventional power models focus on the input and output points of the

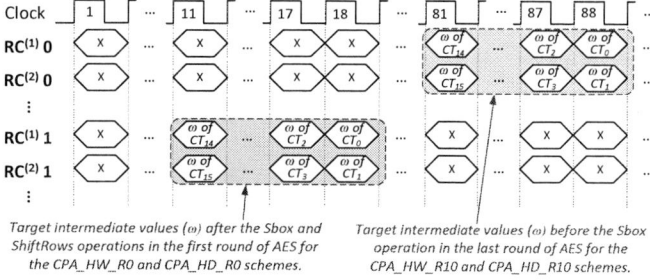

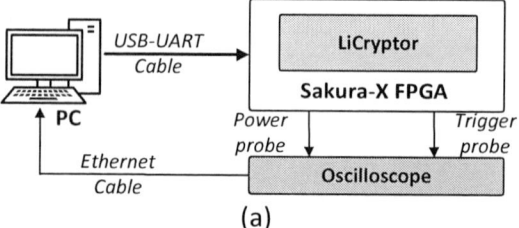

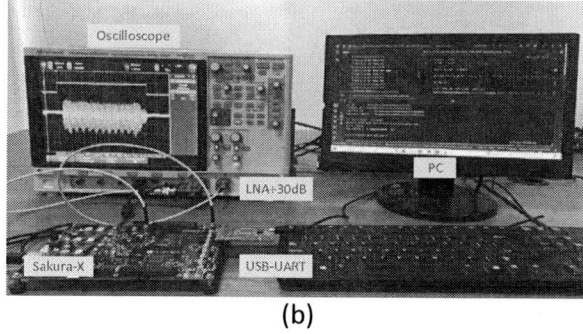

Fig. 4: Illustration of the targeted intermediate values that are used to construct the proposed HD and HW power models.

SBox, where secret key leakage is most likely to occur. To describe these power models, we define the Hamming Distance between two values x and y as $HD(x,y)$, the Hamming Weight of x as $HW(x)$, the SBox transformation of input x as $SBox(x)$, the inverse SBox of input x as $ISbox(x)$, plaintext as PT_j^i, ciphertext as CT_j^i, and keys are the same ($K_j^i = K$).

The LiCryptor architecture, combining pipelining and dual-threading, complicates power analysis attacks by processing two sets of 8 plaintexts simultaneously, increasing the difficulty of isolating key-related information. Although there is no direct data dependency between the two threads, their energy consumption is similar, and the linearity of power models in both HD and HW metrics allows the combined energy model from both threads to be used. This combination accumulates useful energy and improves the SNR, making CPA attacks more effective. As shown in Table II, various power models based on HD and HW metrics are used for CPA attacks on LiCryptor, targeting different intermediate values. In the first round (R0), the ShiftRows operation reorders bytes without changing their values, so the energy consumption after ShiftRows mirrors that of the Sbox operation. Hence, the attack point for the first round is selected at the registers in the RC1 stage after the Sbox and ShiftRows operations. In the last round (R10) of AES, MixColumns is replaced with a No-Operation (NOOP), and the attack targets the intermediate data before the Sbox in the RC0 stage. The attack targets the intermediate values after the Sbox in the first round because they maintain the Sbox's high-entropy characteristics, making them detectable in power analysis. In the last round, the attack targets the values before the Sbox, as MixColumns is skipped, and these values retain predictable patterns that are easier to analyze for key leakage. Since LiCryptor can operate with both one and two threads, there are a total of 8 schemes: two for the HW scheme in the first round (one and two threads), two for the HW scheme in the last round (one and two threads), two for the HD scheme in the first round (one and two threads), two for the HD scheme in the last round (two threads). The targeted intermediate values for the 8 attack schemes are illustrated in Fig. 4, where ω represents the intermediate values corresponding to each ciphertext CT. These intermediate values are analyzed in the power models

Fig. 5: (a) Experimental setup for LiCryptor power measurement on Sakura-X FPGA. (b) Physical devices.

for each of the 8 schemes, which include CPA_HW_R0, CPA_HW_R10, CPA_HD_R0, and CPA_HD_R10, utilizing both single-threaded and dual-threaded configurations. Overall, the HD and HW models we proposed help enhance the effectiveness of CPA-based attacks on LiCryptor's pipelining and dual-threading, enabling more precise key recovery at the correct attack points.

IV. EXPERIMENT AND EVALUATION

A. Experimental Setup

To perform CPA attacks on LiCryptor, we designed an experimental system to collect power traces for side-channel analysis. The block diagram of the experimental setup for power measurement of LiCryptor on FPGA is shown in Fig. 5 (a). It consists of three main components: the control computer (PC) for managing the system and storing data, the FPGA board where the LiCryptor is implemented and runs the cryptographic tasks, and the oscilloscope for measuring the power consumption of the FPGA during the cryptographic operations. Fig. 5 (b) shows the physical implementation of the experimental setup with the actual devices used to measure the power consumption of LiCryptor. The LiCryptor on the Sakura-X FPGA communicates with the control computer via UART and operates at a clock frequency of 50 MHz. A shunt resistor in series with the FPGA measures instantaneous power at the J19 jack, with the signal amplified by +30 dB for efficient analysis. A trigger signal synchronizes the oscilloscope, which is set to a 625 MHz sampling frequency, and collects raw power data for LiCryptor, ensuring compliance with the Nyquist-Shannon criterion.

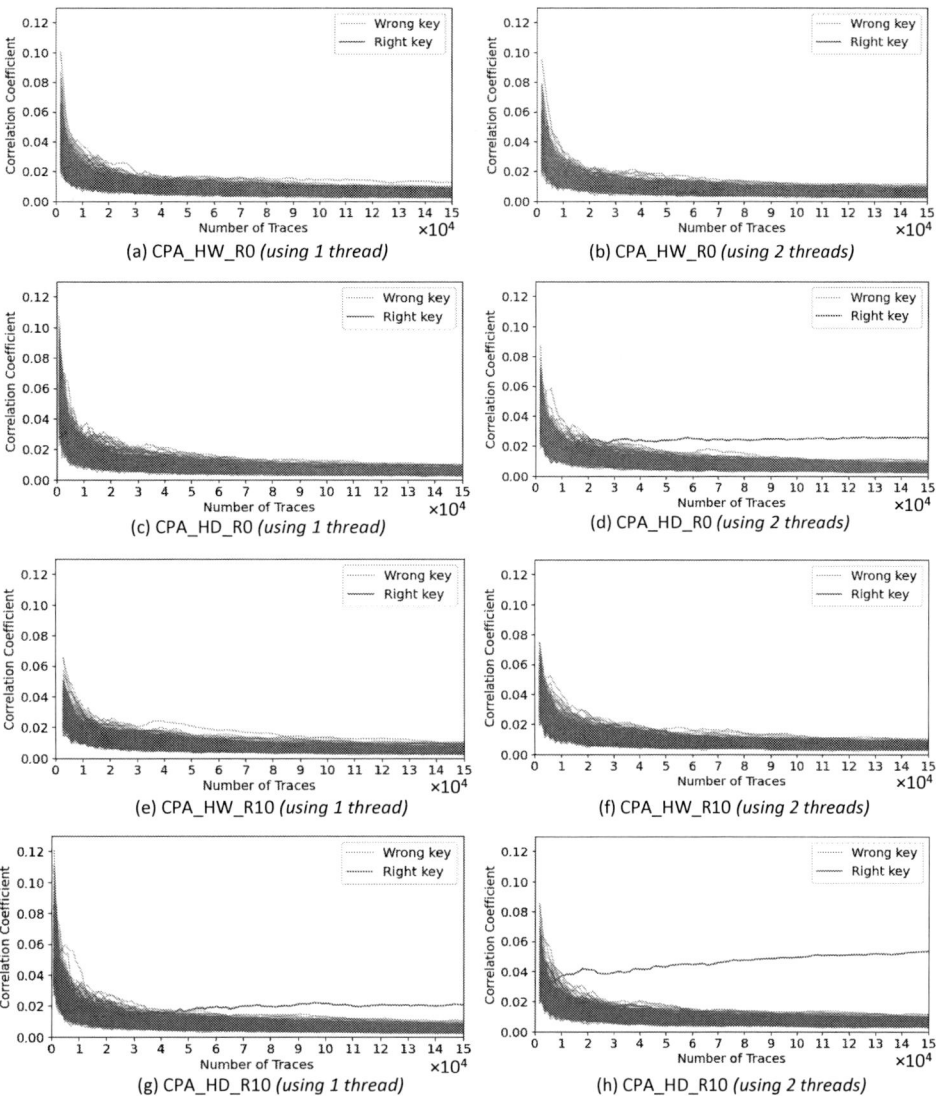

Fig. 6: Correlation Coefficient vs. Number of Traces for 8 HD and HW Power Model Schemes when attacking the first key byte

The control program, developed in Python and running on the PC, automates the process from initialization to power consumption trace collection. The steps are as follows:

- **Step 1:** The control program initializes and configures the oscilloscope, then transfers the contexts to configure LiCryptor on the FPGA to operate in AES-128 mode.
- **Step 2:** The control program switches the oscilloscope to single-trigger mode, sends the same key $K_j^i = K$ and plaintext PT_j^i to LiCryptor for encryption, and triggers the process. Upon successful encryption, the program verifies the ciphertext, collects the power trace, and saves it to the database. If incorrect, the program resends the data for re-execution.

Step 2 is repeated until the desired power traces are collected, then the system returns to the state before *Step 1*.

B. Evaluation

To fully evaluate the effectiveness of the attack methods, we performed the attack on all bytes of the secret key using 150K collected traces. As shown in Fig. 6, the results help determine the number of traces required to reveal the first byte of the secret key. The CPA_HD_R10 attack method outperforms CPA_HD_R0, while the CPA_HW_R10 and CPA_HW_R0 methods show lower effectiveness. Additionally, using the energy from two threads in the proposed model significantly improves attack efficiency compared to using just one thread. Among the attack methods, CPA_HD_R10 with two threading is the most effective, requiring only about 82,100 traces (approximately 54.73% of the total traces) to uncover the full key. In contrast, when using only one thread for the same method, approximately 114,000 traces (around 76% of the total

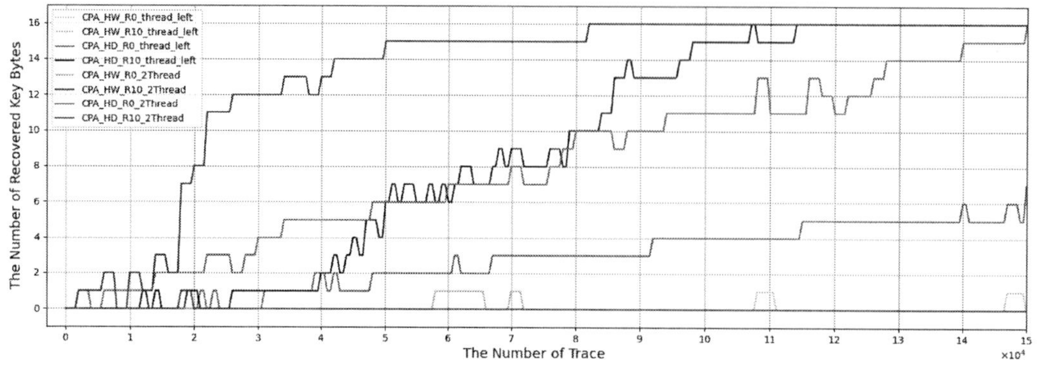

Fig. 7: Mapping of AES-128 with an 8-stage pipeline and dual-threading operation on LiCryptor.

traces) are needed to retrieve the full key.

Based on the analysis of 150K traces, as shown in Fig. 7, three attack methods (CPA_HD_R10 using 1 thread and 2 threads, CPA_HD_R0 using 2 threads) successfully recovered the full AES-128 key with a 100% success rate. CPA_HD_R0 using 1 thread attack method managed to recover 7 out of 16 key bytes, corresponding to a success rate of 43.75%. The remaining attack methods were unable to recover any key bytes, resulting in a 0% success rate. These results demonstrate the significant advantage of using dual-threading and the proposed power models for more effective key recovery in AES-128 when utilizing LiCryptor.

V. CONCLUSION

This paper evaluates LiCryptor's vulnerability to CPA-based SCA attacks, proposing two primary models based on HW and HD. We introduced 8 power model schemes for both single-threaded and dual-threaded pipelined operations. Experimental results show that dual-threaded operation enhances attack effectiveness, reducing the number of traces needed for key recovery. The CPA_HD_R10 model outperformed others in both configurations. Future work will focus on developing a heterogeneous LiCryptor design, analyzing its CPA-based SCA vulnerability, and proposing countermeasures to enhance security and performance for IoT applications.

ACKNOWLEDGMENT

This research was supported by Daiichi-Sankyo "Habataku" Support Program for the Next Generation of Researchers, NAIST Senju Monju Project.

This research was also partially supported by the ASEAN IVO (http://www.nict.go.jp/en/asean_ivo/index.html) Project, "Artificial Intelligence Powered Comprehensive Cyber-Security for Smart Healthcare Systems (AIPOSH)."

REFERENCES

[1] A. Menezes and D. Stebila, "Challenges in cryptography," *IEEE Security & Privacy*, vol. 19, no. 2, pp. 70–73, 2021.

[2] M. Aamir, S. Sharma, and A. Grover, "Chacha20-in-memory for side-channel resistance in iot edge-node devices," *IEEE Open Journal of Circuits and Systems*, vol. 2, pp. 833–842, 2021.

[3] I. Elsadek, S. Aftabjahani, D. Gardner, E. MacLean, J. R. Wallrabenstein, and E. Y. Tawfik, "Hardware and energy efficiency evaluation of nist lightweight cryptography standardization finalists," in *2022 IEEE International Symposium on Circuits and Systems (ISCAS)*. IEEE, 2022, pp. 133–137.

[4] S. Shahabuddin, A. Mämmelä, M. Juntti, and O. Silvén, "Asip for 5g and beyond: Opportunities and vision," *IEEE Transactions on Circuits and Systems II: Express Briefs*, vol. 68, no. 3, pp. 851–857, 2021.

[5] F. Campos, L. Jellema, M. Lemmen, L. Müller, D. Sprenkels, and B. Viguier, "Assembly or optimized c for lightweight cryptography on risc-v?" in *Cryptology and Network Security: 19th International Conference, CANS 2020, Vienna, Austria, December 14–16, 2020, Proceedings 19*. Springer, 2020, pp. 526–545.

[6] G. Sayilar and D. Chiou, "Cryptoraptor: High throughput reconfigurable cryptographic processor," in *2014 IEEE/ACM International Conference on Computer-Aided Design (ICCAD)*. IEEE, 2014, pp. 155–161.

[7] C. Deng, B. Wang, L. Liu, M. Zhu, Y. Wu, H. Li, S. Yin, and S. Wei, "A 60 gb/s-level coarse-grained reconfigurable cryptographic processor with less than 1-w power," *IEEE Transactions on Circuits and Systems II: Express Briefs*, vol. 67, no. 2, pp. 375–379, 2019.

[8] Y. Du, W. Li, Z. Dai, and L. Nan, "Pvharray: An energy-efficient reconfigurable cryptographic logic array with intelligent mapping," *IEEE Transactions on Very Large Scale Integration (VLSI) Systems*, vol. 28, no. 5, pp. 1302–1315, 2020.

[9] T.-D. Vu, H.-L. Pham, T.-H. Tran, and Y. Nakashima, "Flexible and energy-efficient crypto-processor for arbitrary input length processing in blockchain-based iot applications," *IEICE Transactions on Fundamentals of Electronics, Communications and Computer Sciences*, vol. 107, no. 3, pp. 319–330, 2024.

[10] T. S. Duong, H. L. Pham, V. T. D. Le, T. H. Tran, and Y. Nakashima, "Power-efficient and programmable hashing accelerator for massive message processing," in *2023 IEEE 36th International System-on-Chip Conference (SOCC)*. IEEE, 2023, pp. 1–6.

[11] H. L. Pham, V. T. D. Le, T. H. Vu, Y. Nakashima *et al.*, "Licryptor: High-speed and compact multi-grained reconfigurable accelerator for lightweight cryptography," *IEEE Transactions on Circuits and Systems I: Regular Papers*, 2024.

[12] M. Alioto, L. Giancane, G. Scotti, and A. Trifiletti, "Leakage power analysis attacks: A novel class of attacks to nanometer cryptographic circuits," *IEEE Transactions on Circuits and Systems I: Regular Papers*, vol. 57, no. 2, pp. 355–367, 2009.

[13] N.-T. Do, V.-P. Hoang, and C.-K. Pham, "Low Complexity Correlation Power Analysis by Combining Power Trace Biasing and Correlation Distribution Techniques," *IEEE Access*, vol. 10, pp. 17 578–17 589, 2022.

[14] E. Brier, C. Clavier, and F. Olivier, "Correlation power analysis with a leakage model," in *Cryptographic Hardware and Embedded Systems-CHES 2004: 6th International Workshop Cambridge, MA, USA, August 11-13, 2004. Proceedings 6*. Springer, 2004, pp. 16–29.

Data communication security for FANETs using Ascon lightweight cryptography

Huyen-Trang Pham-Thi, Duy-Hieu Bui, Xuan-Tu Tran

VNU Information Technology Institute, Vietnam National University, Hanoi

144 Xuan Thuy street, Cau Giay, Hanoi, Vietnam

Corresponding author's email: hieubd@vnu.edu.vn

Abstract—**Flying Ad-hoc Network (FANET) is a wireless network designed for Unmanned Aerial Vehicles (UAVs). FANET allows many UAVs to connect and exchange data with each other and with ground devices without the need for a fixed network infrastructure. However, data exchange between UAVs in a FANET faces many security risks. This paper proposes a data communication security solution using the lightweight cryptographic standard Ascon designed for resource-constrained devices such as UAVs. The solution is implemented on the STM32WL55JC development kit with an integrated LoRa module, and its performance is evaluated in terms of code size, stack usage, throughput, and power consumption. Experimental results show that this solution protects UAVs against common communication attacks while consuming few resources. With input data ranging from 8 to 245 bytes, Ascon outperforms AES in terms of efficiency. It reduces energy consumption by nearly 50% compared to traditional AES, enabling UAVs to operate more efficiently, extend flight time, and enhance overall system performance in resource-constrained environments.**

Keywords— *FANET, security, lightweight cryptography*

I. INTRODUCTION

Nowadays, Unmanned Aerial Vehicles (UAVs) are widely used in various fields such as agriculture [1], military [2], and transportation [3]. The widespread use of UAVs also brings new challenges, requiring updates to meet different needs and improve operational efficiency. UAV systems can operate individually or in swarms. A single UAV is often used for simple tasks, like aerial photography, filming, and basic data collection. Meanwhile, multi-UAV systems, consisting of multiple UAVs working together, improve efficiency, expand coverage, and optimize costs compared to single UAV systems through coordination and data sharing. This has led to the growing deployment of multi-UAV systems.

The most important issue in multi-UAV systems is communication between UAVs and the ground control station due to high mobility and frequent topology changes. The Flying Ad-hoc Network (FANET) was introduced to address this issue [4]. FANET is a specialized network designed for UAVs, enabling them to communicate with each other without relying on existing infrastructure. FANET is a suitable networking solution for multi-UAV systems, allowing the UAVs to communicate directly and efficiently while supporting complex missions.

However, most commercial UAVs today have many limitations in implementing security mechanisms due to system constraints such as energy, latency, and cost. Therefore, security solutions for UAVs need to be optimized to ensure effectiveness without significantly impacting system performance. Lightweight cryptography is an ideal solution for UAV systems, minimizing resource usage while ensuring security. By using lightweight encryption techniques, UAVs can ensure secure communications and data integrity without heavy computational overhead. Additionally, these cryptographic methods enable UAVs to operate efficiently in resource-constrained environments, ensuring robust protection against cyber threats while preserving battery life and processing power.

One of the recent studies [5] explored lightweight cryptographic algorithms such as Ascon for UAV networks, highlighting their security and computational efficiency advantages. However, that work did not target a specific type of networks and lacked detailed implementation results on real hardware devices. Therefore, in this paper, we propose a communication security solution for FANET networks by applying Ascon to protect the data transmission between nodes, ensuring confidentiality, authentication, and data integrity while maintaining system performance. The proposed solution is implemented on an STM32 microcontroller. Moreover, this paper compares resource usage and energy consumption, demonstrating how the Ascon implementation impacts power usage in UAV systems. Experimental results show that our approach protects the UAV system from specific attacks and offers data on resource and energy usage, which were not addressed in [5].

The remaining parts of the paper are structured as follows. Section II is an overview of FANET and security analysis. Next, the proposed security solutions are described in Section III. Then, Section IV presents our experimental results. Finally, conclusions and future works are given in Section V.

II. FANET AND SECURITY PROBLEMS

A. Introduction to FANET

An ad-hoc network, also known as a self-configuring wireless network, allows devices to connect directly without intermediaries such as routers or access points. The FANET consists of UAVs acting as nodes, connected in an ad-hoc structure, where at least one UAV is linked to a ground control station (GCS) or a satellite. Each UAV must be connected to at least one other UAV. If a UAV cannot complete its task, another UAV in the network can take over the mission. The architecture of FANET is shown in Fig. 1. Due to high mobility and frequent topology changes, maintaining stable connections requires continuous updates on UAV location and efficient energy management. Thanks to its advantages, FANETs are widely applied in real life. For example, FANETs can be used in agriculture to monitor water quality in shrimp farms [6]. Additionally, they can facilitate the exchange of health information to manage disaster victims [7].

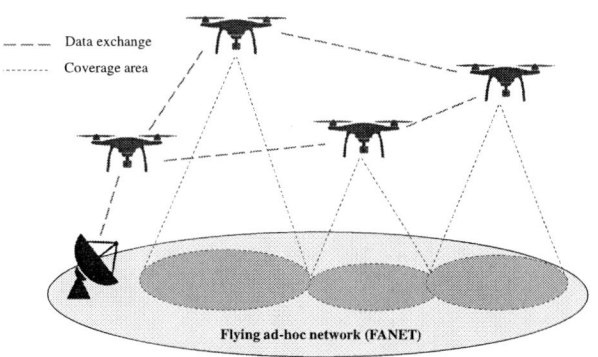

Fig. 1. Architecture of FANET.

Each UAV in FANET utilizes a communication module to exchange data with other nodes within the network. FANET employs wireless technologies such as Wi-Fi [8], LoRa [9], and 4G/5G [10] to ensure fast and stable data transmission between UAVs. Wi-Fi offers high data rates but is limited in range and power efficiency, while 4G/5G provides broad coverage but consumes more energy. In contrast, LoRa stands out due to its long-range capabilities and ultra-low power consumption, making it ideal for UAVs operating in wide-area or energy-constrained environments. This work focuses on the FANETs, which use LoRa as the main communication technology. The structured LoRa packet format proposed by the Skytraxx manufacturer [11] defines how data is transmitted within the FANET. The packet structure has a maximum length of 256 bytes, including a header for packet type and control, source and destination addresses for routing, an extended header for additional control, a 4-byte digital signature for authentication, and a data payload for communication, as illustrated in Fig. 2

Byte	0	1	2	3	4	5	6	7
Stream	Header	Source address			E-Header	Destinations address		
Byte	8	9	10	11	12	13	...	256
Stream	Digital signature				Data			

Fig. 2. The packet format of FANET.

B. Security threats

Communication vulnerabilities are one of the most serious threats to the FANET. These vulnerabilities can arise from various factors, such as insecure communication protocols, lack of encryption or authentication, and inefficient UAV connection management. Attackers can intercept, manipulate, or disrupt data transmission without proper security measures, leading to unauthorized access and data interference.

Moreover, the dynamic structure and high mobility of nodes in the network provide attackers with more opportunities to carry out communication attacks on the system. Some of the most common attacks include eavesdropping, where attackers secretly monitor data transmissions to gather sensitive information without authorization [12]; Man-in-the-Middle (MITM) attacks, which allow unauthorized entities to inject malicious data during transmission [13]; Sybil attacks, where attackers create fake identities to manipulate network behavior [14]. These attacks can lead to loss of UAV control, mission interruptions, or even the disappearance or destruction of the device.

To mitigate communication attacks, UAVs in the FANET must be equipped with security solutions to ensure safe and reliable operation. However, UAVs also limited due to their technical characteristics and operational environment. Firstly, most UAVs have limited resources such as power supply, memory, and processing power. This makes it challenging to implement complex security solutions, as they can impact the operation of UAVs. Additionally, the cost and complexity of security implementation can be a barrier for low-cost commercial UAVs, making the integrating of strong security solutions expensive. Finally, security solutions must have low latency to meet the system's real-time communication requirements. Therefore, UAV security solutions must be optimized to ensure effectiveness without significantly impacting system performance.

C. Lightweight cryptography

One of the methods to protect UAV communication data is to use encryption combined with data authentication. Encryption ensures that only authorized devices can access data, while authentication guarantees data integrity and prevents forgery or tampering. When a UAV sends or receives information, encryption safeguards it from interception, and authentication mechanisms, such as digital signatures or cryptographic hashes, help detect tampering and unauthorized access. UAV systems can maintain secure and reliable communication even in hostile environments by implementing robust traditional encryption protocols like AES or RSA alongside authentication techniques.

However, UAVs often have hardware limitations, including limited processing power, memory capacity, and energy availability. These limitations make it difficult to implement traditional cryptographic methods. While highly secure, advanced encryption algorithms such as AES or RSA, require substantial computational resources, leading to increased processing time and higher power consumption. For example, AES requires multiple rounds of encryption and decryption, each involving complex mathematical operations, making the encryption process time-consuming on UAV systems with limited processing power. This increased computation leads to delays, impacting real-time data transmission and system responsiveness. Moreover, the high energy consumption of these encryption methods drains the UAV's battery more quickly, limiting its flight time and overall efficiency. Due to these challenges, traditional encryption may not be the best choice for UAV systems with strict resource limitations.

A potential solution to the challenges of implementing traditional encryption methods is lightweight cryptography, which is specifically designed for resource-constrained devices. Compared to traditional encryption methods, lightweight cryptography consumes less power, enabling UAVs to extend their operational time without affecting security. One approach to applying lightweight cryptography in UAV communication is Authenticated Encryption with Associated Data (AEAD). This cryptographic technique combines encryption and authentication while allowing additional data to be securely associated with the encrypted message. AEAD can encrypt sensitive UAV data to prevent unauthorized access while also providing authentication to verify the integrity and authenticity of transmitted messages. Additionally, lightweight cryptography algorithms are designed to be highly efficient, ensuring real-time data protection without causing significant latency. Due to its capabilities, lightweight cryptography with AEAD mode can be an ideal choice for securing UAV communication data.

III. PROPOSED METHOD

A. Security requirements and proposed mechanisms

Encryption and authentication techniques can be implemented to enhance security for UAVs data communication. In particular, for long-range, low-power systems using LoRa communication technology, the implemented security solution must meet timing and resource constraints. With limited power sources, commercial UAVs require low-energy security solutions to maximize flight time without affecting performance. Additionally, security solutions not only ensure confidentiality but also guarantee data authenticity and integrity. This helps prevent spoofing attacks or unauthorized modifications. Moreover, the proposed security solution must be designed to maintain real-time efficiency, preventing significant latency that could impact data exchange, especially in time-sensitive missions like surveillance, search and rescue, or delivery services. Therefore, security mechanisms should ensure fast processing speeds, minimal resource consumption, and stable performance.

This paper proposes a security solution using the lightweight cryptographic standard Ascon [15] to encrypt data before transmission. The Ascon algorithm is a symmetric cipher optimized for resource-constrained devices like UAVs, requiring only a single secret key for encryption and decryption. This not only reduces system complexity but also enhances the efficiency of data transmission. Ascon also provides an authenticated encryption mode, ensuring security, data authenticity and integrity. In the proposed system, Ascon is

implemented on each UAV in the FANET network to perform data encryption and authentication before transmission.

The encryption process uses a Pre-Shared Key (PSK) for encryption, authentication, and decryption, reducing key management complexity, system load, and resource usage. While dynamically changing keys enhance security by minimizing the risk of key leakage, they introduce complex management and exchange mechanisms, potentially increasing latency and computational overhead. PSK offers a simpler and more efficient solution, making it ideal for low-latency, resource-constrained networks like FANET, where maintaining efficient communication is important. However, PSK must be carefully protected, as any exposure can lead to data leakage in the whole network. New primitives such as PUF and secure boot can be used to enhance the protection of the secret key.

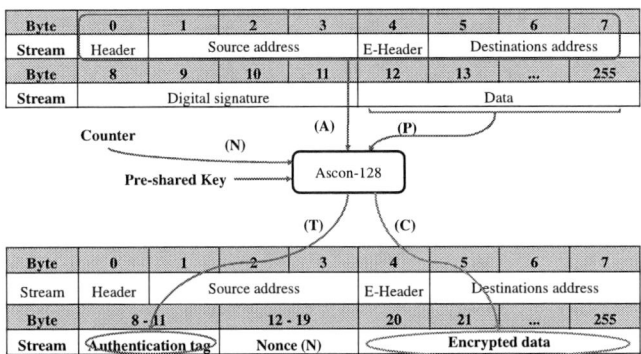

Fig. 4. The packet format of FANET after encryption

Using LoRa communication with Ascon-based data security, the proposed system ensures secure data transmission. It provides an efficient, energy-saving communication method for missions requiring broad coverage and low power consumption. The proposed data communication security mechanism for UAVs is shown in Fig. 3. Sensor data collected by UAVs is processed and encrypted with a PSK. The packet header is used as associated data, and the nonce is a updated counter after each successful transmission. After encryption with Ascon, the output includes the ciphertext and an authentication tag. The digital signature is extracted from the authentication tag. After encryption, the data packet is formatted and transmitted via LoRa. The encrypted packet format is illustrated in Fig. 4. The packet header is left unencrypted to help FANET devices quickly access key routing and control information, improving network efficiency and reducing delay. The nonce is included in each packet to allow the receiver to perform authentication and decryption. The receiving device waits for transmitted packets. The device will verify its authenticity using the digital signature if the packet is successfully received. If the verification is successful, the packet will be decrypted using the pre-shared key (PSK), and the original data will be recovered. If the verification fails, the packet will be discarded, ending the transmission process. This process ensures data security throughout transmission.

979-8-3315-1550-8/25 $31.00 © 2025 IEEE   141

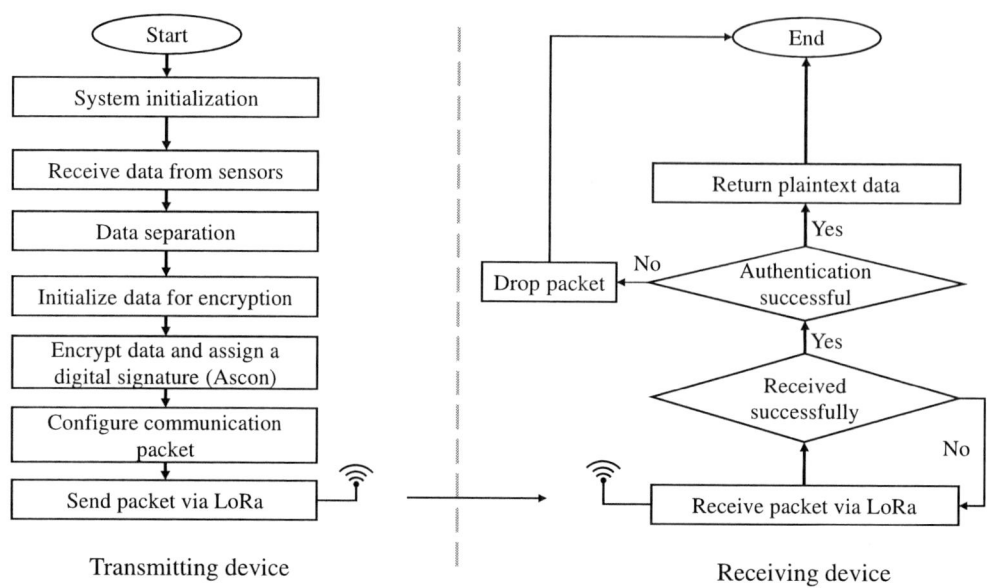

Fig. 3. Flowchart of the proposed communication data encryption algorithm.

The experimental system consists of two STM32WL55JC kits simulating transmitting and receiving devices. Initially, the data is randomly generated using Python software on a computer and then sent to the STM32WL55JC microcontroller via the UART communication protocol. This data includes essential information such as control data, remote configuration, and tracking information for configuration or data transmission. After being sent to the STM32 microcontroller, the received UART commands are analyzed into values that configure communication data packets between devices. Once the data is received and processed, it is encrypted using the Ascon-AEAD algorithm, a lightweight encryption algorithm loaded onto the STM32WL55JC microcontroller. The decryption algorithms were implemented on receiving devices. After receiving an encrypted packet, the receiving device authenticates and decrypts the data using the Ascon-AEAD algorithm.

B. Security analyses

The encryption solution using Ascon with an authenticated encryption mechanism effectively meets security evaluation criteria, including confidentiality, integrity and authenticity. Confidentiality is ensured through a strong encryption mechanism that protects data from leakage. By encrypting the data with a lightweight, highly secure algorithm, Ascon ensures that only authorized parties with the correct decryption key can access the original information. Integrity is maintained through a digital signature. These signatures help detect any changes or tampering that may have occurred during transmission, ensuring that the data remains unchanged from its original state. If any risks are identified, the system automatically rejects the modified packets, preventing the acceptance of malicious data. Additionally, the authentication mechanism

guarantees authenticity by verifying the origin of the data. Digital signatures assure that a valid source generated the received information. This ensures that the data originates from a trusted sender, strengthening the overall security of the communication system.

Using Ascon-AEAD to encrypt data in FANET networks significantly benefits the security of transmitted information between unmanned aerial vehicles (UAVs). Communication data between devices in the network is ensured to prevent information leakage or unauthorized interference during transmission effectively. With the encryption mechanism, the proposed system can ensure that attackers do not eavesdrop on data, preventing eavesdropping attacks. Authentication tags ensure data integrity, drop modified data, and prevent MITM attacks. This mechanism cannot directly prevent Sybil attacks, but it helps mitigate the impact of Sybil attacks. Its encryption and authentication ensure data integrity, making it possible to identify and drop invalid packets.

However, the Pre-Shared Key (PSK) mechanism is a system weakness. Although PSKs are simple and effective, if they leak, all communication data in the network could be intercepted or faked. Besides, these keys are only valid for a limited time. In FANET networks, devices often operate in mobile conditions and exchange data continuously. Therefore, the PSK must be updated frequently to ensure security. Updating keys consumes computational resources and requires a safe way to share new keys between devices. Hence, managing and securely sharing keys between devices is very important.

IV. EXPERIMENTAL RESULTS

The proposed security solution is implemented on the STM32WL55JC hardware platform, which integrates LoRa technology [16]. Additionally, LoRa is utilized to ensure long-range communication, making it suitable for applications in FANET networks. The source code is compiled and optimized through the STM32CubeIDE environment. Once compiled, the program is directly uploaded to the STM32WL55JC microcontroller, ensuring the system operates stably, securely and efficiently. The LoRa transceiver in the system is set to operate at 868MHz with a spreading factor of SF7, a 250 kHz bandwidth, a coding rate of 4/8, and a transmission power of 14dBm.

The performance evaluation criteria include code size, stack usage, throughput and energy consumption. The results are compared with the case of using AES-CCM [17] authenticated encryption mode to evaluate the advantages of the lightweight Ascon cipher over the advanced AES encryption standard in the context of UAV communication over LoRa, where hardware resources and bandwidth are typically limited. At the same time optimization is required to ensure transmission performance and energy efficiency.

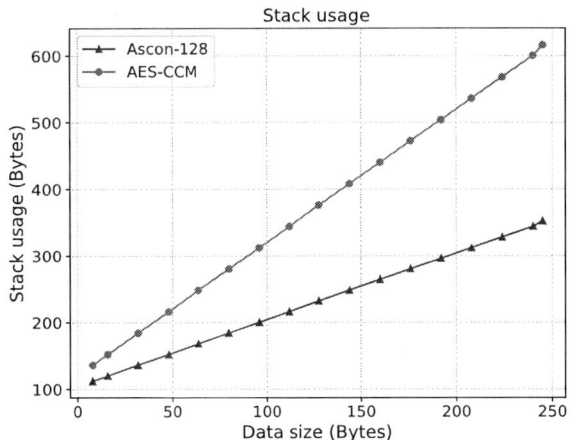

Fig. 6. Stack usage comparison of Ascon-128 and AES-CCM.

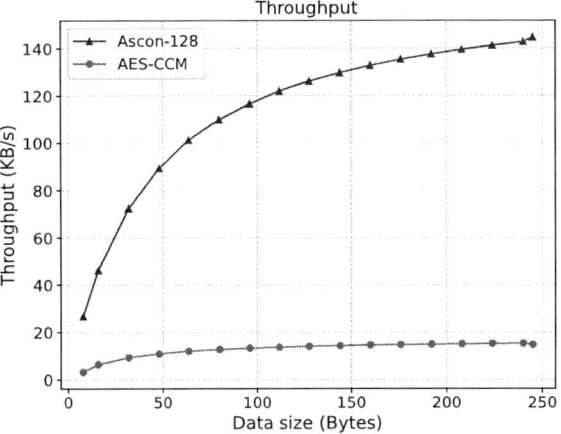

Fig. 7. Throughput comparison of Ascon-128 and AES-CCM..

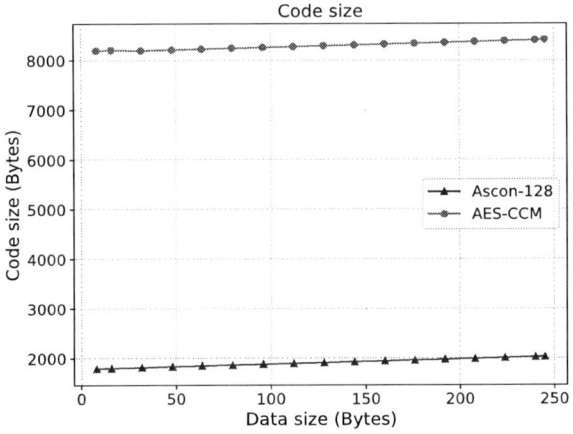

Fig. 5. Code size comparison of Ascon-128 and AES-CCM.

Code size is one of the critical parameters indicating the memory required to store a security solution on systems. Optimizing the size of the security program deployed on UAVs helps reduce memory usage and ensures efficient operation of flight control and sensor communication tasks. The proposed method's code size is shown in Fig. 5, Ascon-128 requires only 2000 bytes, while AES-CCM needs 8000 bytes—four times more. This difference is significant in embedded applications with limited storage area.

Stack usage also reflects the efficiency of memory management in cryptographic algorithms. Monitoring stack usage helps prevent overflow, optimize memory, and maintain system stability, particularly in constrained devices like UAVs. As shown in Fig. 6, Ascon-128 has lower and more stable stack usage than AES-CCM, which becomes more evident as data size increases. This makes Ascon-128 a suitable choice for constrained devices like UAVs.

The security mechanism must quickly process encryption, decryption, and data authentication without disrupting impor-

tant UAV operations. From the graph in Fig. 7, the throughput of Ascon is much higher than AES-CCM, especially as the data size increases. The throughput of Ascon-128 grows sharply, achieving approximately 140 KB/s at 245 bytes, while AES remains stable at 15 KB/s. This difference highlights the faster encryption and decryption of Ascon-128, while the lower throughput of AES-CCM causes delays in real-time applications. Ascon is a suitable solution for FANET networks, where fast, stable, and low-latency communication is essential to support efficient information exchange between nodes.

Energy consumption is crucial for UAV performance, especially in battery-powered systems. As shown in Fig. 8, encryption increases energy usage, but Ascon-128 consumes less energy than AES-CCM. This makes Ascon-128 a more energy-efficient solution, providing higher throughput with lower energy consumption, making it ideal for UAVs requiring high performance and low energy use.

After testing with 240 input data bytes, the results in Table I show that Ascon-128 is better in every evaluation criterion. Specifically, Ascon-128 has a code size that is 75.9%

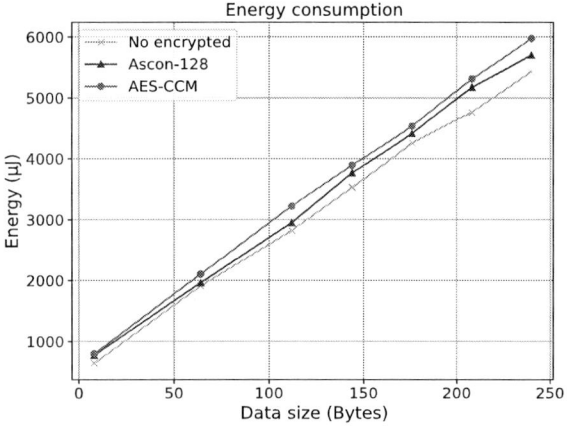

Fig. 8. Energy consumption comparison of Ascon-128 and AES-CCM.

smaller and uses 42.67% less stack memory, significantly saving memory, making it especially suitable for IoT devices or embedded systems with limited resources. In addition, the throughput of Ascon-128 is 838.7% higher than that of AES-CCM, effectively meeting the demands of applications that require high-speed data transmission. Moreover, Ascon-128 consumes 50.14% less energy than AES-CCM. Thanks to these advantages, the proposed solution is a suitable choice for the communication security challenges faced by UAVs while also meeting the constraints of limited resources.

TABLE I
PERFORMANCE COMPARISON BETWEEN AES-CCM AND ASCON-128

Parameters	AES-CCM	Ascon-128	Improvement
Code size	8400B	2024B	Reduced by 75.9%
Stack usage	600B	344B	Reduced by 42.67%
Throughput	15.58 KB/s	146.29 KB/s	Increased by 838.70%
Energy consumption	540.904 μJ	269.68 μJ	Reduced by 50.14%

V. CONCLUSION

This work proposes a security mechanism for data communication in FANET using Ascon lightweight cryptography, implemented on the STM32 microcontroller. The implementation results are evaluated to have better performance than a traditional encryption. Ascon-128 achieves a 75.9% reduction in code size, 42.67% less stack memory usage, 838.7% higher throughput, and 50.14% lower energy consumption compared to the traditional AES-CCM implementation. These significant improvements show that the security solution based on Ascon not only ensures data protection but also optimizes system resources, making it suitable for the requirements of UAV systems.

Ascon-AEAD offers strong encryption and authentication, but key management in the FANET remains challenging. Future research could focus on using Physical Unclonable Functions (PUFs) for secure key management, which would reduce the risk of key leakage, enhance the robustness of the FANET against cyber threats, and improve the overall security and efficiency of UAV communications in dynamic environments.

ACKNOWLEDGMENT

This work has been supported by Vietnam National University, Hanoi (VNU) under Project No. QG.24.88 (LoPoRISC).

REFERENCES

[1] J. Kim, S. Kim, C. Ju, and H. I. Son, "Unmanned aerial vehicles in agriculture: A review of perspective of platform, control, and applications," *IEEE Access*, vol. 7, pp. 105 100–105 115, 2019.

[2] ABCnews, "Ukraine and Russia launch overnight drone attacks amid missile strike tensions," 2024. [Online]. Available: https://abcnews.go.com/International/ukraine-russia-launch-overnight-drone-attacks-amid-missile/story?id=116174858

[3] H. Menouar, I. Guvenc, K. Akkaya, A. S. Uluagac, A. Kadri, and A. Tuncer, "UAV-enabled intelligent transportation systems for the smart city: Applications and challenges," *IEEE Communications Magazine*, vol. 55, no. 3, pp. 22–28, 2017.

[4] A. Chriki, H. Touati, H. Snoussi, and F. Kamoun, "FANET: Communication, mobility models and security issues," *Computer Networks*, vol. 163, p. 106877, 2019.

[5] A. Patel and A. K. Cherukuri, "Analysis of lightweight cryptography algorithms for UAV-networks," *arXiv preprint arXiv:2504.04063*, 2025.

[6] A. G. Orozco-Lugo, D. C. McLernon, M. Lara, and S. A. R. Zaidi, "Monitoring of water quality in a shrimp farm using a FANET," *Internet of Things*, vol. 18, p. 100170, 2022.

[7] A. Crețu, C. Avram, D. Radu, B. Parrein, A. Aștilean, and C. Domuța, "Health information exchange for management of disaster victims using FANET," in *6th International Conference on Advancements of Medicine and Health Care through Technology; 17–20 October 2018, Cluj-Napoca, Romania*, S. Vlad and N. M. Roman, Eds. Singapore: Springer Singapore, 2019, pp. 201–205.

[8] A. Guillen-Perez, R. Sanchez-Iborra, M.-D. Cano, J. C. Sanchez-Aarnoutse, and J. Garcia-Haro, "Wifi networks on drones," in *2016 ITU Kaleidoscope: ICTs for a Sustainable World (ITU WT)*. IEEE, 2016, pp. 1–8.

[9] W. D. Paredes, H. Kaushal, I. Vakilinia, and Z. Prodanoff, "Lora technology in flying ad hoc networks: A survey of challenges and open issues," *Sensors*, vol. 23, no. 5, 2023.

[10] M. F. Khan and K. lim Alvin Yau, "Route selection in 5G-based flying ad-hoc networks using reinforcement learning," *2020 10th IEEE International Conference on Control System, Computing and Engineering (ICCSCE)*, pp. 23–28, 2020.

[11] Naviter. (2019) How it works: Fanet+. [Online]. Available: https://naviter.com/2019/09/how-it-works-fanet/

[12] H.-N. Dai, H. Wang, H. Xiao, X. Li, and Q. Wang, "On eavesdropping attacks in wireless networks," 2016, pp. 138–141.

[13] M. Conti, N. Dragoni, and V. Lesyk, "A survey of man in the middle attacks," *Commun. Surveys Tuts.*, vol. 18, no. 3, p. 2027–2051, Jul. 2016.

[14] A. Vasudeva and M. Sood, "Survey on sybil attack defense mechanisms in wireless ad hoc networks," *Journal of Network and Computer Applications*, vol. 120, pp. 78–118, 2018.

[15] C. Dobraunig, M. Eichlseder, F. Mendel, and M. Schläffer, "Ascon v1.2: Lightweight authenticated encryption and hashing," *J. Cryptol.*, vol. 34, no. 3, Jul. 2021.

[16] STMicroelectronics, "STM32WL55JC Datasheet," 2020. [Online]. Available: https://www.st.com/resource/en/datasheet/stm32wl55jc.pdf

[17] M. J. Dworkin, "Recommendation for block cipher modes of operation: The cmac mode for authentication," 2016.

Author Index

Yassin Abdullah	79	Khang Nguyen Minh	103
Krishna Baishnab	109	Nguyen D. Minh	19
Bui Ngoc Thanh Binh	91	Yasuhiko Nakashima	55, 133
Saroj Biswas	1, 109	Le Nguyen Nhat Nam	67
Dong Bui	13	Adhiraj Nandy	109
Duy-Hieu Bui	61, 139	Sourav Nath	109
Thanh-Dat Bui	25	Ngo Minh Nghia	97
Trong-Tu Bui	7	Hoang Anh Vy Ngo	13
Bofan Chen	79	Viet N. D Ngo	31
Chukhu Chunka	1	Bao Chau Pham Ngoc	73
Anh-Vu Dinh-Duc	73	Kha Duy Thai Ngoc	97
Nguyen Dung	97	Duy-Cong Nguyen	115, 127
Quan Doan Duy	103	Ha Thi Viet Nguyen	37
Nilüfer Ertekin	43	Hoa Quang Nguyen	115, 127
Koushik Guha	109	Hoai Son Nguyen	83
Nguyen Viet Ha	91	Hoang-Long Nguyen	127
Muhammad Hashim	79	Manh Long Nguyen	115
Long Pham Hoang Ho	49	Phuong Nguyen	115
Thien-Duy Ho	61	Quoc Minh V. Nguyen	7
Chi Phuong Hoang	19	Tung-Bach Nguyen	61
Kha Manh Hoang	37	Van-Tinh Nguyen	133
Van-Phuc Hoang	133	Xuan Bach Duy Nguyen	73
Cuong Huynh	31, 49	Thuat Nguyen-Khanh	121
Huu-Thuan Huynh	25	My Nguyen-Le-Ha	121
Tan Phat Huynh	115	Linh Nguyen-Thi-Thuy	19
Thi-Minh-Tuyen Huynh	25	Luu Nguyen-Van	19
Atiya Khan	1	Vishnu Padmakumar	109
Dang Le Khoa	97	Cong-Kha Pham	25, 37
Tran Nguyen Tuan Kiet	91	Hoai Luan Pham	55, 133
Tuan Anh La	83	Huy-Duc Pham	25
Wen Cheng Lai	87	Thai-Bao Pham	121
Duc-Hung Le	7	Huyen-Trang Pham-Thi	139
Nhu Thai Le	83	Giang Phan	115
Tri-Hieu Le	127	Vinh Truong Quang	103
Trung-Khanh Le	7	Xuan Thanh-Pham	37
Vu Trung Duong Le	55, 133	Hang Le Thi	83
Quan Le-Trung	121	Dat Hoang Tran	13
Wen Lei	43	Diem Thi Tran	67
Zhennan Li	79	Tuan-Kiet Tran	25
Zhiqun Li	79	Tuan-Phong Tran	61
Weiwen Lin	79	Van Duy Tran	55
Van-Tuan Luu	133	Van-Toan Tran	127
Duong Van Minh	115	Xuan-Tu Tran	61, 139

Quang-Kien Trinh	127	Luan Van-Thien	121
Nguyen Van Trung	133	Thanh Cong Vu	83
Nguyen Thi Xuan Uyen	97	Tuan Hai Vu	55
Lam Thien Van	49	Yan Yao	79